Fluid Mechanics in Channel, Pipe and Aerodynamic Design Geometries 1

...To our Family

Depy
Andreas
Nikos
Giannis
Lilian

...as without their support none of this would have ever been possible for us

Engineering, Energy and Architecture Set

coordinated by
Lazaros E. Mavromatidis

Volume 2

Fluid Mechanics in Channel, Pipe and Aerodynamic Design Geometries 1

Christina G. Georgantopoulou
George A. Georgantopoulos

iSTE

WILEY

First published 2018 in Great Britain and the United States by ISTE Ltd and John Wiley & Sons, Inc.

ISTE Ltd
27-37 St George's Road
London SW19 4EU
UK

www.iste.co.uk

John Wiley & Sons, Inc.
111 River Street
Hoboken, NJ 07030
USA

www.wiley.com

MIX
Paper from responsible sources
FSC® C013604

Library of Congress Control Number: 2018932715

British Library Cataloguing-in-Publication Data
A CIP record for this book is available from the British Library
ISBN 978-1-78630-139-0

Contents

Preface

This book presents an extended and detailed analysis of both the flow phenomena in closed and open channels and the flows around solid bodies. It comprises two volumes. This book is a specialized resource for those students, engineers and researchers who want to focus on the industrial applications of flows and study the fascinating world of internal and external flow phenomena.

We have both had extensive experience in teaching, studying and researching fluids since the completion of our respective PhD theses. We felt that it was time to write about the practical and analytical aspects of flow applications, all of which can be applied in industrial flows, to support researchers, engineering students and industrial engineers in the field of fluids in order to optimize their work in "flows".

For the first author, the "fluids direction" began in the early stages of her PhD thesis study in Computational Fluid Dynamics in 1998 at the National Technical University of Athens. The second author's knowledge of the fluids' path is very extensive, obtained from more than 45 years of studies and work involved in his PhD thesis and further research work at the University of Patras, as well as through his position as Professor of Aerodynamics at the Hellenic Air Force Academy, spanning more than 35 years.

We have both gained substantial experience in Fluid Mechanics research through numerous publications, presentations at international conferences, academic textbook authoring, teaching through international experiences and collaborations. However, we felt that more should be offered to the Fluid Mechanics community, and hence this book.

Although we both have experience in writing for academic textbooks, this is our first publication that caters to international students, researchers and engineers, considering the industrial phenomena that are met in international industries and we have tried to present most of the applications in flows inside or around bodies. This book is based on books written previously by us on Fluid Mechanics and on Aerodynamics, but for the first time our work focuses on the practical aspects of industrial internal and external flows.

Christina, the first author, offers this textbook to the Bahrain polytechnic engineering students and all the industrial delegates who have worked with her in "flows" for many years. She also wishes to express her appreciation for her colleagues, namely Payal Modi, for the thousands of hours of constructive discussions and collaborations in fluids aspects, to Lazaros E. Mavromatidis for his support during the publishing procedure, to her father George who has been her mentor for all these years and to Stephanie Sutton and Amerissa Kapela for their continuing support with the quality of the academic English language. Additionally, George, the second author, wishes to share his more than 40 years of experience in fluids with the fluids community around the world and support them in their "flows" work as best he can.

We both have a special sentimental feeling for this book in that we are extremely proud that we have been able to write, publish and offer it to you, hoping that it will really support you in your fluids journey. We have both worked on fluids with a passion not only for our students, but also to honor our colleagues around the world. We are equally happy to say that the Fluid Mechanics community has been served by the same family for more than 40 years. We hope that we will be physically and mentally healthy to continue to serve our students and support our colleagues in the fluids aspects in the future.

We hope that you will enjoy this book and be engaged with the fascinating world of flows.

Christina G. GEORGANTOPOULOU
George A. GEORGANTOPOULOS
February 2018

Introduction

I.1. Introduction

Fluid Mechanics consist of two main categories. The first one refers to the quantitative and qualitative analysis and study of fluids in motion, the velocity or acceleration as well as the forces exerted by nature. The second category analyzes the physical forces that are developed on solid–fluid interfaces, where the solids represent the containers. The first category can be called Theoretical Fluid Mechanics, while the second one is called Applied Fluid Mechanics.

This book primarily presents the aspects and problems of internal and external flows, including certain fundamental principles of fluids.

To develop an extended study of Applied Fluid Mechanics problems, mathematical modeling and analysis is considered necessary. On the other hand, the empirical or experimental investigation of fluid phenomena only provides us with certain measurements and information about individual cases, and it is often difficult to generalize our conclusions. Hence, the appropriate way to study fluid flows is to investigate the related phenomena with a combined analytical–computational and experimental approach in order to improve step by step the proposed fluid theories or solutions.

Industrial engineers have raised various issues related to the main assumption that all fluids are considered to be ideal. In order to overcome these issues, every technological problem is considered to be an individual one, resulting in a lack of theoretical background. Year after year, a huge gap has been created between theoretical and practical hydrodynamics researchers, which exists even today. This book bridges this gap between various industrial flows, and an attempt has been made to present a common strategy. The flows inside pipes or channels as well as the flows around bodies are considered to be real life applications, setting the appropriate theoretical background simultaneously.

I.2. Fluid Mechanics sections

The Fluid Mechanics study comprises fluid motion and fluid balance. During the last decades, it has evolved in two major directions. Theoretical Fluid Mechanics includes the mathematical exploitation of fluid phenomena, and Technical Fluid Mechanics includes the applications of mechanical engineering, aeronautics, shipbuilding and meteorology. Technical Fluid Mechanics is considered an applied science, and hence it is often referred to as Applied Fluid Mechanics, which includes the possible solutions of fluid problems and the explanation of natural phenomena. Moreover, it aims to produce numerical predictions or experimental validation for direct practical applications.

Classic Fluid Mechanics can be derived from various areas according to the mechanical condition or fluid properties. The categories presented in Table I.1 are based on the motion of fluids as well as on compressibility, where the density varies according to the fluid condition.

Fluid mechanics	Fluids at rest	Fluids in motion
Hydrodynamics (ρ=ct)	Hydrostatics	Hydrodynamics
Aeromechanics ($\rho \neq$ct)	Aerostatics	Aerodynamics

Table I.1. *Fluid Mechanics categories*

I.3. Systems of units

I.3.1. *Definitions and general considerations*

Units are fundamental for physics, especially for all the applied sciences such as mechanical engineering. The number without units means absolutely nothing for Fluid Mechanics, as it represents a natural quantity such as pressure, velocity or force.

Historically, various systems of units have been developed according to the theoretical principle demands or to practical applications. In most countries (not including the USA), the metric system is the official system of measurement, which is accepted by both scientists and engineers. The International System of Units (SI) was defined and established at the 11th General Conference on Weights and Measures, where more than 36 countries accepted it to be the most complete and appropriate one, including the USA. Since then, the USA has made huge progress in introducing SI units to engineering. For example, many NASA laboratories use SI units for their technical research results, and the AIAA (American Institute of Aeronautics and Astronautics) also supports the SI in its research papers.

Therefore, students who want to study engineering have to know both unit systems. The following table presents the corresponding basic units in both systems based on the theory that all the derived units at the metric system can be produced by the base ones.

Base quantity	**SI**	**BS**
Length	Meter (m)	Foot (ft)
Time	Second (s)	Second (sec)
Mass	Kilogram (kg)	Pounds of mass (lb_m) or slug
Temperature	Celsius (°C)	Fahrenheit (°F)
Absolute temperature	Kelvin (K)	Rankine (R)

Table I.2. *Base units in SI and BS (British system)*

As we have just mentioned, the derived units can be produced by the base units following the nature of interrelationships or the basic formulas with the need for adding any conversion factor, as in the following, using Newton's law:

$$F = m \times \alpha$$
$$(1\ \text{newton}) = (1\ \text{kilogram})(1\ \text{meter/second}^2) \qquad [I.1]$$

Thus, we further confirm the definition of Newton as the force that is required to accelerate a mass of 1 kg at a rate of 1 m/s^2. Similarly, the ideal gas constant for air (R=287 J/(kg·K) can also be expressed in the following way:

$$R = 1716\frac{\text{J}}{(\text{kg})(\text{K})} = 287\cdot\frac{\text{N}\cdot\text{m}}{(\text{kg})(\text{K})} = 287\frac{\text{kg}\cdot\text{m}}{\text{s}^2}\cdot\frac{\text{m}}{(\text{kg})(\text{K})} = 287\frac{\text{m}^2}{(\text{s}^2)\cdot(\text{K})} \qquad [I.2]$$

The BS is also a consistent system, and the same procedure can be followed for the derived quantities:

$$F = m \cdot a \tag{I.3}$$

$$(1\text{ pound}) = (1\text{ slug})(1\text{ foot/second}^2)$$

$$R = 1716 \frac{\text{ft} \cdot \text{lb}}{(\text{slug})(\text{R})} = 1716 \frac{\text{ft}^2}{(\text{s}^2)(\text{R})} \tag{I.4}$$

However, more systems of units are not consistent; therefore, it is necessary to use a factor in order to produce the required conversion as shown below. These systems have been used in the past by engineers but have often not been convenient to be applied:

$$F = (1/g_c) \times m \times a \tag{I.5}$$

$$g_c = 32.2 \quad (\text{lb}_\text{m}) \cdot (\text{ft})/(\text{s}^2) \cdot (\text{lb}_f)$$

$$\begin{array}{cccc} F = & (1/g_c) & & m \times a \\ \uparrow & \uparrow & \uparrow & \uparrow \\ \text{lbf} & (1/32.2)\,\text{lb}_\text{m} & & \text{ft}/\text{s}^2 \end{array} \tag{I.6}$$

The various temperature units are of high importance. We often denote absolute temperature by *T*, where the minimum temperature value can be zero. Kelvin (K) and Rankine (R) are the absolute temperature units, where 0 R = 0 K indicates the temperature at which all the molecular motion theoretically stop. In addition, the relationships among the temperature units are:

$$0°\text{F} = 460\,\text{R} \tag{I.7}$$

$$0°\text{C} = 237\text{k} = 32°\text{F} \tag{I.8}$$

It is worth mentioning that the temperature *T* in the ideal gas equation of state (equation [I.9]) is absolute:

$$p = \rho R T \tag{I.9}$$

where p is the pressure, ρ is the density of gas and the other symbols are defined as above.

I.3.2. *Definitions and fundamental units in fluids*

Natural quantity	Units	Symbol
Force	Newton	N = kg m/s^2
Energy	Joule	J = N m
Power	Watt	W = J/s

Table I.3. *Units of common quantities in physics and fluids*

10^{-6}	micro	μ
10^{-3}	milli	m
10^{3}	kilo	k
10^{6}	mega	M

Table I.4. *Common metric prefix in SI*

Length *l*	
1 in	25.4 mm
1 ft	0.3048 m
1 yd	0.9144 m
1 mile	1.6093 km
Area *S*	
1 in^2	645.16 mm^2
1 fr^2	0.0929 m^2
1 yd^2	0.8361 m^2
1 mile^2	2.590 km^2
1 acre	4046.9 m^2
Volume *v*	
1 in^3	16387 mm^3
1 ft^3	0.02832 m^3
1 UK gal	0.004546 m^3
1 US gal	0.003785 m^3

Table I.5. *Length, area and volume conversion factors*

Mass *m*	
1 kg	10^3 g
1 oz	28.352 g
1 lb	453.592 g
1 cwt	50.802 kg
1 ton (UK)	1016.06 kg
Density *ρ*	
1 lb/ft^3	16.019 kg/m^3
1 lb/UK gal	99.776 kg/m^3
1 lb/US gal	119.83 kg/m^3
Force *F*	
1 dyne	10^{-5} N
1 poundal	0.1383 N
1 lb-f	4.4482 N
1 kg-f	9.8067 N
1 ton-f	9.9640 kN
Viscosity *μ*	
1 poise (1 g/cm sec, 1 dyn sec cm^2)	0.1 N sec/m^2
1 lb/ft sec (1 poundal sec/ft^2)	1.4882 N sec/m^2
1 lb/ft hr (1 poundal hr/ft^2)	0.4134 mN sec m^2
Pressure *p*	
1 bar (10^5 dynes/cm^2)	10^5 N/m^2
1 atm (1 kg-f/cm^2)	98.0665 kN/m^2
1 atm (standard)	101.325 kN/m^2
1 psi (1 lb-f/in^2)	6.8948 kN/m^2
1 psf (1 lb-f/fn^2)	47.880 N/m^2

Table I.6. *Conversion factor mass, density, force, viscosity and pressure*

Energy *E*	
1 erg	10^{-7} J
1 ft poundal	0.04214 J
1 ft lb-f	1.3558 J
1 cal (international table)	4.1868 J

1 Btu	1055.06 J	
1 hph	2.6845 MJ	
1 kwh	3.6 MJ	
Power *P*		
1 erg/sec	10^{-7} W	
1 hp (British)	745.70 W	
1 hp (metric)	735.40 W	
1 ft lb-f/sec	1.3558 W	
1 Btu/hr	0.2931 W	
Surface tension σ		
1 dyne/cm (1 erg/cm^2)	10^{-3} J/m^2	
Moment of inertia *M*		
1 lb.ft^2	0.04214 kg m^2	
Momentum *J*		
1 lb-fit/sec	0.1383 kg.m/sec	
Specific temperature *c*		
1 Btu/lb°F (1 cal/g.°C)	4.1868 kJ/kg.°C	
Heat transfer coefficient *h*		
1 Btu/h.ft^2.°F	5.6783 W/m^2.K	
Thermal conductivity *K*		
1 Btu/h.ft.°F	1.7307 W/m.K	
Water (18°C and air properties (STP))		
	Water	*Air*
Density (kg/m^3)	10^3	1.3
Viscosity (N sec/m^2)	10^{-3}	1.7×10^{-5}
Specific heat (KJ/kg.K)	4	1
Thermal Conductivity (W/m.K)	0.6	0.024

Table I.7. *Other conversion factors*

1

Fundamental Principles in Fluids

1.1. Introduction

The most important characteristics of fluids are distinguished under shear conditions. As we have already mentioned, every fluid is incapable of maintaining its static balance under shear stress conditions.

As a matter of fact, many objects, even solids, show fluid properties if high-grade shear stresses are developed. In addition, some real fluids show a viscosity that is so high under certain conditions that they behave more like solids than fluids. However, the definition that has been given is adequate for fluids despite the occurrence of some extreme cases. In general, fluids are comprised of liquids and gases.

Figure 1.1 shows the shear stresses in both solids and fluids. A tangential force F is applied to the upper surface of each object, thereby developing shear stresses (F/S). When the shear stresses are applied to a solid body, deformation is developed, which is depicted by its change in shape and the angle *da.*

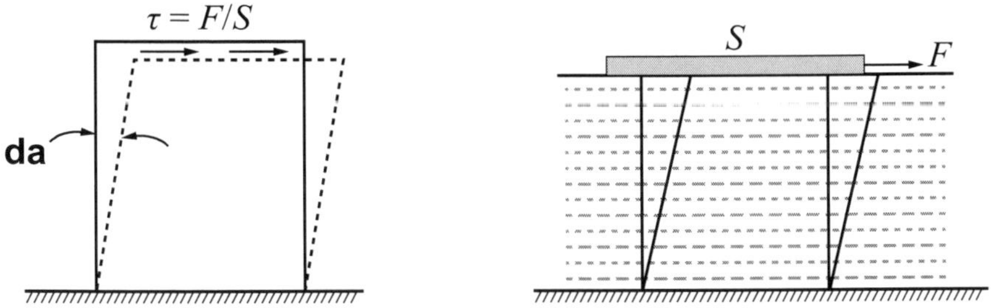

Figure 1.1. *Shear stresses in solids and fluids*

However, if the same stresses are applied to fluids, deformation will not be permanent. Deformation is developed only during the stress application. If some objects sometimes behave like solids and other times like fluids, according to the size of the applied stresses, then they are called *plastics*, for example the paraffin. This type of material is deformed like solids if the shear stresses are less than a certain value τ_0, and flows like fluids if the stresses are $\tau < \tau_0$. The official name of this type of material is *Bingham* plastic.

1.2. Incompressible and compressible fluids

According to the variation in physical properties, fluids can be classified as *incompressible* or *compressible.* Under certain conditions of static balance, some fluids show minimum density variation even at high pressure values. This type of fluid is called incompressible.

Consequently, incompressible fluids are fluids with constant density regardless of the applied pressure value. However, this definition is theoretical or can be considered as an assumption, as real fluids are not able to maintain their density totally constant under various pressure conditions. The density always shows a certain variation even if it is limited. If the variation is close to zero, then we can assume the fluids to be incompressible and neglect the aforementioned variation. For gases, this can be assumed only if the temperature is also constant.

Most liquids can be assumed to be incompressible. The study of incompressible fluids under static balance conditions is called hydrostatics. If the density of fluids cannot be considered as constant, then these fluids are called *compressible.* Although most liquids can be assumed to be incompressible, the parameter that determines the distinction between incompressible and compressible in air flows is the *Mach* number (*M*). If $M < 0.4$, the air flow can be considered to be incompressible, while if $M > 0.4$, the flow is considered to be compressible.

1.3. Fluid properties

The main characteristics of fluids are called fluid properties. As these characteristics may vary according to the space or time, we will define these as characteristics of a fluid particle. Their definition will be presented on a point within a fluid or within a flow field.

1.3.1. *Density* (ρ)

Fluid density (ρ) is defined as the mass *dm* of a fluid particle per volume *dv* unit. Consequently, if we want to define the density of the fluid in a point *P* of a flow field, we may write the following expression:

$$\rho(P) = \lim_{dv \to 0} \frac{dm}{dv}. \qquad [1.1]$$

The SI unit of density is kg/m^3. Density depends on both the location of point *P* and the time:

$$\rho = \rho(x, y, z, t). \qquad [1.2]$$

It can also be varied according to the pressure and temperature of the fluid.

If the density of a fluid remains constant regardless of the other property's variation, then the fluid is called incompressible. In the case of liquids, we may assume that their density remains constant even after high pressure or temperature variation. This assumption cannot be made for gases, and hence they are called *compressible* fluids. The density of most gases is proportional to the pressure and irreversibly proportional to the temperature.

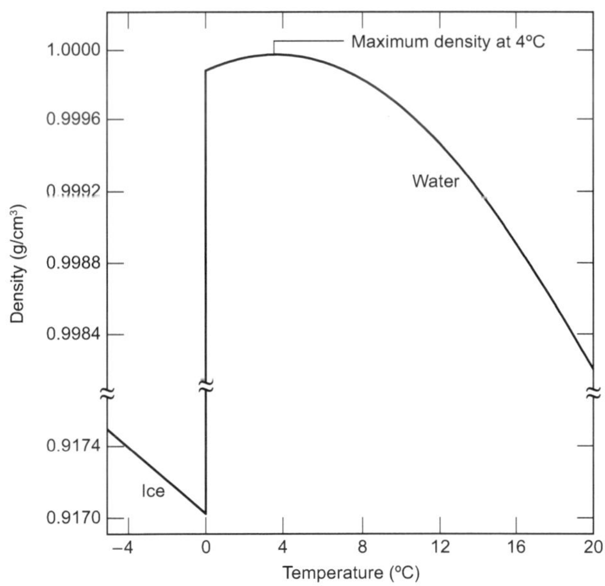

Figure 1.2. *Density of water variation according to the temperature*

As the density is defined per volume unit, the *temperature* and *pressure* are the main conditions for its variation. In addition, the volume of the fluid and consequently the density are affected by the possible *dissolved substances*.

Temperature variation affects fluid density. As the temperature is increased, the volume of the body is increased (expansion) and consequently the density has lower values. There is an exception in the case of water at the temperature range of 0°C–4°C: the density is decreased when the temperature is increased in this range of values, and thus it reaches a minimum value (Figure 1.2).

T (°C)	ρ (kg/m^3)	T (°C)	ρ (kg/m^3)	T (°C)	ρ (kg/m^3)	T (°C)	ρ (kg/m^3)
0	999,839	7	999,901	30	995,647	65	980,502
1	999,898	8	999,848	35	994,032	70	977,771
2	999,940	9	999,781	40	992,215	75	974.850
3	999,964	10	999,699	45	990,213	80	971,799
4	999,972	15	999,099	50	988,037	85	968,621
5	999,964	20	999,204	55	985,696	90	965,321
6	999,940	25	997,045	60	983,200	100	958,365

Table 1.1. *Density of pure water according to various T values*

The expansion is more intense in gases than in liquids. For instance, the expansion coefficient of oil is four times lower than air's coefficient, and the water expansion coefficient is 17 times higher (Table 1.1). Consequently, if the temperature is increased from 20 to 30°C under constant pressure, the volume of the air will be increased by 3.3%, the volume of the oil by 0.9% and the volume of the water by approximately 0.2%. Table 1.1 presents the density of pure water for *T* ranging from 0 to 100°C. As can be easily seen, the maximum density is achieved at 4°C.

In the case of gases under conditions far below the dew point, the density as well as its variation in relation to the *T* and *p* can be calculated using the following state equation for gases (equation [1.4]):

$$p \cdot v = n \cdot R_w \cdot T = \frac{m}{m_r} \cdot R_w \cdot T \Rightarrow$$

$$\Rightarrow p = \rho \cdot R \cdot T \qquad [1.3]$$

$$\rho = \frac{p}{R \cdot T} \qquad [1.4]$$

where p is the pressure, T is the absolute temperature and R is the gas constant.

The gas constant R can be defined as the ratio of the universal gas constant over the molar mass of the gas:

$$R = \frac{R_w}{m_r} \qquad [1.5]$$

where $R_w = 8,314\,\mathrm{J/(mole \cdot K)} = 8,314\,\mathrm{J(Kmole \cdot K)}$.

In the case of air as a fluid that comprises two main components (21% oxygen and 79% nitrogen), the molar mass is:

$$m_r = 32 \cdot 0.21 + 28 \cdot 0.79 = 29\,\mathrm{g/mole} = 29\,\mathrm{kg/Kmole} \qquad [1.6]$$

and then $R = 287.1/\mathrm{kgK}$.

The *pressure* also significantly affects the density of fluids. If the temperature remains constant, a double value of pressure implies a double value of density. Concerning gas vapor, the relationship between pressure and density remains the same. For example, if the temperature of a vapor quantity is equal to 150°C (423 K) and the pressure to 1 bar (100 kPa), the density will be ρ = 0.590 kg/m^3 (if the state equation is applied, the results would be ρ = 0.512 kg/m^3). If the pressure is doubled, the density will be increased by 77%.

Unlike gases, the effect of pressure on the density of liquids is minimum. Hence, in most of the practical or technical applications in liquids, we assume that the density is constant despite possible pressure variation. As already mentioned, if the density remains constant when the pressure varies, the fluids are called incompressible, whereas these are called compressible if the density also varies. *In fact, compressible fluids are gases, as liquids are assumed to be incompressible.*

The *dissolved substance* in a mixture increases the fluids' density. When a solid substance is dissolved in a solvent, the mass is increased without any corresponding increase in the volume. This occurs due to the chemical relationship between the liquid and the dissolved substance, which can be explained by the dissolution theory and mechanism.

%	0°C	10°C	25°C	40°C	60°C	80°C	100°C
1	1,007.5	1,007.1	1,004.1	999.1	990.0	978.5	965.1
2	1,015.1	1,014.4	1,011.1	1,005.9	996.7	985.2	971.9
4	1,030.4	1,029.2	1,025.3	1,019.8	1,010.3	998.8	985.5
8	1,061.2	1,059.1	1,054.1	1,048.0	1,038.1	1,026.4	1,013.4
12	1,092.4	1,083.5	1,083.6	1,077.0	1,066.7	1,054.9	1,042.0
16	1,124.2	1,120.6	1,114.0	1,106.9	1,096.2	1,084.2	1,071.3
20	1,156.6	1,152.5	1,145.3	1,137.7	1,126.8	1,114.6	1,101.7
24	1,190.0	1,185.6	1,177.8	1,169.7	1,158.4	1,146.3	1,133.1
26	1,207.1	1,202.5	1,194.4	1,186.1	1,174.7	1,162.6	1,149.2

Table 1.2. *Sea water density*

This phenomenon receives great importance in the case of water. Natural water contains many dissolved solid substances such as salts. Therefore, the density of the natural water is slightly higher than the one presented in Table 1.1. For the temperature range of 0–20°C, the density of natural water is equal to 1,000 kg/m^3. Sea water, which contains high concentrations of sodium chloride, has a higher density range (1,025–1,028 kg/m^3). Table 1.2 presents the density of sea water analytically according to both the temperature and the concentration.

If we mix two liquids with densities ρ_1 and ρ_2, the *density value of the mixture* will be between these two values, depending on the mixture ratio. The mass of the solution will be $m = m_1 + m_2$, while the volume $v \leq v_1 + v_2$. If we assume that $v \cong v_1 + v_2$, we are able to estimate the solution's density. However, the precise density estimation can be retrieved by the rating table or graphs.

The measurement of density can be achieved by specific instruments, which will be studied in the next chapter.

1.3.2. *Specific weight (γ)*

Specific weight (γ) is defined as the weight per volume unit:

$$\gamma = \frac{dw}{dv} \qquad [1.7]$$

where w is the weight of the body and v is the volume.

As can be easily seen according to the w and ρ definition, the specific weight can be calculated as follows:

$$\gamma = \rho \cdot g \tag{1.8}$$

where ρ is the density of the fluid and g is the acceleration of gravity.

The SI unit of specific weight is N/m^3.

Table 1.3 presents the density and specific weights for various fluids at 20°C, $p = 101.3$ KPa, $g = 9.807$ m/s^2.

Fluid	ρ (kg/m^3)	γ (KN/m^3)	Fluid	ρ (kg/m^3)	γ (KN/m^3)
Water (pure)	998.2	9.79	Pentane	630	5.71
Natural water	1,000	9.81	Hexane	659	6.46
Sea water	1,025	10.05	Heptane	684	6.71
Ether	708	6.94	Octane	703	6.89
Alcohol	789	7.74	Decane	730	7.16
Aceton	792	7.77	Hexadecane	774	7.59
Benzene	879	8.62	Petrol	700–750	6.86–7.36
Aniline	1,022	10.0	Kerosene	780–820	7.65–8.04
Chloroform	1,489	14.6	Diesel	810–860	7.94–8.43
Carbon tetrachloride	1,590	15.6	Crude oil	900–1,000	8.83–9.81
Sulfuric acid	1,834	28.0	Mineral oil	880–940	8.63–9.22
Mercury	13,550	133.0	Olive oil	910–920	8.92–9.02

Table 1.3. *Density and specific weight of various fluids*

1.3.3. *Relative density*

Sometimes, especially in relation to industrial applications, the relative density (*RD*) or specific gravity (*SG*) is used instead of density. *RD* is a dimensionless quantity, which expresses the relationship between the densities of two fluids where the second one is chosen as a reference material:

$$RD = \frac{\rho}{\rho_f} \tag{1.9}$$

where ρ is the density of the fluid and ρ_f is the density of the reference material at a certain temperature. For liquids, the reference material is water and ρ_f is often equal to $1{,}000\ \mathrm{kg/m^3}$ at 4°C. In the case of gases, the corresponding reference density is the density of the air for T = 293 K and p = 101.3 KPa, which is $\rho_f = 1.205\ \mathrm{kg/m^3}$. As the relative density is a dimensionless quantity, it is advantageous for most engineers, who thus prefer it to the density or specific weight. In Figure 1.3, various relative densities of liquid mixtures are depicted.

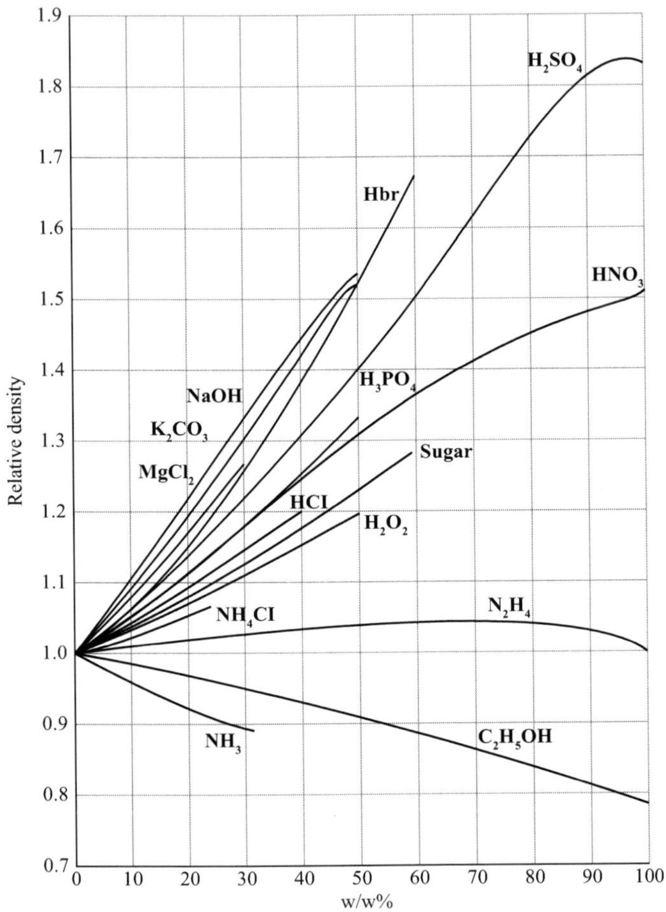

Figure 1.3. *Relative densities of liquid mixtures*

For non-homogeneous fluids, as some types of mixtures, the density definition differs. The concentration in this case is not steady, and mass motion is produced among various fluids' points regardless of the fluidity. This process which is called diffusion will be continued until a homogeneous mixture is formed. The concentration and density are connected according to their definition. Concentration *A* of a mixture component is defined as the mass per volume unit. The mass concentration or the partial density ρ_a is often used. The mixture density is equal to the sum of the mass concentration of the components:

$$\rho = \rho_\alpha + \rho_\beta + ... + \rho_n = \Sigma\rho_n \quad [1.10]$$

where ρ_i are the densities of the *n* mixture components.

According to the data given, the mixture density can also be calculated as follows:

$$\rho = \frac{m}{v} = \frac{\rho_1 n_1 + \rho_2 n_2 + ... + \rho_i n_i}{n_1 + n_2 + ... + n_i} \quad [1.11]$$

where the symbols are defined as above.

Special density calculation is applied for the atmospheric air case, which is a mixture of oxygen, carbon dioxide, nitrogen oxide and noble gases containing water vapor simultaneously. The vapor precedence is not constant and depends on the pressure and temperature conditions. In these cases, the density of the air is given by:

$$\rho_0 = \rho l - 0,337\varphi\frac{p_a}{p} \quad [1.12]$$

where ρ is the density of the dry air, φ is the relative humidity, p is the pressure and p_a is the saturation pressure of water.

According to *API gravity* standards, the relative density of a fluid is defined as the ratio of the fluid density to the density of water in 60°F. Moreover, according to American Petroleum Institute standards, this ratio can be calculated based on the value of API gravity (API) as follows:

$$RD(60) = \frac{141.5}{131.5 + API}. \quad [1.13]$$

The aforementioned formula can also be used for the API gravity calculation, which is known as the relative density value.

The relative density can be calculated at a different temperature using the following formula:

$$RD(T) = \frac{RD(60)}{1 + a(T - 60)} \tag{1.14}$$

where a is equal to:

$$a = \exp(0.0106 \cdot API - 8.05) \tag{1.15}$$

The *Baume scale* is also used by the industry for various fluids with different density values of the water (lower or higher). The corresponding relationship between the Baume scale and relative density is presented in Table 1.4.

Baume Degrees	**RD for fluids with** $< \rho_{water}$	**RD for fluids with** $> \rho_{water}$
0	–	1.000
10	1.000	1.074
20	0.933	1.160
30	0.875	1.261
40	0.823	1.381
50	0.778	1.526
60	0.737	1.706
70	0.700	1.933
80	0.669	2.231
90	0.636	–
100	0.609	–

Table 1.4. *Baume scale values*

Finally, according to the Twaddell scale, the Twaddell degrees T_w and the *RD* of the fluid are connected by the following relationship:

$$RD(60) = \frac{T_w}{200} + 1 \qquad [1.16]$$

1.3.4. *Pressure*

The fluid pressure is the same in all directions, which is developed on every fluid surface. Consequently, the *pressure* is defined as the ratio of the force *F* that a fluid exerts on a surface *S* over this surface:

$$p = \frac{F}{s}. \qquad [1.17]$$

If the force at a point is not uniform, then the pressure can be defined as:

$$p = \lim_{\delta S \to 0} \frac{\delta F}{\delta S} = \frac{dF}{dS} \qquad [1.18]$$

where the δS includes the point *P*.

Hydrostatics pressure is defined as the pressure which is equal to the weight of a vertical liquid column extended to the free liquid surface with a cross-sectional area equal to 1. The pressure head (h) is defined as the ratio of the pressure by the weight of the liquid:

$$h = \frac{p}{\gamma} \qquad [1.19]$$

The difference in pressure between two points in a liquid which are located in different positions and heights can be defined as follows (Figure 1.4):

$$|p_1 - p_2| = (h_1 - h_2)\gamma \qquad [1.20]$$

$$|p_1 - p_3| = (h_1 - h_3)\gamma \qquad [1.21]$$

where γ is the specific weight, p_i is the pressure at different points and h_i are the heights.

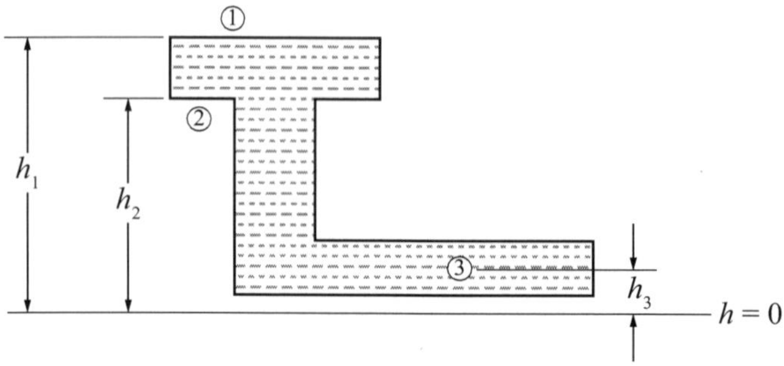

Figure 1.4. *Pressure heads*

Two main types of pressure are distinguished: static and dynamic. *Static pressure* is defined as the pressure which is created by the fluid's molecular forces. Static pressure obtains a value irrespective of whether the fluid is moving or not. The *dynamic pressure* which is originated by the forces exerted by the fluid during its motion can be calculated as:

$$p_d = \frac{1}{2}\rho V^2 \tag{1.22}$$

where p_d is the dynamic pressure, ρ is the density of the fluid and V is the velocity.

The sum of static and dynamic pressure is the *total pressure.*

1.3.5. *Compressibility*

Compressibility is a characteristic property of fluids as the meter of the volume variation under the influence of external forces. The volume variation is a function of the fluids' elasticity, which is expressed in a similar way by Hook's law in solids. Consequently, a linear function pressure–volume variation is applied as:

$$\Delta_p = -E\frac{\Delta v}{v_0} \tag{1.23}$$

where E is the *meter of elasticity (bulk modulus)* and $\Delta v/v_0$ is the volume variation due to the pressure variation Δp. The minus (−) sign means that the volume variation is negative.

As given by equation [1.23], the meter of elasticity is reversed to the compressibility (high meter of elasticity means low compressibility), and hence the compressibility is expressed by $1/E$. In addition, as the ratio $\Delta v/v_0$ is dimensionless, the meter of elasticity units are pressure units.

The compressibility of liquids is very low. For example, the bulk modulus of water is equal to 20,000 kP/cm^2, which means that if we increase the pressure of the water volume v_0 for 1 kP/cm^2, the volume variation will be approximately 1/20,000–0.00005. Most liquids show similar behaviors, and hence they are assumed to be incompressible.

In the case of gases, the bulk modulus E is equal to the initial pressure p_0 of a certain gas quantity v_0 when the process is isothermal (the isothermal process is called a gas process where the pressure and the volume are varied but the temperature remains constant). Consequently, we have to prove that in gases, if $T = const$, then $E = p_0$.

In practical applications, the gas flow is considered to be incompressible, if the variation of the specific gravity is less than 0.08 ($\frac{\Delta\rho}{\rho_0} < 0.08$), which means, for example, that in the case of an aircraft motion inside the atmosphere, its velocity should not exceed 140 m/s, that is, the *Mach number* (M = V/a, where V is the velocity of the air and a is the speed of sound) should be less than 0.4. However, the air flow is often considered to be incompressible if the velocity is less than 100 m/s. Obviously, as the velocities of most aircrafts are much higher than this value, it is important to consider the compressibility in aerodynamics.

1.3.6. *Viscosity*

Fluids can be easily deformed under the influence of shear stresses. The deformation velocity variation is connected with shear stress, which can be applied to all fluids. In addition, during this continuing deformation, the fluid develops a relating *resistance*, which is expressed in terms of shear stresses. The type and the tension of the fluids' resistance depend on its relative motion as well as every velocity variation. This property is called *viscosity*.

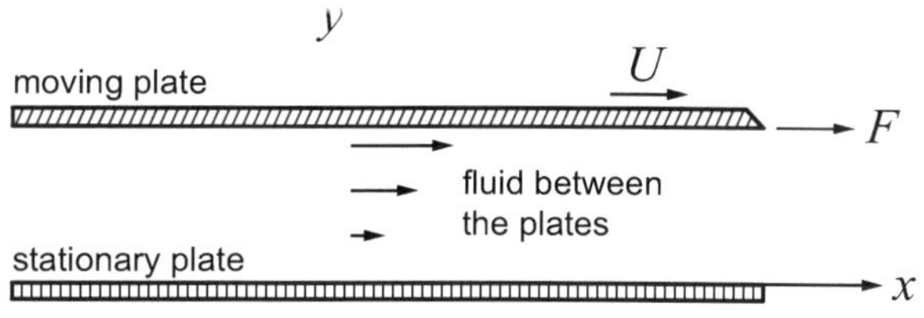

Figure 1.5. *Fluid between infinite plates*

In order to investigate the relationship between viscosity and shear stresses, we assume a relating experiment. Let us assume two large, flat, thin and parallel surfaces (plates) located in a very short distance between them as it can be seen in Figure 1.5. The gap between the plates has been filled with a fluid. The below surface remains constant while the above one is moving with constant velocity U under the influence of a force F. During this experiment, we observe that the fluids' molecules, which are attached to the below surface remain motionless due to the intermolecular forces, while the molecules attached with the above plate are moving with the same plate velocity U. Consequently, a velocity distribution is observed along the y-axis as $u = u(y)$. For $y = 0$, we have $u = 0$ and for $y = h$ (where h is the maximum distance between the plates) the velocity is equal to U ($u = U$). Due to the infinite length of the plates, the velocity is independent of the x coordinate (Figure 1.6).

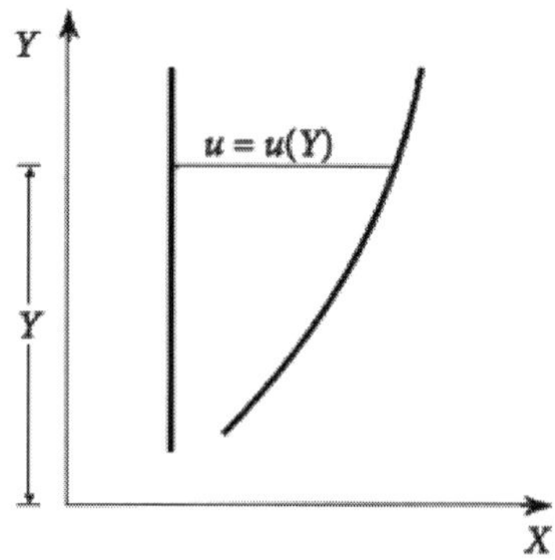

Figure 1.6. *Velocity distribution between the plates*

It is worth mentioning that the velocity U of the above plate remains constant despite the continuing force exertion F. Therefore, we conclude that the fluid exerts a similar force in the opposite direction. This force is responsible for the development of *shear stresses*.

We also assume the 2D motion of differential fluids' rectangle section, with dimensions dx and dy. We also assume that one of its edges is moving with a velocity equal to u, while the other one with $u + du$ as the velocity is a function of y. Due to the relative motion between the two edges AB and CD, the new position of the $ABCD$ rectangle after dt time will be $A'B'C'D'$, as shown in Figure 1.7. Due to this difference in velocity, the rectangle is deformed, while the meter of this deformation constitutes the small angle $d\varphi$.

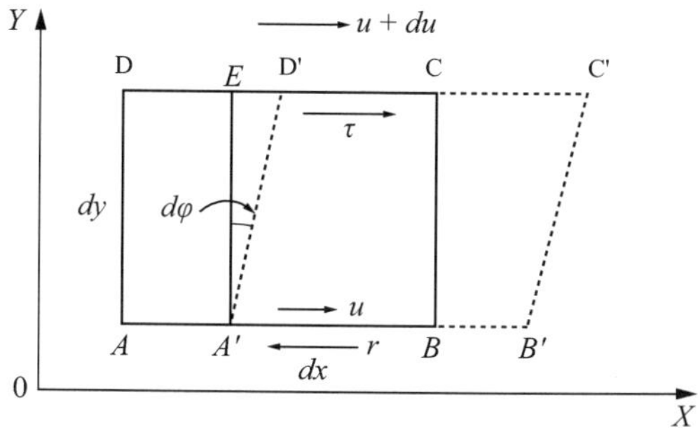

Figure 1.7. *Deformation of the rectangular fluid particle*

The proportionality of the shear stresses (τ) to the deformation velocity variation can be experimentally proved as follows:

$$\tau = \mu \frac{d\varphi}{dt} \quad [1.24]$$

which means:

$$\tau = \mu \frac{du}{dy} \quad [1.25]$$

where μ is the ratio coefficient, called *dynamic viscosity* or *viscosity*.

Equation [1.25] expresses the experimental *Newton's law* of the internal friction or viscosity.

Consequently, the viscosity units in SI are:

$$\mu = \frac{\mathrm{N \cdot s}}{\mathrm{m}^2} \quad [1.26]$$

Moreover, in BS:

$$m = \frac{\ell bz \cdot \mathrm{s}}{\mathrm{ft}^2}. \quad [1.27]$$

as

$$1\,\mathrm{Pa} = 1\,\mathrm{N/m^2}.$$

Viscosity can be expressed in SI units as follows:

$$\mu = \text{Pa} \cdot \text{s}. \tag{1.28}$$

Kinetic viscosity is defined as:

$$\nu = \frac{\mu}{\rho} \tag{1.29}$$

where ν is the kinetic viscosity, μ is the dynamic viscosity and ρ is the fluid density.

The meaning of kinetic viscosity was introduced by Maxwell in 1560. It is used for the incompressible fluid study to simplify the calculations and the fluid equations. The SI unit of kinetic viscosity is m^2/s.

Two popular units of kinetic viscosity are connected by the SI units as follows:

$$1\ \text{Stoke} = \text{cm}^2/\text{s} = \left(\frac{1}{100}\right)^2 \text{m}^2/\text{s} = 10^{-4}\,\text{m}^2/\text{s}. \tag{1.30}$$

$$\text{In addition, } 1\,\text{Poise} = \frac{\text{dyn} \cdot \text{s}}{\text{cm}^2} \tag{1.31}$$

The kinetic viscosity in liquids shows similar dependence on the dynamic viscosity according to the temperature as the density variation is minimum when T is varied. However, in gases, the ν is increased rapidly, due to the μ raise and the reduction of density ρ. In Table 1.2, some kinetic and dynamic viscosity values are presented for water and air.

Temperature T (°C)	**Water**			**Air**		
	ρ	$\mu x\ 10^6$	$\nu x\ 10^6$	ρ	$\mu x\ 10^6$	$\nu x\ 10^6$
–20	–	–	–	0.00270	0.326	1.22
–10	–	–	–	0.00261	0.338	1.30

Table 1.5. *Viscosity values for water and air*

In Table 1.6, various values of kinetic viscosity are provided for T = 20°C.

As already mentioned, the viscosity is influenced by all the other fluid properties such as the pressure and temperature. However, changes in μ and ν depending on the pressure are negligible, except for some extreme pressure changes. The viscosity depends on the temperature (T): *when T is increased, the viscosity of liquids decreases but the viscosity of gases increases.*

Fluid	μ 10^3 kg/ms	ν 10^6 m²/s	Fluid	μ 10^3 kg/ms	ν 10^6 m²/s
Air	0.018	14.5	Mercury	1.56	0.12
Ammonia	0.22	0.36	Kerosene	1.92	2.4
Benzene	0.65	0.74	Ethylene glycol	21.4	19.1
Carbon tetrachloride	0.967	0.61	Oil SAE 10	104.	120
Water	1.00	1.00	Oil SAE 30	290	325
Alcohol	1.20	1.52	Glycerol	1,490	1,182

Table 1.6. *Kinetic viscosity values for various fluids*

Some formulas that are used to calculate the viscosity dependence of the temperature are as follows.

For gases:

$$\mu = \mu_0 + at - \beta t^2 \qquad [1.32]$$

where, for air, the corresponding values are:

$$\mu_0 = 0.000017,\ a = 0.00000056,\ \beta = 1{,}189 \times 10^{-9}. \qquad [1.33]$$

For liquids:

$$\mu = \mu_0 \frac{1}{1 + at - \beta t^2} \qquad [1.34]$$

where μ is the liquid viscosity at T°C in poise, μ_0 is the liquid viscosity at *0*°C in poise and α and β are constants for the specific liquid. For water, these constants are equal to:

$$\mu_0 = 1.79 \times 10^{-3} \text{poise } \mu\varepsilon \; a = 0.03368 \; \kappa\alpha\iota \; \beta = 0.000221 \quad [1.35]$$

Specifically for oils and lubricants (petroleum products), a different clarification is used by the SAE (Society of Automotive Engineers) in the USA, according to the value of the kinetic viscosity at a certain temperature (usually 0°F or 210°F). Assuming that:

$$1\,\text{cSt} = 10^{-2}\,\text{St} = 10^{-6}\,\text{m}^{-2}\,\text{s}10^{-1} = 1\,\text{mm}^2\,\text{s}^{-1} \quad [1.36]$$

various kinetic viscosity values in the SAE are presented in Table 1.7:

SAE	Kinetic Viscosity range in cST
5 W	v_{max} = 869 at 0°F
10 W	1303 < v < 2606 cST at 0°F In the same category, all of the oils with v < 1,303 cST at 0°F and v = 4.18 cST at 210°F are included
20 W	2606 < v < 10423 cST at 0°F In the same category, all of the oils with v < 1,303 cST at 0°F and v > 5.73 cST at 210°F are included
20	5.73 < v < 9.62 cST at 210°F
30	9.62 < v < 12.94 cST at 210°F
40	12.94 < v < 16.77 cST at 210°F
50	16.77 < v < 22.68 cST at 210°F
80	3257 < v < 21711 cST at 210°F In the same category, all of the oils with v < 3,257 cST at 0°F and v > 6.66 cST at 210°F are included
90	14.24 < v < 25.04 cST at 210°F In the same category, all of the oils with v > 25.04 cST at 210°F and v < 162,900 cST at 0°F are included
140	25.04 < v < 42.7 cST at 210°F
250	v > 42.7 cST at 210°F

Table 1.7. *Kinetic viscosity SAE values*

The kinetic viscosity of oils is often determined according to the Saybolt scale in seconds. Consequently, according to the SAE indication, the Saybolt values in seconds (s) are:

SAE	Saybolt (s)	Temperature (°C)
10	90–120	130°F
20	120–185	-II-
30	185–255	-II-
40	>255	-II-
40	<80	210°F
50	80–105	-II-
60	105–125	-II-
70	125–150	-II--II-

Table 1.8. *Kinetic viscosity Saybolt values*

The *viscosity index (VI)* indicates the grade of fluidity of each oil at a certain temperature range. It expresses the relative variation in kinetic viscosity according to the temperature which depends on the composition of the oil mixture. For example, the oils based on paraffin often have low rates of viscosity change, and hence their VI value is set as 100. On the contrary, the oils based on naphtha have high rates of viscosity change, and hence their VI value is fixed as 0.0. Consequently, the VI of an oil X is defined according to the above two values using the following formula:

$$V.I = \frac{S_{100/0} - SX_{100}}{S_{100/0} - S_{100/100}} \qquad [1.37]$$

where:

VI is the viscosity indicator of oil X;

SX_{100} is the Saybolt indications in seconds, at 100°F, for the kinetic viscosity of X;

$S_{100/0}$ is the Saybolt indication in seconds for the kinetic viscosity of oil A at 100°F, which has *VI* = 0, but at 210°F has the same VI with oil X;

$S_{100/100}$ is the Saybolt indication in seconds for the kinetic viscosity of oil B in 100°F,which has *VI* = 100, but at 210°F has the same VI with oil X.

From equation [1.79], it can be observed that the lower the VI of an oil, the higher the rate of change in kinetic viscosity for every temperature change.

1.3.7. *Specific volume*

Density is defined as the mass per unit volume. The reversible size is quite often used in Fluid Mechanics. We define as Specific Volume *U* of a fluid the volume per unit mass, which is:

$$U = \frac{1}{\rho} \tag{1.38}$$

Applying the state equation, we obtain:

$$p = \rho RT = \frac{1}{U} RT \tag{1.39}$$

or

$$pU = RT \tag{1.40}$$

The SI unit for the specific volume is m^3/kg.

1.4. Surface tension

In every liquid molecule, cohesive forces are exerted by the neighboring ones while their resultant is equal to zero (0). However, if the molecules are located at the free surface of the liquid, these forces are only exerted by the ones which are below them and, in this case, the resultant force is not equal to zero (0). In addition, the interfaces between liquids and gases, gases and solids or liquids and solids behave as a film under tensile. *Consequently, surface tension is defined as the force per length unit, which is applied on the fluid surface in a vertical direction.* We may also say that the surface tension is the work which is demanded in order for new molecules to be transferred from the interior to the fluid surface per unit area.

Let us assume a film of soap solute ABCD and a force F, which is exerted vertical to the B, and that the film area is expanded to B'C'. If δW is the work that has been consumed for the area expansion, then (Figure 1.8):

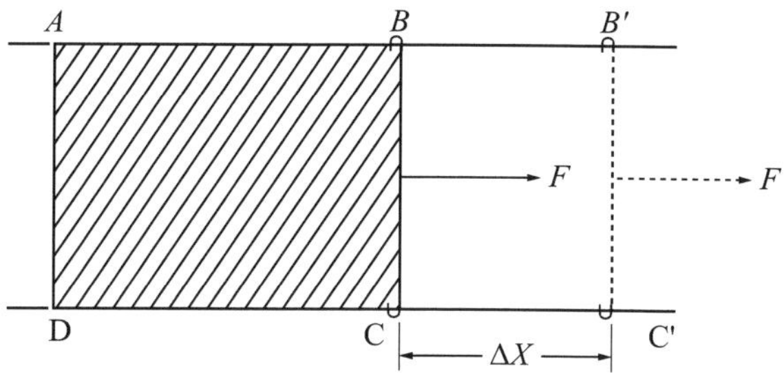

Figure 1.8. *Fluid's area extension*

$$\delta W = F\delta x \qquad [1.41]$$

If $\delta S = l\delta x$ is the film area variation for each edge, then $2\delta S = 2l\delta x$ will be the total area variation and if σ is the surface tension and $l = AD = BC = B'C'$, then:

$$\sigma = \frac{\delta W}{2\delta S\ell} = \frac{\delta W}{2\ell\delta x} = \frac{F}{2\ell} \qquad [1.42]$$

Therefore,

$$\sigma = \frac{F}{2\ell} \qquad [1.43]$$

1.5. Surface tension applications

Water drop: let us assume a water drop as shown in Figure 1.9.

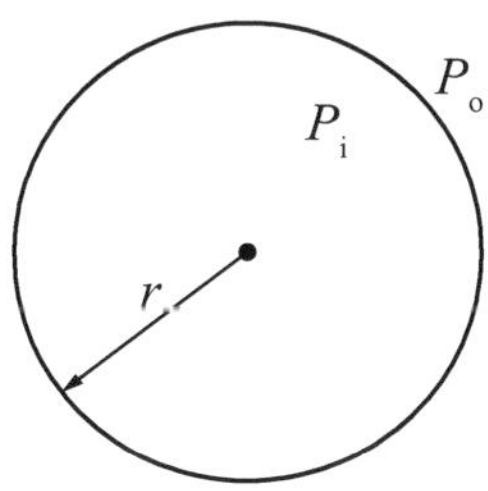

Figure 1.9. *Water drop*

If p_i is the pressure inside the drop, p_0 the pressure outside of the drop and r the drop radius, we may write:

$$p_i - p_0 = \frac{2\sigma}{r} \qquad [1.44]$$

where $p_i > p_0$ and σ is the surface tension.

Liquid bubble: let us assume a bubble of a soap solute (Figure 1.10), which presents two interfaces: (a) air–solute and (b) solute–air.

a) For the first interface, we may write:

$$p_i - p_0 = \frac{2\sigma}{r_1} \qquad [1.45]$$

where r_1 is the radius of the area.

b) For the second interface, we apply (Figure 1.10):

$$p_i - p_0 = \frac{2\sigma}{r_2} \qquad [1.46]$$

where r_2 is the radius of the area and all the other symbols are defined as above.

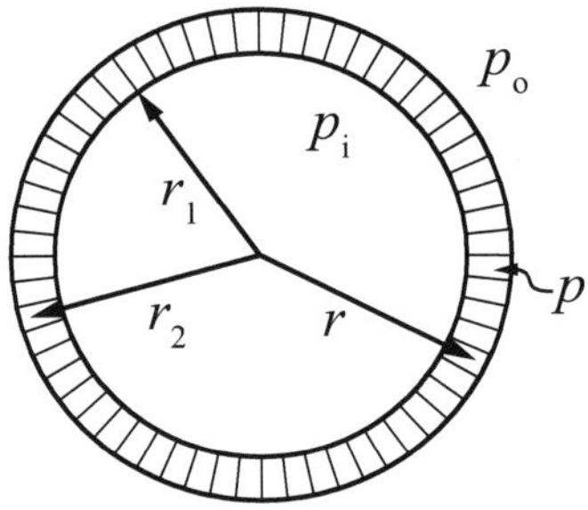

Figure 1.10. *Soap solute bubble*

Using equations [1.42] and [1.44], we obtain:

$$p_i - p + p - p_0 = \frac{2\sigma}{r_1} + \frac{2\sigma}{r_2} \qquad [1.47]$$

In addition, due to $r_1 \approx r_2 \approx r$, equation [1.45] becomes:

$$p_i - p_0 = \frac{4\sigma}{r} \qquad [1.48]$$

The surface tension units are $\text{CGS:1 dyn/cm} = 1\,\text{erg/cm}^2$.

The surface tension phenomenon is fundamental concerning a fluid's flow through a small bore under a drop's shape. When the liquid is exerted by a bore, as, for example, chiffon's bore, it is held at the lower edge, if its quantity is small enough, due to the surface tension development. When its quantity is increased, it is detached and falls as a drop. It is obvious that the size of the drops depend on the surface tension value. The higher the surface tension, the greater the size of the liquid drops.

1.6. Capillarity effect

In the case of fluid–solid contact, it is observed that the liquid sometimes drenches the solid and sometimes does not. In both cases, the free liquid surface does not remain horizontal and flat, but becomes a curve. These phenomena are called capillarity effects, as they were first observed in low-diameter capillary tubes and develop due to the shear forces between liquid and solid molecules.

The capillarity effect is depicted in Figure 1.11, where θ is the contact angle between the liquid and the solid, which is formed due to the tangent line at the contact point and the tube's wall.

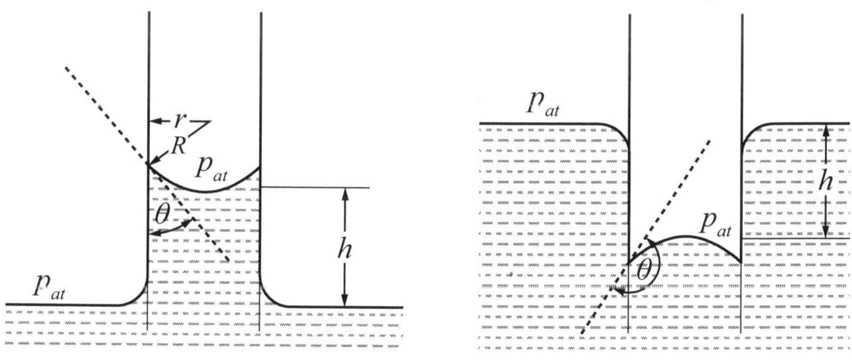

Figure 1.11. *Capillarity effect*

The angle θ can be acute or obtuse according to the position of liquids (Figure 1.11). The height h can be calculated using the following formula:

$$h = \frac{2\sigma \cos\theta}{\rho \cdot g \cdot r} \quad [1.49]$$

where h is the capillary rise or drop, σ is the surface tension, θ is the contact angle, ρ is the liquid density, g is the acceleration of gravity and r is the tube radius.

1.7. Newtonian and non-Newtonian fluids

The fluids that obey Newton's law, which is expressed as:

$$\mu = \frac{\tau}{\frac{du}{dy}} \quad [1.50]$$

are called Newtonian fluids. Two more categories of fluids that do not obey Newton's law are ideal fluids (or Pascal's fluids) and non-Newtonian fluids.

Ideal fluids, where $\mu = 0$, have developed no resistance to flow under any shear stresses.

Newtonian fluids obey Newton's law and μ has various values according to the fluidity of each fluid.

Non-Newtonian fluids obey a nonlinear relationship between the shear stresses τ and the deformation rate $\frac{du}{dy}$ and, consequently, do not obey Newton's law. In general, for the non-Newtonian fluids, the following is applied:

$$\mu = f\left(\frac{du}{dy}\right) \quad [1.51]$$

which means that the viscosity is a function of the deformation rate.

In the case of non-Newtonian fluids, the viscosity μ depends not only on the fluid's molecule forces, but also on the shear stresses and the shear velocity γ as given by the following formula:

$$\dot{\gamma} = \frac{du}{dy} \quad [1.52]$$

which means:

$$\dot{\mu} = \mu\left(\tau, \frac{du}{dy}\right) \quad [1.53]$$

or:

$$\dot{\mu} = \mu(\tau, \dot{\gamma}) \quad [1.54]$$

where all the symbols are defined as above.

Two popular fluid models for the non-Newtonian fluids are Bingham and Ostwald–de Waele models.

The Bingham model is related to various pastes (such as toothpaste), which describes the fluids behavior when, for certain values of shear stresses τ_{yx} less than the fluid stress value τ_0, they behave as solids. The relationship for τ_{yx} is:

If $|\tau_{yx}| > \tau_0$, then

$$\tau_{yx} = \pm\tau_0 - \mu\frac{du}{dy}. \quad [1.55]$$

However, if $|\tau_{yx}| < \tau_0$, then

$$\frac{du}{dy} = 0. \quad [1.56]$$

The Ostwald model is expressed by the following formula:

$$\tau_{yx} = \hat{\mu}\frac{du}{dy} \quad [1.57]$$

where

$$\bar{\mu} = \mu_1 \left| \frac{du}{dy} \right|^{n-1} \qquad [1.58]$$

and is called *apparent viscosity*.

Fluids composition (**% of weight**)	$\mu_1\,(\frac{Ns^n}{m^2})$	N (–)
33%, Hydrated Lime	7.182	0.171
4%, Hydrated Paper pulp	20.01	0.575
54%, Cement in water	2.509	0.153
0.67%, Hydrated Cellulose	0.304	0.716
1.5%, Hydrated Cellulose	3.127	0.554
3%, Hydrated Cellulose	9.288	0.566

Table 1.9. *Apparent viscosity values*

The viscosity μ_1 only depends on the molecules' properties and if $n = 1$ and $\mu_1 = \mu$. The fluid is Newtonian. The exponential Ostwald–de Waele flow law can be applied in various fluids according to the n value. In Table 1.9, some μ_1 and n values for fluids can be found.

The relationship between shear stresses and shear velocity is depicted in Figure 1.12 for Newtonian and non-Newtonian groups of fluids.

The Bingham line inclination is quite similar to the Newtonian fluid's one. The Ostwald–de Waele relationship is divided into pseudo-plastics ($n < 1$) and broadly fluids ($n > 1$). As shown in Figure 1.12, the broad fluids present solid intention in high-inclination angles of fluid velocity. Some examples of these fluids are the solid-concentrated dispersions in fluids, as the sand, the quartz and the metallic oxides in water. In pseudo-plastics, the shear stresses are reduced if the shear velocity is increased. In the case of bearing lubrication, the lubricant is assumed to be pseudo-plastic after the addition of some organic substances such as colloid polymers (their molecules comprise 10^3–10^9 atoms).

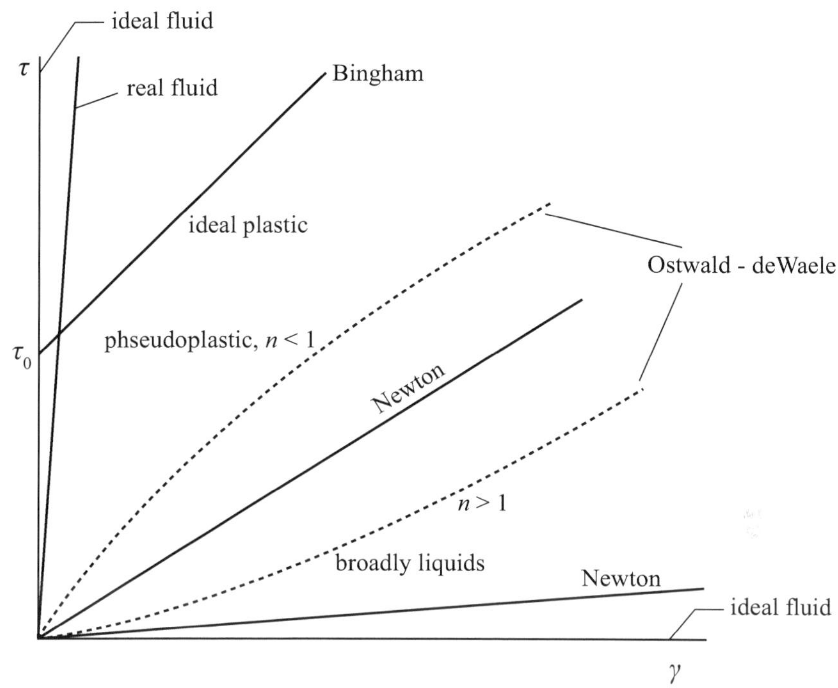

Figure 1.12. *Shear stresses and shear velocity*

Last but not least, one more difference between Newtonian and non-Newtonian fluids is related to the *relaxation time* t_r. It is defined as the time needed for a fluid reaction when the fluid is suddenly stressed under shear forces. The relation time for the Newtonian fluids is much shorter than the corresponding one in non-Newtonian fluids.

1.8. Vapor pressure

The liquid molecules abandon the free surface of the liquid, and create a cloud covering all the surface. This cloud is increased or decreased according to the molecules' motion direction (Figure 1.13). The pressure of this cloud is called *vapor pressure*.

If the space above the free surface is limited, the vapor pressure value varies according to the temperature. At the time the vapor pressure is equal to the free surface pressure, the liquid boils even in ambient temperatures. In Figure 1.14, the vapor pressure is depicted against the water temperature.

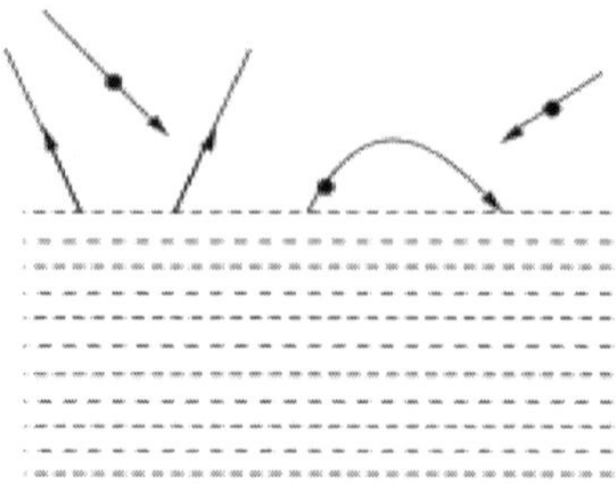

Figure 1.13. *Vapor pressure*

The partial vapor pressure p_v in water depends on the temperature and the relative humidity as well. The vapor pressure can be calculated as follows:

$$pv(\theta_a) = 6.05 \cdot 10^{-4} \cdot RH \cdot \{[(\theta_a + 7.066) \cdot \theta_a + 908.88] \cdot \theta_\alpha + 9567\}\ \mathrm{Nm}^{-2} \quad [1.59]$$

where RH is the relative humidity in precedence (%), θ_a is the air temperature and p_v is the partial vapor pressure in N/m^2.

In Figure 1.15, the relationship between the partial vapor pressure and the temperature is presented for three different values of the relative humidity: 40%, 50% and 60%.

If during a liquid movement inside pipes or under any internal flow, the liquid pressure is equal or lower than the vapor pressure, then air bubbles start to appear. This phenomenon is called *cavitation*, which can cause high damage in turbines, pump blades or other similar applications.

1.9. Cavitation

In water flows with a high velocity, low instant pressure values are developed, so at this point, the lower instant pressure is lower than the vapor pressure. This occurs only for this point. Therefore, the water is transformed in a gas state and small air bubbles are developed. If these air bubbles are moved towards a higher pressure area, they get transformed into a gas state and instantly release very high pressure, which are responsible for the solid boundary damage even if the material is characterized by high strength. This phenomenon is called cavitation. Rayleigh (1916) was the first to study and analyze the meaning and importance of cavitation in order to explain the damages on ship propellers.

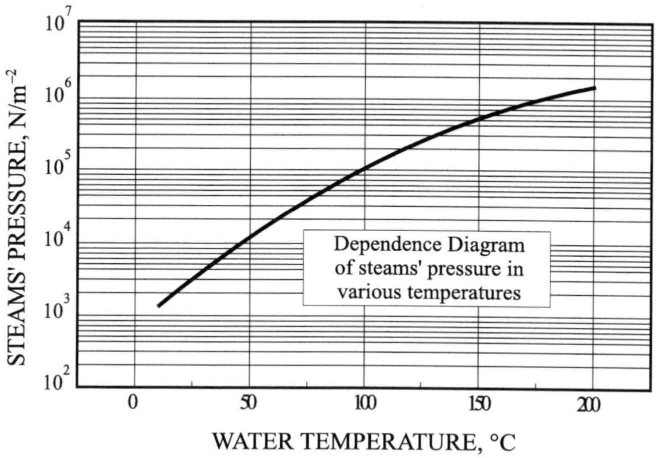

Figure 1.14. *Vapor pressure diagram for water*

Cavitation is developed in water pumps, in ship propellers and in high-speed transportation pipes. Recently, the cavitation effect has been applied in various workshops (such as cement cutting), where a very thin line with high speed is collided to the solid body surface which is for the cutting process. In addition, a scientific effort has been made to succeed nuclear fusion caused by the cavitation effect in the water when the sound wave is produced inside it. In the barrier spillways, the cavitation effect occurs, where the water falls to the free surface up to 300 m and its velocity is more than 20 m/s, creating low pressures and multiple air bubbles of cavitation.

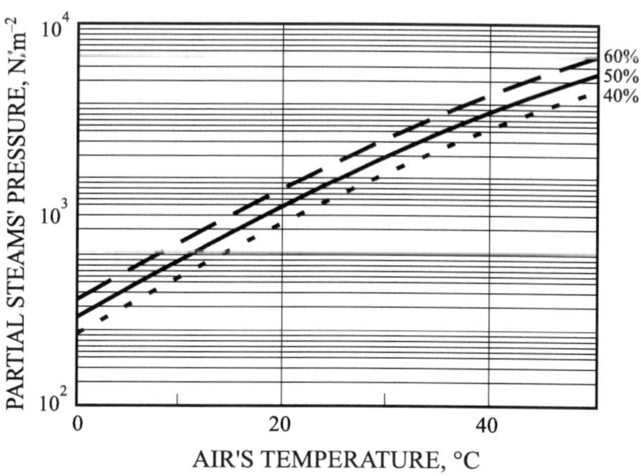

Figure 1.15. *Partial vapor pressure according to the relative humidity of the air*

It is worth mentioning that the negative pressure value development is not necessary in order for the cavitation to be produced. For example, if we assume water flow inside a straight pipe with velocity equal to 17 m/s and with a depth of 3 m, cavitation is developed very close to the bottom (1.6 mm). In this case, the pressure values are positive.

Obviously, the damage is progressive and not instantaneous. However, according to the applications, various regulations have to be applied to avoid the cavitation effect. For example, the smooth surface of a weir is designed to delay the cavitation effect.

The possibility of the cavitation effect to occur can be estimated by calculating the cavitation number c_a using the following formula:

$$c_a = \frac{2(p_0 - p_v)}{\rho V_m^2} \qquad [1.60]$$

where:

– p_0 is the absolute pressure on the solid bound. It applies that $p_0 = p_a + p_h$, with p_h being the average static pressure and p_a being the atmospheric pressure;

– p_v is the vapor pressure;

– ρ is the density of water;

– g is the acceleration of gravity;

– V is the local velocity.

1.10. Formulae

1) Density, ρ:

$$\rho = \frac{m}{\mathrm{v}} (\mathrm{kg/m^3})$$

where m is the mass and v is the volume.

2) Relative density, RD:

$$RD = \frac{\rho}{\rho_f}$$

where ρ is the density of the fluid and ρ_f is the density of the reference fluid.

i)

$$RD(60) = \frac{141.5}{131.5 + API}$$

where API = API (American Petroleum Institute) scale values.

ii)

$$RD(T) = \frac{\rho_D(60)}{1 + \alpha(T - 60)}$$

iii)

$$RD(60) = \frac{Tw}{200} + 1$$

where T_w = Twaddell scale values at 60°F.

3) Mixture density, ρ_m:

$$\rho = \frac{m}{\mathrm{V}} = \frac{\rho_1 n_1 + \rho_2 n_2 + \ldots + \rho_i n_i}{n_1 n_2 + \ldots + n_i}$$

where ρ_i are the densities of the mixture components n_i.

4) Density of dry air, ρ_v:

$$\rho_v = \rho(1 - 0.377\varphi = \frac{p_d}{p}$$

where ρ is the density of dry air, φ is the relative humidity, p is the pressure and p_a is the saturation pressure of water.

5) Specific weight, γ:

$$\gamma = \frac{dw}{dv}$$

where w is the weight of the body and v is the volume.

6) Specific weight and density:

$$\gamma = \rho \cdot g$$

where ρ is the density of the fluid and g is the acceleration of gravity.

7) Static pressure, p:

$$p = \frac{F}{S} (\mathrm{N/m^2})$$

where F is the force and S is the surface.

8) Dynamic pressure:

$$p_d = \frac{1}{2}\rho V^2$$

where p_d is the dynamic pressure, ρ is the density of the fluid and V is the velocity.

9) Total pressure, p_{tot}:

$$p_{tot} = p + p_d$$

where p is the static pressure and p_d is the dynamic pressure.

10) Mach number, M:

$$M = \frac{V}{\alpha}$$

where V is the fluid velocity and a is the speed of sound.

11) Viscosity, μ:

$$\tau = \mu = \frac{\partial u}{\partial y} \quad (m = Kgr / \mathrm{m} \cdot \mathrm{s})$$

where τ is the shear stress and $\frac{\vartheta u}{\vartheta y}$ is the variation in the fluid's deformation velocity.

12) Kinetic viscosity, v:

$$v = \frac{\mu}{\rho}$$

where v is the kinetic viscosity, μ is the dynamic viscosity and ρ is the fluid density.

13) State equation of ideal gases:

$$p = \rho RT$$

where p is the pressure, ρ is the density and R is the gas constant.

14) Specific volume, U:

$$U = \frac{1}{\rho}(\mathrm{m}^3 / \mathrm{kg})$$

where ρ is the density of the fluid.

15) Sound speed, a:

$$a = \sqrt{\gamma RT}$$

where $\gamma = C_p / C_v$ is the specific heat under constant pressure and volume, respectively, R is the gas constant and T is the temperature.

16) Compressibility:

$$\Delta p = -E\frac{\Delta_v}{V_c}$$

where E is the bulk modulus, Δ_v / v_v is the relative volume difference and Δ_p is the pressure difference.

17) Viscosity variation according to the temperature:

i) Gases

$$\mu = \mu_0 + \alpha t - \beta t^2$$

where μ_0 is the gas viscosity at 0°C in poise, t is the temperature in poise and a and β are gas constants.

ii) Liquids

$$\mu = \mu_0 = \left(\frac{1}{1+\alpha t+\beta t^2}\right)$$

where μ is the liquid's viscosity in poise at $t°C$, μ_0 is the liquid viscosity in poise at 0°C, *a* and *β* are liquid constants and t is the temperature in °C.

18) Viscosity index:

$$V.I. = \frac{S_{100/0} - SX_{100}}{S_{100/0} - S_{100/100}}$$

where SX_{100}, $S_{100/0}$ and $S_{100/100}$ are the Saybolt viscometer indications for the X oil.

19) Surface tension:

$$\sigma = \frac{F}{2\ell}$$

where F is the force and l is the increase in length of the surface side.

20) Shear velocity:

$$\dot{\gamma} = \frac{du}{dy}$$

where $\frac{du}{dy}$ is the angular deformation according to the time unit.

21) Shear stresses in non-Newtonian fluids (Bingham model):

$$\tau_{yx} = \pm \tau_0 - \mu \frac{du}{dy}$$

where τ_0 is the flow tense, m is the viscosity and $\frac{du}{dy}$ is the angular deformation according to the time unit.

22) Shear stresses in non-Newtonian fluids (Ostwald model):

$$\tau_{yx} = \hat{\mu} \frac{du}{dy}$$

where $\hat{\mu}$ is the apparent viscosity and μ_1 is the viscosity depending on the molecular fluid forces. If $n = 1$ and $\mu = 1$, the fluid is Newtonian.

23) Internal energy (Boltzmann theory):

$$E = \frac{1}{2} K T$$

where $K = 1.3805 \cdot 10^{-25}$ J/K and T is the absolute temperature.

24) Thermodynamic processes for gases:

i) T = const

$$p_1 \mathrm{v}_1 = p_2 \mathrm{v}_2$$

ii) ρ = const

$$\frac{p_1}{T_1} = \frac{p_2}{T_2}$$

iii) p = const

$$\rho_1 T_1 = \rho_2 T_2$$

where ρ is the density, p is the absolute pressure and T is the absolute temperature.

iv) Adiabatic process

$$\frac{T_1}{T_2} = \left(\frac{\mathrm{v}_2}{\mathrm{v}_1}\right)^{\gamma-1}$$

and

$$\frac{T_1}{T_2} = \left(\frac{p_1}{p_2}\right)^{\frac{\gamma-1}{\gamma}}$$

where $\gamma = c_p/c_v$, with c_p,c_v being the specific heat under constant pressure and volume, respectively.

25) Cavitation number:

$$c_\alpha = \frac{2(p_o - P_v)}{\rho V_m^2}$$

where p_o is the absolute pressure on the solid bound, with $p_o = p_\alpha + p_n$, where p_n is the average static pressure and p_α is the atmospheric pressure, p_v is the vapor pressure, ρ is the density of the water and V is the local velocity.

1.11. Questions

1) How are fluids defined?

2) How are incompressible and compressible fluids defined?

3) How is the density of fluids defined and on what does it depend?

4) How many measurement scales of density do we use in the chemical oil industry and which ones?

5) How do temperature differences affect the density of fluids?

6) How do pressure differences affect the density of fluids?

7) How is the relative density of fluids defined?

8) Under certain humidity values, how can the density of air be calculated?

9) How is the specific weight of fluids defined?

10) How are dynamic pressure and total pressure defined?

11) How is the compressibility of fluids defined?

12) Prove the relationship between the pressure and density difference based on the continuity equation.

13) How is viscosity defined in fluids? What are the SI units of viscosity?

14) How is kinetic viscosity defined and what are its SI units?

15) How does viscosity depend on temperature in liquids and gases? Explain its dependence.

16) What is a viscosity index and where is it used mostly?

17) How is specific volume defined and what are its SI units?

18) Explain the capillarity effect.

19) What are Newtonian and non-Newtonian fluids?

20) How is the shear velocity of a non-Newtonian fluid defined?

21) Depict the qualitative comparison diagram of a Newtonian and a non-Newtonian fluid.

22) What is the relaxation time between Newtonian and non-Newtonian fluids?

23) How is vapor pressure defined?

24) Describe the cavitation phenomenon and explain the parameters that affect this mechanism.

1.12. Problems with solutions

1) The atmospheric pressure π and the density of air at a point of a Boeing 747 wing are 1.10×10^5 N/m^2 and 1.20 kg/m^3, respectively. What is the temperature at that point?

Solution:

From the state equation, we obtain:

$$T = \frac{p}{R \cdot \rho} = \frac{1.10\times10^5\,\text{N/m}^2}{(1.20\,\text{kg.m}^3)[287\,j(\text{kg})(K)]} = 319\,\text{K}.$$

2) The high pressure air chamber of a supersonic tunnel has a volume equal to 1,000 ft^3. If the pressure of the air is 30 atm and the temperature 530 R, what is the air mass in slugs and lb_m?

Solution:

As we know:

$$\rho = \frac{p}{R \cdot T}$$

If we substitute, we estimate the air density to be:

$$\rho = 0.0698\,\text{slug}/\text{ft}^3.$$

Consequently, the air mass is given by:

$$m = \rho \cdot v$$

$$m = 69.8 \text{ slugs}$$

$$m = 2{,}248 \text{ lb}_\text{m}.$$

3) For what temperature values in Kelvin and Rankin do two Fahrenheit and Celsius thermometers indicate the same temperature value?

Solution:

$$°F = °\text{C}$$

$$°F = 1.8°\text{C} + 32.$$

$$°F = °\text{C} = -40.$$

$$K = °\text{C} + 273.$$

$$K = 233$$

$$R = °\text{F} + 460$$

$$R = 420$$

4) A supersonic tunnel is operated due to six air chambers. The total volume of these tanks is 153 m^3 and the pressure is equal to 4,137,000 Pa at 38°C. Estimate the total air mass inside the air chambers.

Solution:

$$p = \rho RT \Leftrightarrow \rho = \frac{p}{R \cdot T} \Leftrightarrow \frac{4{,}137{,}000}{287 \cdot 311}\, Kgr/\text{m}^3 \Leftrightarrow \rho = 46.35\, Kgr/\text{m}^3$$

Consequently, the mass of air will be:

$$m = \rho v = 7{,}091.4 \text{ kg}$$

5) The volume of a high-pressure tank is equal to 1,000 ft^3. If the pressure of the air is 30 atm at 530 R, what is the mass value of the air in slugs and lb_m?

Solution:

We know that:

$$1 \text{ atm} = 2{,}116 \text{ lb/ft}^2.$$

Consequently,

$$p = (30)\,(2{,}116) \text{ lb/ft}^2 = 6{,}348 \times 10^4 \text{ lb/ft}^2$$

As we know that:

$$p = \rho RT \Rightarrow \rho = p / RT$$

The density of air will be:

$$\rho = \frac{6,348 \times 10^4 \text{ lb / ft}^2}{1,716 \text{ ft J b /(slug)(R)(530 R)}} 6.98 \times 10^{-2} \, slug / \text{ft}^3$$

Finally, the total mass is:

$$M = \rho \cdot \text{v} = (6.98 \times 10^{-2} \text{ slug / ft}^3)(1,000 \text{ ft}^3) = 69.8 \text{ slug}$$

As 1 slug = 32.2 lb_m.

The mass in lb_m is:

$$m = (6.98) \cdot (32.2) = 2.248 \text{ lb}_\text{m}.$$

6) Air flow of high velocity in a wind tunnel has 0.3 atm pressure and a temperature of $-100°\text{C}$. What is the air density and the specific volume?

Solution:

$$1 \text{ atm} = 1,01 \times 10^5 \text{ N/m}^2$$

Thus, $p = (0.3)(1.01 \times 10^5) = 0.303 \times 10^5 \text{ N/m}^2$.

The temperature –100°C is not the absolute temperature.

Thus, T = –100 + 273 = 173 K

Since $\rho = p / RT \Rightarrow \rho = \dfrac{0.303 \times 10^5 \text{ N/m}^2}{[287 \text{ J/kg} \cdot \text{K}](173 \text{ K})} = 0.610 \text{ kg/m}^3$

$$U = \frac{1}{\rho} = \frac{1}{0.610} = 164 \text{ m}^3\text{/kg}$$

NOTE.– It is worth reminding that:

$$1 \text{ atm} = 2,116 \text{ lb / ft}^2 \text{ and } 1 \text{ atm} = 1.01 \times 10^5 \text{ N/m}^2$$

7) The tank of the super-compressor of a supersonic wind tunnel has a volume v = 28.3168 m^3. If the air pressure of the tank is p = 30 atm and the temperature is T = 299.44 K, what is its mass?

Solution:

The air mass m is given by the relation:

$$\rho = \frac{m}{\mathrm{v}} \Rightarrow m = \rho \mathrm{v}. \tag{1}$$

The density ρ is given by the relation:

$$\rho = \rho RT \Rightarrow \rho = \frac{p}{RT}. \tag{2}$$

Converting atm into N/m^2: we have that 1 atm = 1.01×10^5 N/m^2; thus,

$$p = (0.3)(1.01 \times 10^5) = 0.303 \times 10^5\,\mathrm{N/m^2}.$$

Replacing in (2) we have:

$$\rho = \frac{30.3 \times 10^5\,\mathrm{N/m^2}}{\left[287 \dfrac{J}{\mathrm{kgK}}\right] \cdot 2{,}940.44\,\mathrm{K}} = 35.85\,\mathrm{kg/m^3}.$$

Replacing the (1) and we have: m = 35.85 kg/m^3·28.3168 m^3 = 101,667 kg.

8) Two homo-axial cylinders have a large length L, with regard to their radii R_1 and R_2. The space between them is filled with fluid of cohesion μ. The internal cylinder remains steady, while the external rotates with constant moment M. Calculate the angular velocity ω of the external cylinder.

Solution:

The velocity of the fluid will be approximately a function of the radius r from the axis of the cylinders, to the surface of the external cylinder. Therefore, it will be $V = \omega R_2$. We use the relation

$$\tau = \mu \frac{\partial V}{\partial r},$$

where the shear stress τ is found from the known relation

$$\tau = \frac{F}{S} = \frac{M}{r(2\pi r L)}.$$

When the force $F = M/r$ is applied on the surface $2\pi rL$, combining these relations, we have:

$$\frac{M}{2\pi r^2 L} = \mu \cdot \frac{\partial V}{\partial r}.$$

The integration of the last one takes place easily, that is:

$$\frac{M}{2\pi L}\int_{R_1}^{r}\frac{dr}{r^2} = \mu\int^{v}\cdot\partial V \quad \text{or} \quad \frac{M}{2\pi L}\left(\frac{1}{R_1}-\frac{1}{r}\right) = \mu V$$

Therefore, the linear velocity of the rotating fluid is

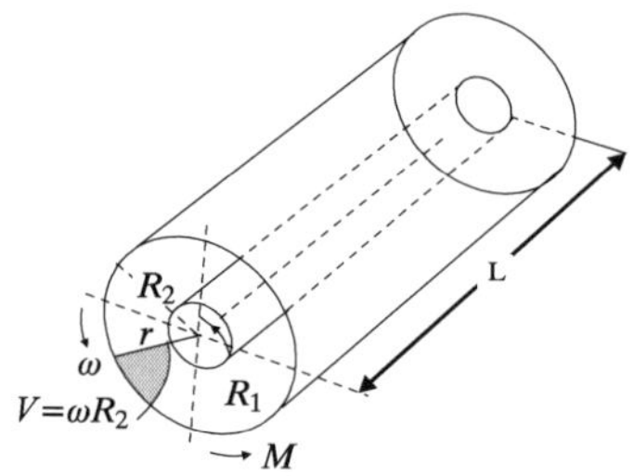

$$V(r) = \frac{M}{2\pi\mu L}\left(\frac{1}{R_1}-\frac{1}{r}\right),$$

and the angular velocity of the external cylinder is

$$\omega = \frac{M}{2\pi\mu L R_2}\left(\frac{1}{R_1}-\frac{1}{r}\right),$$

since we have $V(R_2) = \omega R_2$.

9) A flat flagstone moves on another at a velocity of 30 m/s, with oil between them. The thickness of the oil between them is D = 3 mm and the oil has a density of 0.75 kg/m^3 and a viscosity (μ) of 0.7×10^{-9} kg/m s. Calculate:

i) the kinematic viscosity;

ii) the shear stress on the top and bottom of the flagstone;

iii) the directions of these stresses.

Solution:

i) The kinematic viscosity is

$$v = \frac{\mu}{\rho} \frac{0.7 \times 10^{-3}}{0.75} = 0.93 \times 10^{-3}\,\mathrm{m^2/s}$$

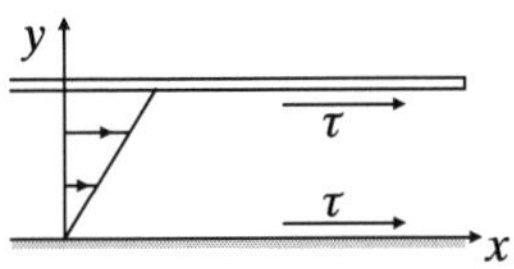

ii) The shear stress on the above flagstone is

$$\tau = \mu \frac{\partial u}{\partial y}.$$

However, from the similarity of the triangles, it is $\partial u / \partial y = u / D$, so:

$$\tau = \mu \frac{V}{D} = 0.7 \times 10^{-3} \times \frac{0.3}{0.003}\,\mathrm{N/m^2} = 0.07\,\mathrm{N/m^2}$$

The shear stress on the bottom flagstone is equal to a quantity with 0.07 $\mathrm{N/m^2}$.

iii) The directions of the shear stresses are shown in the figure.

10) Calculate the variation of the volume 1 $\mathrm{m^3}$ of water in normal conditions, when the pressure is increased by $100 \dfrac{Kp}{\mathrm{cm}^2}$ psi when the elasticity constant is *E*=300,000 psi.

Solution:

From the expression [1.31], the variation of the pressure due to the compressibility is

$$\Delta p = -E \frac{\Delta}{\mathrm{v}_0} \Rightarrow$$

$$\Rightarrow \Delta p, \mathrm{v}_0 = -E \Delta \mathrm{v} \Rightarrow \Delta r = -\frac{\Delta p}{E} =$$

$$= -\frac{1\,\text{m}^3 \times 100\,\text{Kp/cm}^2}{\frac{300,000}{14.7}\text{Kp/cm}^2} = -\frac{100 \times 14.7}{300,000}\text{m}^3 =$$

$$= \frac{-1.47 \times 10^{-3}}{3 \times 10^5}\text{m}^3 = -4.9 \times 10^{-3}\,\text{m}^3 \text{ or } 4{,}900\text{ cm}^3$$

thus the volume became smaller by $4.9 \times 10^{-3}\,\text{m}^3$ or 4,900 cm^3.

11) Calculate the density of air (m_{mol} = 29) at 60°F and at an absolute pressure of 180 lb-f/in^2 in the FPS system.

Solution:

We first calculate the value of the constant *R* in standard conditions (*P* = 1 atm and *T* = 0°C) using the FPS system.

$$P = 1\,\text{atm} = 1.013 \times 10^6\text{ dynes/cm}^2 = 1.013 \times 10^6 \times 10^{-5}\,N / \left(\frac{1}{30.48}\text{ft}\right)^2 =$$

$$= \frac{1.013 \times 10^6 \times 10^{-5}}{(1/30.48)^2} \times \frac{1}{0.1393}\text{Pounda l/ft}^2 =$$

$$= \frac{10.13}{0.1383} \times (30.48)^2\,\text{Pounda l/ft}^2 = 68{,}048.285\,\text{Pounda ls/ft}^2$$

$$\rho = 0.9\,\text{gr/cm}^3 = 0.9 \times \frac{1}{453.6}\text{lb} / \left(\frac{1}{28{,}316.8}\right)\text{ft}^3 = 56.18\,\text{lb/ft}^3$$

so the kinematic viscosity is

$$v = \frac{\mu}{\rho} = \frac{0.6045\text{ ps/ft}^2}{56.18\text{ lb/ft}^3} = 0.01076\text{p.ft.s/lb} = 0.01076\,\text{ft}^2/\text{s}.$$

12) A fluid fills the space between two parallel flat flagstones that refrain from each other at a distance *y* = 0.1 ft. What is the viscosity μ of the fluid if for the transportation of a flagstone at constant velocity *u* = 0.1 ft/s (when the other one remains still) a force *F* = 27.64 dynes is needed? The moving flagstone has a surface area *S* = 1 ft^2.

Solution:

As already known, in this case, the viscosity is calculated from the relation

$$\tau = \mu \frac{du}{dy} \quad \text{or} \quad \frac{F}{S} = \mu \frac{u}{y} \tag{1}$$

since the distance between the two flagstones is small and the μ alters linearly after the y.

From (1), we have:

$$\mu = \frac{F \cdot y}{u \cdot S} = \frac{27.64 \times 0.1}{0.1 \times 1} \text{dynes.s/ ft}^2 = \frac{27.64\,\text{dynes.s}}{(30.48\,\text{cm}^2)} = 0.0298\,\text{poise}$$

13) The space between two homo-axial cylinders of height h = 2 ft and radii r_i = 0.4 ft and r_0 = 0.5 ft is filled with a fluid. What is the viscosity of this fluid if, for the rotation of the internal cylinder with a frequency v equal to 60 rot./min, the momentum required is equal to 0.463(lb-f) ft?

Solution:

The 0°C corresponds to 32°F (Fahrenheit) and the molar volume is:

$$\text{v}_{mol} = 22.4 \,{}^{\text{lit}}\!/_{\text{mol}} = 22,400\,\text{cm}^3/\text{mol} =$$

$$= 22,400 \left(\frac{1}{30.48} \text{ft} \right)^3 / \left(\frac{1}{453.6} \right) \text{lb} - \text{mol} = 359\,\text{ft}^3 / \text{lb} - \text{mol}.$$

Therefore, the value of the constant R is

$$R = \frac{P\text{v}_{\text{mol}}}{T} = \frac{68,048.285 \times 359}{32} \frac{P / \text{ft}^2 \times \text{ft}^3 / \text{lb} - \text{mol}}{°\text{F}}$$

$$= 763,416.7 \frac{\text{Pounda ls.ft}}{\text{lb / mol.°F}}.$$

The absolute pressure is

$$P_{\alpha\pi} = 180\,\text{lb} - f / in^2 = 180 \times 32.2\,\text{pounda ls} \left(\frac{1}{12} \text{ft} \right)^2 =$$

$$= 180 \times 32.2 \times 144\,\text{p / ft}^2 = 834,624\,\text{p / ft}^2,$$

and so the density of the air in the given conditions will be:

$$\rho = \frac{P_{\alpha\pi} m_{\text{mol}}}{RT} = \frac{834,524 \times 29}{763,416.7 \times 60°} \text{lb}/\text{ft}^2 = 0.528\,\text{lb}/\text{ft}^3.$$

14) The viscosity μ and the density ρ of an oil is μ = 9 poise and ρ = 0.9 g/cm^3, respectively. What is the kinematic viscosity of the oil in the (i) CGS and (ii) FPS systems?

Solution:

i) As already known, 1 poise = 1 dyne.s/cm^2, so the kinematic viscosity v of the oil is

$$v = \frac{\mu}{\rho} = \frac{9\,\text{dynes.s/cm}^2}{0.9\,\text{gr/cm}^3} = 10\frac{\text{gr cm/s}^2\text{.s/cm}^2}{\text{gr/cm}^3} = 10\,\text{cm}^3/\text{s} = 10\,\text{Stokes}$$

ii) In the FPS system, we have

$$\mu = 9\,\text{dynes.s/cm}^2 = 9 \times 10^{-5}\,\text{N.s}/\left(\frac{1}{30.48}\text{ft}\right)^2 =$$

$$= 9 \times 10 - 5 \times \frac{1}{0.1383}\text{p.s} \times 30.48^2/\text{ft}^2 = 0.6045\,\text{p.s/ft}^2$$

15) If the velocity distribution in a flat surface is given by the relation $u = \frac{2}{3}y - y^2$, where u is the velocity in meters per second and y is the distance in meters from the surface, calculate the shear stress τ for y = 0 and y = 0.15 m. It is given that the viscosity of the fluid is 8.63 poises.

Solution:

Since

$$u = \frac{2}{3}y - y^2 \Rightarrow \frac{du}{dy} = \frac{2}{3} - 2y \quad (1)$$

Therefore, for

$$y = 0 \Rightarrow \left(\frac{du}{dy}\right)_{y-0} = \frac{2}{3} - 2(0) = \frac{2}{3} = 0.667 \quad (2)$$

Also for

$$y = 0.15 \Rightarrow \left(\frac{du}{dy}\right)_{y=0.15} = \frac{2}{3} \cdot 2(0.15) = 0.667 - 0.30 = 0.367 \tag{3}$$

Also, it has been given that $\mu = 8.63$ poise

$$= \frac{8.63}{10} \text{SI units} = 0.863 \text{N/m}^2 \tag{4}$$

However, the shear stress τ is given from the relation

$$\tau = \mu \frac{du}{dy} \tag{5}$$

Thus,

i) for $y = 0$, the shear stress τ is

$$\tau = \mu \left(\frac{du}{dy}\right)_{y=0} = 0.863 \times 0.667 = 0.576 \text{ N/m}^2. \textit{Ans}$$

and

ii) for $y = 0.15$ m, relation (5) with relations (3) and (4) gives

$$\tau_{0.15} = \mu \left(\frac{du}{dy}\right)_{y=0.15} = 0.863 \times 0.367 = 0.3165 \text{ N/m}^2. \textit{Ans.}$$

1.13. Problems to be solved

1) The rotation boost of a vertical axis takes place from a horizontal flagstone of diameter 100 mm, which is adjusted to the bottom edge of the axis and is divided by its motionless reception with oil of thickness 0.25 mm. If the axis rotates with 1,000 rotations per minute and the viscosity of the oil is $0.3–10^{-9}$ kg/m·s, calculate the power that is absorbed by the internal friction with oil.

2) A flat surface of length 2.0 m and a width of 0.10 m is dragged horizontally on another level. If the gap between these two levels is 0.01", which is filled with a fluid of relevant density 0.9 and kinematic viscosity $\nu = 1.05 \times 10^{-3}$ ft^2 / s, calculate the force that is required to drag the flat surface at a velocity of 0.5 m/s.

3) A horizontal circular surface of diameter 20" rotates, with rotation axis, the vertical axis of its symmetry, with a velocity of 120 RPM, as shown in the following figure:

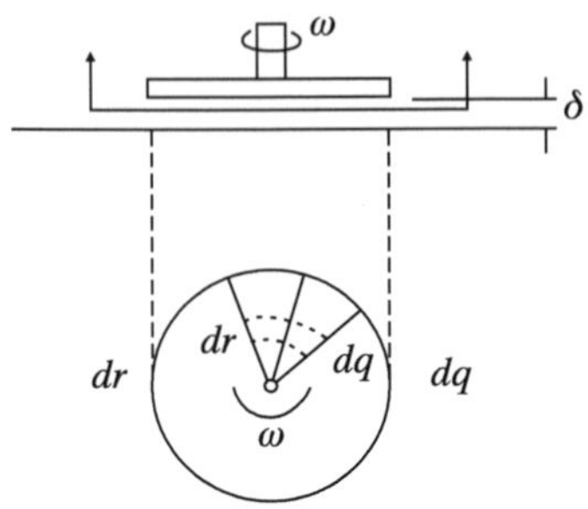

If the distance between the flagstone and the circular periphery is 0.005", and if it has been filled with mineral oil SAE 10W-30, at temperature 150°C, calculate the torque moment that is necessary to maintain the motion if the centrifuge (radiant) motion of the mineral oil is considered to be negligible. Finally, calculate the percentage rise in torque moment when the simple SAE 30 is used instead of the SAE 10W-30. For the solution of the problem, make use of the following table, which presents the values of viscosity for various types of mineral oil according to their temperature conditions. Calculate the value of torque moment in SI units.

Temperature		Viscosity, μ (kg/m - s)			
°F	°C	SAE 50	SAE 30	SAE 10W-30	SAE 10W
20	6.67	10.43	3.26	0.79	0.60
40	4.44	3.28	1.04	0.34	0.25
60	15.56	1.12	0.45	0.15	0.11
80	26.67	0.48	0.13	0.09	0.063
100	37.78	0.22	0.097	0.055	0.037
120	48.89	0.12	0.060	0.037	0.022
140	60.00	0.072	0.037	0.025	0.016
160	71.11	0.045	0.025	0.016	0.011
180	82.22	0.030	0.018	0.012	0.0083
200	93.33	0.022	0.012	0.0094	0.0063
220	104.44	0.015	0.0090	0.0075	0.0049
240	115.56	0.011	0.0071	0.0062	0.0042

Table 3. *Viscosity table of various mineral oil lubricants*

4) Consider a low velocity flight of a spacecraft bus, which lies in the landing condition. If the air pressure and the temperature at the snout of the bus is 1.2 atm and 300 K respectively, calculate the density and the specific volume.

5) Consider that 1,500 lb_m of air is placed inside a 900 ft^3 tank and the air temperature inside the tank is uniform at 70°F. Calculate the air pressure inside the tank in atmospheres.

6) Consider that at a point of the wing of a passenger supersonic Concorde, the air temperature is –10°C and the pressure is 1.7×10^4 N/m^2. Calculate the density at this point.

7) At some point of an area examined at a supersonic wind tunnel, consider that the pressure and the temperature of air are 0.5×10^5 N/m^3 and 240 K, respectively. Calculate the specific volume.

8) A wind tunnel operates from six big tanks of compressible air. The tanks have a total volume of 153^3 and a pressure of 4.137×10^6 N/m^2 at 38°C. What is the mass contained in the tank? Calculate in kg and slugs.

9) Inside a tin with glycerin, a small sphere is allowed to drop. Show graphically and with a free evaluation (i) the relation between the sphere's velocity and the time and (ii) the relation between the acceleration and the time.

10) A drop of oil while falling in the air with constant velocity travels 0.5 m in 0.15 s. (i) Which forces are applied on the sphere? (The buoyancy is considered to be negligible.) (ii) If the coefficient of the air's viscosity is $\mu = 1.82\times10^{-4}$ poise, what is the weight of the sphere (specific weight of oil $\gamma_\lambda = 0.9\, gr/\text{cm}^2$, g = 10 m/s^2?

11) A flat plate that abstains from another motionless plate 0.025 mm is moving with a velocity u = 60 cm/s under the influence of a force F = 2 N/m^2, which increases the above velocity. Calculate the viscosity of the fluid between these two flagstones.

12) Calculate the kinematic viscosity of some oil of density 981 kg/m^3. The shear stress of the oil at a point is 0.2452 N/m^2 and the variation in the velocity at this point is 0.2 m/s.

13) The velocity distribution of a fluid that moves on a flagstone is given from the relation $u = \frac{1}{4}y - y^2$, where u is the velocity in meters per second (u = m/s) and y is the distance in meters from the flagstone (y = m). Calculate the shear stress for y = 0.15 m. The viscosity of the fluid is μ = 8.6 poise.

2

Hydrostatics

2.1. Introduction

Every fluid that is in balance applies forces on osculated surfaces or the vessel that contains the fluid. These forces are categorized into external forces and hydrostatic forces.

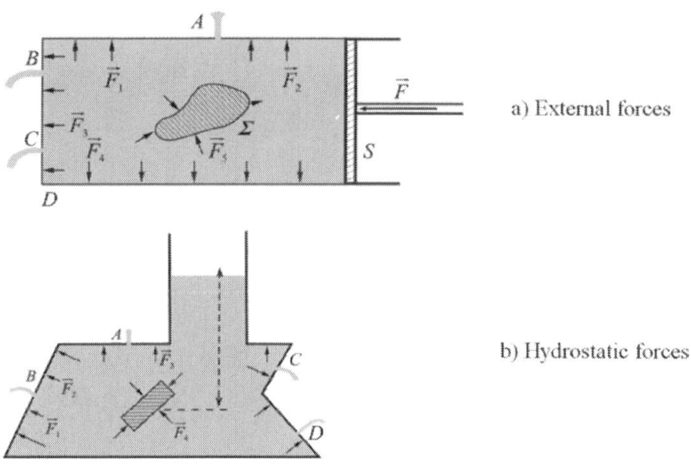

Figure 2.1. *External and hydrostatic forces*

External forces are forces that a fluid applies on the osculated surfaces when the fluid is forced due to an external source, e.g. a plunger, as shown in Figure 2.1.

Let us suppose that the fluid in the container D (Figure 2.1(a)) has no weight. When we push the plunger S with a force $\vec{F}$, the fluid applies forces

F_1, F_2, F_3, F_4, F_5, etc. on the walls of the container D and on the surface of the body Σ. The forces $\vec{F}_1$, $\vec{F}_2$, $\vec{F}_3$, $\vec{F}_4$, $\vec{F}_5$, etc. are external forces that arise from the force $\vec{F}$ which is applied on the fluid. The external forces are vertical on the surfaces. Indeed, if we open the holes A, B, C, we will note that the fluid veins that are formed are vertical on the surfaces of the holes (Figure 2.1(a)).

Hydrostatic forces are forces that the fluid develops due to its weight on the attached surfaces. Therefore, these forces are a result of its weight. *Hydrostatic forces are vertical on the surfaces on which they are applied.*

Indeed, if we open the holes A, B, C, D (Figure 2.1(b)), we will note that the fluid veins that are formed are vertical near the base of the holes.

2.2. Basic law of hydrostatic pressure

If the hydrostatic force $\vec{F}$ is applied by the fluid on a very small surface ΔS at point A (Figure 2.2), then at point A, the pressure p is produced:

$$p = \frac{F}{\Delta S} \qquad [2.1]$$

The pressure p, which is produced at point A by the hydrostatic force $\vec{F}$, is called the *hydrostatic pressure at this point.*

In general, *hydrostatic pressure* at a point of a fluid is the pressure produced on it by the hydrostatic force applied at this point. The hydrostatic forces produce hydrostatic pressures due to the weight of the fluid.

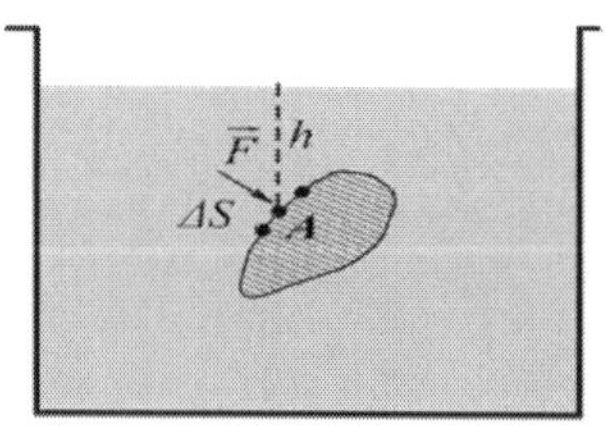

Figure 2.2. *Definition of hydrostatic pressure*

Let us assume an open tank that contains a still fluid. We consider its vertical column where its base is AA', and it has a surface area S and is located away from its free surface at a distance h (Figure 2.3). The weight of the fluid is applied vertically on the horizontal surface *AA'*. This means that the weight *W* of the column has a base with surface area *S* and height *h*.

Therefore, the weight *W* of the fluid produces this *hydrostatic pressure* on the surface *AA'*:

$$p = \frac{W}{S} \tag{2.2}$$

If the specific weight of the fluid is γ and the volume of the column is *v*, then the following relationships are applicable:

$$W = \gamma \cdot \mathrm{v} \tag{2.3}$$

$$\mathrm{v} = S \cdot h \tag{2.4}$$

From the above relationships, we have:

$$p = \frac{W}{S} = \frac{\gamma \cdot \mathrm{v}}{S} = \frac{\gamma \cdot S \cdot h}{S} = \gamma \cdot h$$

and

$$p = \gamma \cdot h \tag{2.5}$$

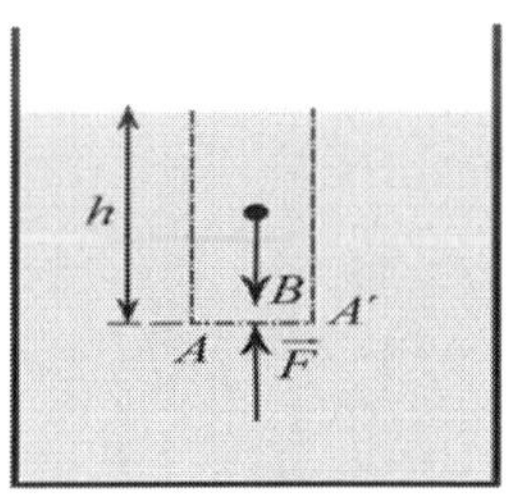

Figure 2.3. *Liquid in open tank*

Relationship [2.5] expresses *the basic law of hydrostatics, which states that the hydrostatic pressure p is equal to the specific weight γ of the fluid multiplied by the vertical distance h of the point, from the free surface of the fluid.*

If the density of the fluid is ρ and the numerical value of the acceleration of gravity is *g*, then the following relationship is applicable:

$$\gamma = \rho \cdot g \qquad [2.6]$$

Consequently, the basic law of hydrostatics is expressed as follows:

$$p = \rho \cdot g \cdot h \qquad [2.7]$$

By solving relationship [2.7] for *h*, we have:

$$h = \frac{p}{\rho \cdot g} \qquad [2.8]$$

where *h* is the *pressure head.*

The basic law of hydrostatics states that: *the total pressure (p_{tot}) at a point in the fluid that is in balance is equal to the sum of the external pressure (p_{ex}) of the fluid and the hydrostatic pressure (p_{hyd}) at this point.* The total pressure in *A* (Figure 2.3) is the sum of the pressure applied by the force $\vec{F}$ as well as the hydrostatic pressure. That is:

$$p_{tot} = p_{ex} + p_{hyd} = \frac{F}{S} + \gamma \cdot h \qquad [2.9]$$

In a fluid that lies in balance in a container, shown in Figure 2.4, we consider two points *A* and *B*, which are located away from its free surface at distances (h_1) and (h_2), respectively, with $h_1 < h_2$. Therefore, we have:

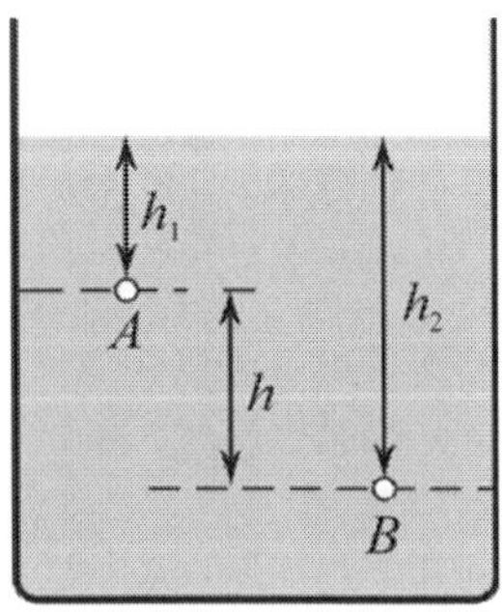

Figure 2.4. *Fluid at rest*

The pressure p_A at point *A* is:

$$p_A = \gamma \cdot h_1 + p_{ex} \qquad [2.10]$$

where $\gamma \cdot h_1$ is the hydrostatic pressure on A and p_{ex} is the external pressure on the fluid.

The pressure p_B at point B is:

$$p_B = \gamma \cdot h_2 + p_{ex} \qquad [2.11]$$

where $\gamma \cdot h_2$ is the hydrostatic pressure on B.

The difference in pressures between points A and B is

$$\Delta p = p_B - p_A = \gamma \cdot h_2 + p_{ex} - (\gamma \cdot h_1 + p_{ex}) \Rightarrow$$

$$\Delta p = \gamma \cdot h_2 - \gamma \cdot h_1 \Rightarrow$$

$$\Delta p = \gamma \cdot (h_2 - h_1) \Rightarrow$$

$$\Delta p = \gamma \cdot h \qquad [2.12]$$

where $h = h_2 - h_1$ and γ is the specific weight of the fluid.

From relationship [2.12], it can be observed that the difference in pressure, when we examine two points of a fluid that is in balance, *does not depend on the external pressure of the liquid*. From the same relationship, it can also be observed that the difference in pressure, when we examine two points of a fluid that is in balance, is equal to the difference in their hydrostatic pressures, and that all the points of the fluid with the same pressure are located on the same horizontal level.

Combining the above, we have:

$$\Delta p = p_B - p_A = \gamma (h_2 - h_1) \qquad [2.13]$$

If $p_B = p_A$, we will have $p_B - p_A = 0$

we have:

$$0 = \gamma (h_2 - h_1) \qquad [2.14]$$

As $\gamma \neq 0$, we get:

$$h_2 - h_1 = 0 \text{ and } h_2 = h_1. \tag{2.15}$$

Finally, at all the points of a horizontal level inside a fluid that is in balance, the same pressure is applied. We have:

$$\Delta p = p_B - p_A = \gamma(h_2 - h_1) \tag{2.16}$$

If A and B are points of the same horizontal level, then we have:

$$h_2 - h_1 \text{ and } h_2 - h_1 = 0 \tag{2.17}$$

Since $\gamma \neq 0$, we get:

$$p_B - p_A = 0 \text{ and } p_B = p_A \tag{2.18}$$

Relationship [2.16] constitutes the *basic theorem of hydrostatics*, which states that:

> *The difference in the total pressures between two points A and B of a fluid that is in balance is equal to the specific weight γ of the fluid, multiplied by the vertical distance between the two points.*

2.3. Law of communicating vessels

If we assume communicating fluid containers of different shapes and sizes, shown in Figure 2.5, it can be seen that when the fluid is in balance, the free surfaces of the fluid in all the vessels lie on the same horizontal level.

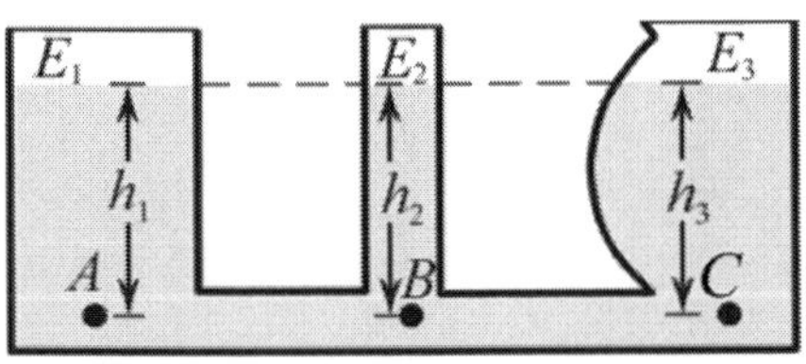

Figure 2.5. *Communicating vessels*

Let us assume three points A, B and C of the fluid that lie on the same horizontal level. As the fluid is in balance and points A, B and C lie on the same horizontal level, the following relationships are applicable:

$$p_A = p_B = p_\Gamma \quad [2.19]$$

If the pressure applied on the free surfaces E_1, E_2, E_3 of the fluid is p_{atm} and the distances of points *A*, *B*, *C* from these surfaces are h_1, h_2 and h_3, , respectively, then the following relationships are applicable:

$$p_A = p_{atm} + \gamma \cdot h_1 \quad [2.20]$$

$$p_B = p_{atm} + \gamma \cdot h_2 \quad [2.21]$$

$$p_C = p_{atm} + \gamma \cdot h_3 \quad [2.22]$$

From the above relationships, we get:

$$p_{atm} + \gamma \cdot h_1 = p_{atm} + \gamma \cdot h_2 = p_{atm} + \gamma \cdot h_3$$

$$h_1 = h_2 = h_3 \quad [2.23]$$

From the above relationship, it can be observed that the free surfaces are far away from the horizontal level on which *A*, *B* and *C* lie, so they are on the same horizontal level.

Relationship [2.23] is a common fluid property that is called the *law of communicating vessels*, which indicates that: *the free surfaces of the fluid contained in vessels that communicate with each other and are in balance, regardless of the vessel's shape, lay on the same horizontal level.*

Communicating vessels find their applications in the city's water distribution systems or fountains (Figure 2.6), or in artesian wells.

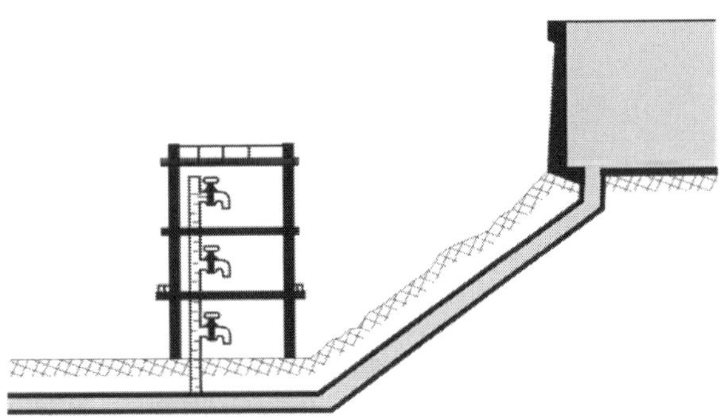

Figure 2.6. *Water distribution pipe system*

2.4. Forces applied by fluids on flat surfaces

2.4.1. *Forces applied on the horizontal bottom of a vessel*

As the bottom of the vessel in Figure 2.7 is horizontal, each of its points has the same hydrostatic pressure:

$$p = \gamma \cdot h \quad [2.24]$$

where γ is the specific weight of the fluid and h is the distance of the horizontal bottom from the fluid's free surface.

As the hydrostatic pressure at different points of the horizontal bottom is the same, the force F that the fluid applies on the horizontal bottom is:

$$F = p \cdot S \quad [2.25]$$

where S is the surface area of the horizontal bottom.

From relationships [2.24] and [2.25], the following relationship arises:

$$F = p \cdot S = \gamma \cdot h \cdot S$$

$$F = \gamma \cdot h \cdot S \quad [2.26]$$

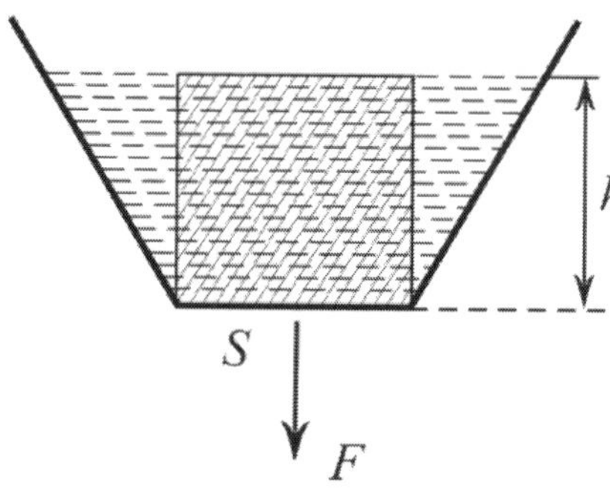

Figure 2.7. *Fluid in a vessel*

From equation [2.26], it arises that the force F does not depend on the shape of the vessel and the total weight of the contained fluid, but on the specific weight (γ) of the fluid, the area (S) of the horizontal bottom and the distance (h) of the bottom from the free surface of the fluid.

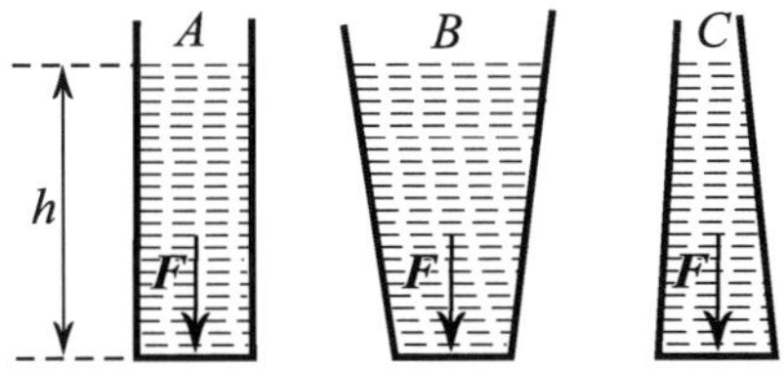

Figure 2.8. *Various types of vessels*

If the three vessels *A*, *B* and *C* (Figure 2.8) of equal bottom size (*S*) contain the same fluid (γ) with the same height (*h*), then we can see that at their bottom surfaces, same forces are applied $F(F = \gamma hS)$, but the weights W_A, W_B and W_C of the fluid that they contain are different.

The fact that the force applied by the fluid on the bottom of the vessel *C* is higher than the weight of the fluid constitutes the *hydrostatic paradox*.

Finally, in relationship [2.26], the $(h \cdot S)$ shows the volume (v) of a column that has a height (*h*) and surface area (*S*), that is:

$$\mathrm{v} = h \cdot S \qquad [2.27]$$

From relationships [2.26] and [2.27], we have:

$$F = \gamma \cdot h \cdot S = \gamma \cdot \mathrm{v} \qquad [2.28]$$

The weight W' of the fluid column that has a specific weight (γ) and volume (v) is:

$$W' = \gamma \cdot \mathrm{v} \qquad [2.29]$$

From relationships [2.28] and [2.29], we have:

$$F = W' \qquad [2.30]$$

Therefore, the force *F* applied by the fluid at the horizontal bottom of the vessel, where it balances, is equal to the weight W' of a vertical fluid column, with base (*S*) of the bottom and height (*h*), the distance from the bottom up to the free surface of the fluid. It is worth mentioning that we have not taken into account the atmospheric pressure.

2.4.2. Forces applied on the flat side walls of a vessel

If we assume point A of the flat side wall of a vessel (Figure 2.9), a hydrostatic pressure is produced, which is given by the following relationship:

$$p = \gamma \cdot h \quad [2.31]$$

where γ *is* the specific weight of the fluid and h the distance of point A from the free surface of the fluid.

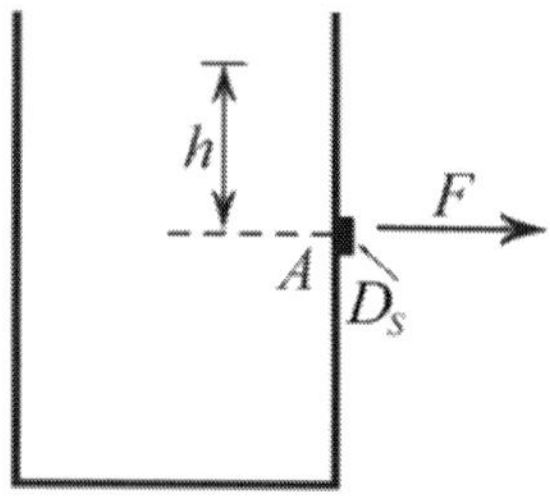

Figure 2.9. *Force applied on the side wall of a vessel*

Therefore, at every point of the flat wall, the fluid applies a force F vertical to the wall.

The value of this force is:

$$F = p \cdot \Delta S = \gamma \cdot h \cdot \Delta S \quad [2.32]$$

where ΔS is the area of an infinitesimal surface of the wall at point A.

From relationship [2.32], it arises that the numerical value of the forces applied by the fluid at various points of the flat side wall increases when the distance of the points from the fluid's free surface increases. Of course, the forces that the fluid applies at various points of the flat wall are parallel between them, as they are all vertical to it.

Therefore, in order to find the force that the fluid applies on the whole flat wall of the vessel, we must compose the forces F_1, F_2, F_3, etc. applied by the fluid at all of the points of the flat wall (Figure 2.10).

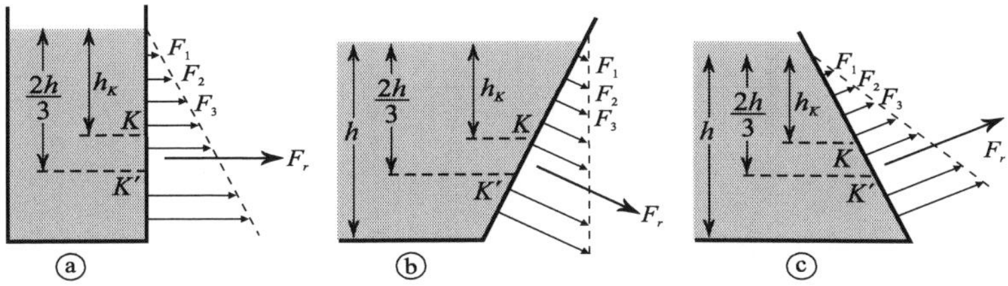

Figure 2.10. *Forces applied on the vessel's wall*

The resultant F_r of the forces that the fluid applies on the whole flat wall is vertical to it and its numerical value is given by the relationship:

$$F_r = \gamma \cdot h_K \cdot S \qquad [2.33]$$

where S is the area of the surface of the flat wall which is immersed in the fluid and h_K is the distance of the center of gravity from the free surface.

The application point K' of the force F_r is called the *center of pressure*. In general, it lies under the center of gravity K of the wall's surface that is immersed.

If the wall's surface is a *rectangular parallelepiped*, then the center of pressures K' is far away from the free surface of the fluid at a distance $h_{K'}$:

$$h_{K'} = \frac{2}{3} \cdot h \qquad [2.34]$$

where h is the distance of the bottom's surface from the free fluid's surface.

In this case, the center of gravity K of the wall's surface that is immersed is far away from the fluid's free surface at a distance equal to:

$$h_K = \frac{h}{2} \qquad [2.35]$$

If the flat wall is vertical, then F_r is horizontal (Figure 2.10(a)).

If the flat wall is inclined, then the force is directed downwards or upwards (Figures 2.10(b) and (c)), respectively. The barriers are built in such a way that their thickness increases according to the depth because the force applied on the walls increases, respectively.

2.5. Forces applied by fluids on curved surfaces

We consider a curved surface AB at the bottom of a fluid that is in balance, as shown in Figure 2.11.

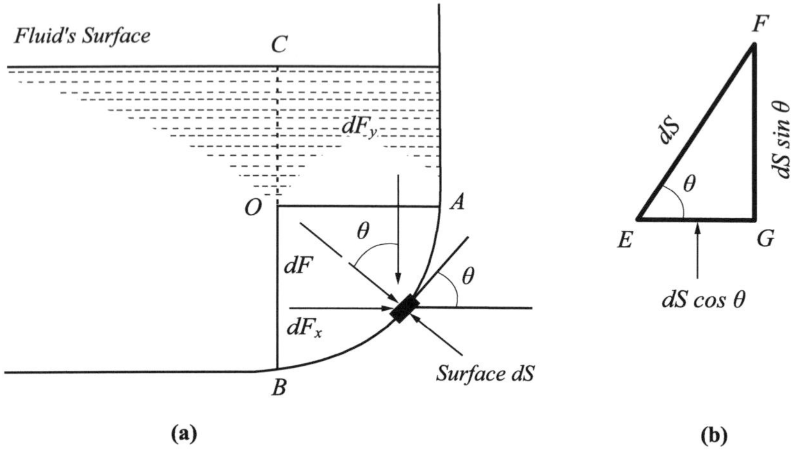

Figure 2.11. *Fluid vessel with curved surfaces*

Let us assume that dS is an elementary surface at depth h from the free surface of the liquid. Then, the pressure applied on the surface dS by the fluid is:

$$p = \rho \ g \ h \qquad [2.36]$$

Moreover, the pressure's dynamic dF is:

$$dF = p \cdot dS = \rho \ h g \cdot dS \qquad [2.37]$$

This force dF is vertical to the surface. Therefore, the total force of the pressure on the curved surface will be:

$$F = \int \rho \ g h dS \qquad [2.38]$$

In this case, the direction of the force on the small surface dS is not in the same direction, but varies from point to point. Therefore, the integration of relationship [2.38] for the curved surface is not possible. For this purpose, we analyze the force dF in two components dF_x and dF_y regarding the x and y axes, respectively.

The total force in the x and y directions, Fx and Fy, respectively, can be found by integrating the force dF. Thus, the total force on the curved surface will be

$$F = \sqrt{F_x^2 + F_y^2} \qquad [2.39]$$

and the slope of the resultant force F on the horizontal level is:

$$\tan\theta = \frac{F_g}{F_x} \qquad [2.40]$$

By analyzing the force dF given by relationship [2.39] in the x and y directions, we have:

$$dF_x = dF\sin\theta = \rho\, g\, h\, dS\sin\theta \qquad [2.41]$$

and

$$dF_y = dF\cos\theta = \rho\, h\, h\, dS\cdot\cos\theta \qquad [2.42]$$

Therefore, the total force in the x and y directions is:

$$F_x = \int dF_x = \int \rho\, g\, h\, dS\sin\theta \qquad [2.43]$$

and

$$F_y = \int dF_y = \int \rho\, g\, h\, dS\cos\theta.$$

From Figure 2.11(b), we have:

$$EF = dS$$

$$EG = dS\sin\theta$$

$$EG = dS\cos\theta$$

Therefore, in equation [2.43], the $dS\sin\theta = FG$ is a vertical projection on the surface dS and the relationship $\rho g\int h\, dS\sin\theta$ expresses the total force of the pressure multiplied by the curved surface on the vertical level, and thus F_x is the total force of the pressure on the curved surface on the vertical level.

Moreover, $dS\cos\theta = EG$ is the horizontal projection of dS, and therefore $hdS\cos\theta$ is the volume of the fluid contained in the elementary area from the free surface of the fluid.

Therefore,

$$\mathrm{v}_{o\lambda} = \int h\,dS\cos\theta \qquad [2.44]$$

is the total volume contained between the surface column and the free surface of the fluid.

Consequently,

$$W_{o\lambda} = \rho\,g\int h\,dS\cos\theta \qquad [2.45]$$

is the total weight of the fluid that pushes the curved surface. Therefore,

$$F_y = \rho\,g\int h\,dS\cos\theta \qquad [2.46]$$

is the fluid's weight applied on the curved surface.

2.6. Archimedes' principle

Let us consider a body Σ, which is of a prismatic shape, (Figure 2.12) and is immersed in a fluid of specific weight γ. The forces F_3 and F_4 that the fluid applies on the body's lateral areas cancel each other out. The force F_1 that the fluid applies on the body's top base (which is horizontal) is vertical, which acts downward and has a numerical value:

$$F_1 = p_1 \cdot S = \gamma \cdot h_1 \cdot S, \qquad [2.47]$$

where:

p_1 is the hydrostatic pressure on the top base of the body;

S is the area of the base of the body;

γ is the specific weight of the fluid;

h_1 is the body's top base distance from the fluid's free surface.

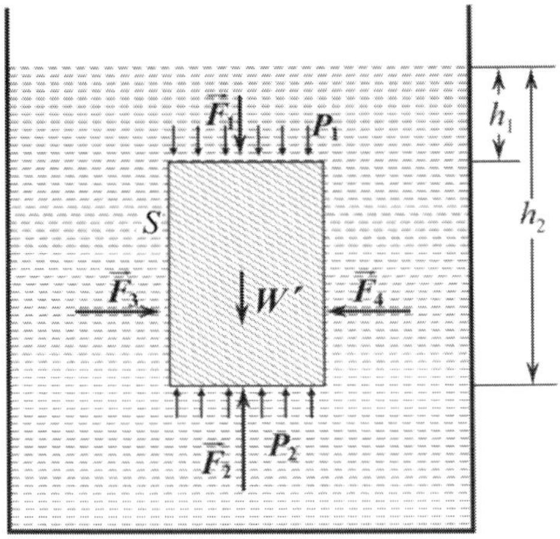

Figure 2.12. *Body immersed in a fluid*

The force F_2 applied by the fluid on the body's bottom base (which is horizontal) is vertical upwards and has the value:

$$F_2 = p_2 \cdot S = \gamma \cdot h_2 \cdot S, \qquad [2.48]$$

where p_2 is the hydrostatic pressure on the body's bottom base and h_2 is the body's bottom base distance from the free surface of the fluid.

Since the forces F_1 and F_2 lie on the same vertical level and have opposite directions, their resultant, which is the body's buoyancy, is:

$$A = F_2 - F_1 \qquad [2.49]$$

From the above relationships, we obtain the following equations:

$$A = F_2 - F_1 = \gamma \cdot h_2 \cdot S - \gamma \cdot h_1 \cdot S = \gamma \cdot S\,(h_2 - h_1) \text{ and}$$

$$A = \gamma \cdot h \cdot S \qquad [2.50]$$

where $h = (h_2 - h_1)$ is the height of the prismatic body.

The volume v of the body, and therefore the volume of the outplaced fluid, is:

$$\mathrm{v} = h \cdot S \qquad [2.51]$$

From relationships [2.50] and [2.51], we obtain the following relationship:

$$A = \gamma \cdot \mathrm{v} \qquad [2.52]$$

The weight W' of the outplaced fluid is:

$$W' = \gamma \cdot \mathrm{v} \qquad [2.53]$$

From relationships [2.52] and [2.53], we obtain the following relationship:

$$A = W' \qquad [2.54]$$

which means that the buoyancy's numerical value is equal to the weight of the outplaced fluid. Consequently:

1) the buoyancy's direction is vertical (since the buoyancy is the resultant of F_1 and F_2, which are also vertical);

2) the buoyancy's direction is bottom-up (since the force that is higher than $F_1 = (\gamma \cdot h_2 \cdot S > \gamma \cdot h_1 \cdot S)$ has a bottom-up direction);

3) the buoyancy's numerical value is equal to the numerical value of the weight of the outplaced fluid [2.54].

Therefore, when a body is totally immersed in a fluid that is in balance, at every point of the body's surface, a vertical force is applied by the fluid. This force is called *buoyancy* and this principle is called *Archimedes' principle.*

Center of buoyancy is called the application point of the body's buoyancy and is the same with the fluid's center of gravity that is outplaced from the body. It only coincides with it when the body is homogeneous and totally immersed in the fluid.

2.7. Consequences of Archimedes' principle

2.7.1. *Fully immersed body*

When a solid body is fully immersed in a fluid, two forces are applied on it: its weight W and its buoyancy A. Under the effect of the forces A and W, the body is oriented in such a way that its center of gravity and its buoyancy center are on the same vertical level (Figure 2.13).

The body's motion when it is let free depends on the *resultant of the forces* A *and* W.

We may distinguish the following cases:

a) When the buoyancy A of the body is *opposite* to its weight W, it means that:

$$A = -W. \quad [2.55]$$

However, we also know that:

$$A = W.$$

Therefore, their resultant F is:

$$F = W - A = 0 \Rightarrow$$

$$F = 0. \quad [2.56]$$

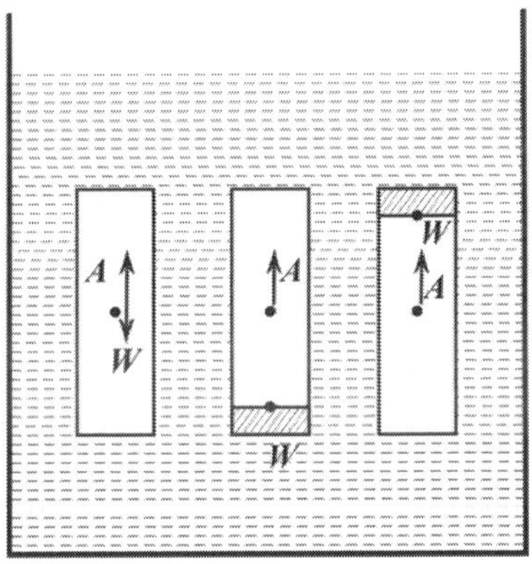

Figure 2.13. *Forces applied on a fully immersed body*

Therefore, in this case, the resultant of the forces applied on the body is zero, and thus the body will balance in any position inside the fluid.

We also distinguish the following categories of balance:

i) If the center of gravity lies *under* (Figure 2.14) its buoyancy center, then the body's balance is steady (*stable equilibrium*), because with a small displacement, restoring forces on it are developed.

ii) If the body's center of gravity lies above (Figure. 2.14) its buoyancy center, then the body's balance is not steady (*unstable equilibrium*), because with a small displacement, inversion forces on it are developed.

iii) If the body's center of gravity *coincides* with the buoyancy center, then the body's balance is in a *neutral equilibrium*.

b) When the numerical value of the body's buoyancy A is *smaller* than the numerical value of its weight W (they definitely lie on the same vertical and have opposite direction), it means that:

$$W = A.$$

Then, their resultant F is given by:

$$F = W = -A$$

$$F = W - A = const \neq 0 \qquad [2.57]$$

Therefore, in this case, the resultant F of the forces that are applied on the body is vertical, and acts in a downward direction and has a constant numerical value. Therefore, the body sinks with a constant acceleration, so long as the fluid has no other resistance.

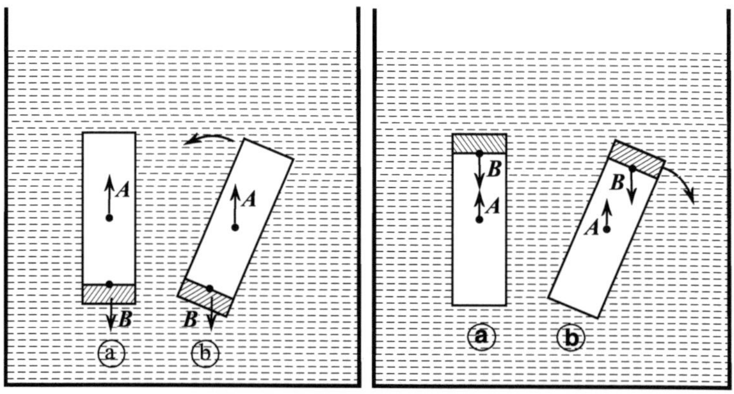

Figure 2.14. *Stable and unstable equilibrium*

c) When the numerical value A of the body's buoyancy is *larger* than the numerical value W of its weight (for certain A and W lie on the same vertical and have opposite directions), it means:

$$W < A,$$

then their resultant F is:

$$F = A - W$$

$$F = A - W = \text{const} \neq 0 \quad [2.58]$$

Therefore, in this case, the resultant F of the forces A and W that are applied on the body is vertical, has upward direction and constant numerical value. The body lifts inside the fluid with the effect of the resultant, until it reaches the fluid's free surface. Then a part of the body's volume comes out of the fluid, the buoyancy A is reduced and it becomes *opposite* to the body's weight W and finally, the solid body floats on the free surface of the fluid.

2.7.2. *Partially immersed (floating) body*

If only a part of the body is immersed and the body is in balance, then we say that the body is floating. A body floats when the numerical value of the body's buoyancy is equal to the numerical value W, its weight, and the body's center of gravity and buoyancy lie on the same vertical axis.

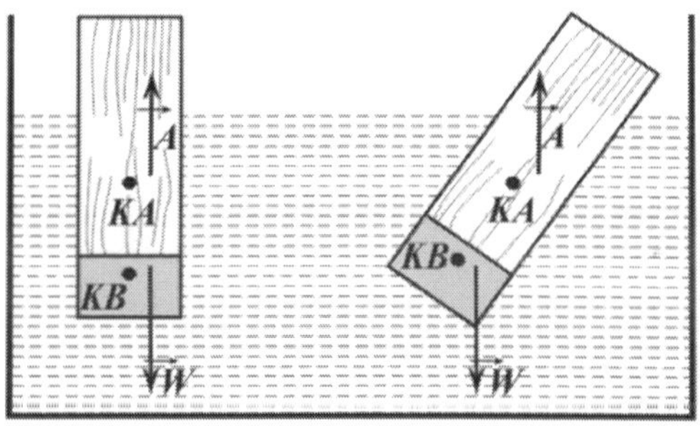

Figure 2.15. *Center of gravity under body's center of buoyancy*

If the body's center of gravity is under its center of buoyancy, then the body's balance is always stable (Figure 2.15). If the body's center of gravity is under its buoyancy center and we divert the body a little from its balancing position, the two forces A and W form a pair which bring the body back to its initial position, which means that this balance (floating) is stable as well.

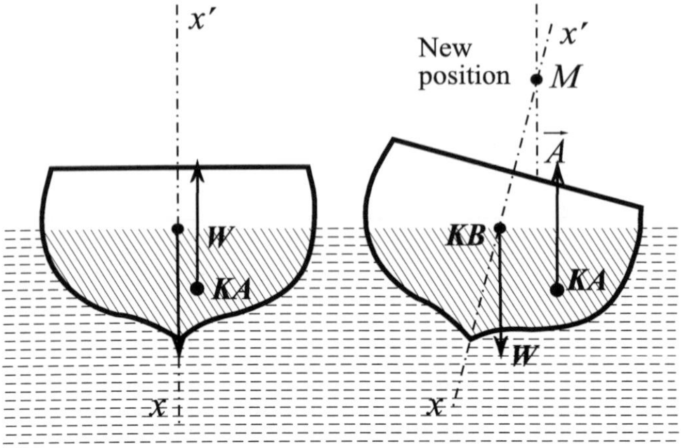

Figure 2.16. *Center of gravity over body's center of buoyancy*

If the body's center of gravity (Figure 2.16) lies over the center of buoyancy, the balance is stable if, in any diversion of the body from it, the metacenter lies over the body's center of gravity. The balancing axis of a floating body is the straight line XX′ (Figure 2.16) that crosses from the center of gravity and the buoyancy center when the body is in balance.

Metacenter M of a body is called the intersection point of the body's balancing axis XX′ and the vertical one that crosses from the body's buoyancy center. From above, it can be observed that the metacenter's position varies for different diversion angles of the body in its balancing position. Therefore, the balancing axis of the body is intersected by the vertical one that crosses the buoyancy's center and it depends on the diversion's angle of the body in its balancing position.

2.8. Formulae

1) Pressure, *p*:

$$\boxed{p = \frac{F}{S}}$$

where F is the vertical force on the surface and S is the surface area.

2) Absolute pressure, p_{abs}:

$$p_{abs} = p_{\alpha tm} + p_{vac}$$

where p_{atm} is the atmospheric pressure and p_{vac} is the vacuum pressure.

3) Vacuum pressure, p_{vac}:

$$p_{vac} = p_{\alpha tm} - p_{abs}$$

where p_{atm} is the atmospheric pressure and p_{abs} is the absolute pressure.

4) Basic law of hydrostatics:

$$p = \rho \cdot g \cdot h$$

where ρ is the fluid density, g is the acceleration of gravity and h is the distance from the free surface.

5) Basic theorem of hydrostatics:

$$\Delta p = \gamma \cdot h$$

where γ is the specific fluid weight and h is the distance from the free surface.

6) Law of communicating vessels:

$$h_1 = h_2 = h_3$$

where h is the height from the free surface of the communicating vessels.

7) Total force of the pressure on a curved surface:

$$F = \int \rho g h dS$$

where ρ is the fluid density, g is the acceleration of gravity, h is the depth from the fluid's free surface and dS is the elementary surface of the curved surface.

8) Archimedes' principle:

$$A = \gamma \cdot \mathrm{v}$$

where A is the buoyancy of the fluid, γ is the specific weight of the fluid and v is the volume of the displaced fluid.

2.9. Questions

1) What are *pressure* and *hydrostatic pressure*?

2) What are *hydrostatic forces*?

3) How is the *basic law of hydrostatics* stated verbally and mathematically?

4) How is the *basic theorem of hydrostatics* stated verbally and mathematically?

5) What is the *law of communicating vessels*?

6) What is *hydrostatic paradox*?

7) What is the *law of Archimedes* for fluids?

8) What do we call *buoyancy's center* and with what does it coincide within a solid body inside the fluid?

9) What do we call a *body's metacenter* and when do we have a *stable*, an *unstable* and an *uninteresting* balance?

10) Calculate the *pressure* applied on a *curved surface* from a fluid that is in a calm state.

2.10. Problems with solutions

1) Point A lies at some depth inside a fluid that has a specific weight. How much pressure is applied at point A if the atmospheric pressure is $p_{atm} = 1033.6\ p/\mathrm{cm}^2$ (Figure (1))?

Solution:

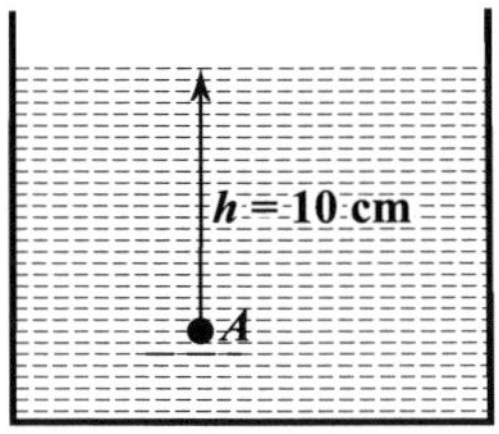

Figure (1)

At every point of the fluid's free surface, atmospheric pressure is applied (p_{atm}). According to Pascal's Law, this pressure is also applied at point *A*.

Moreover, at point *A*, hydrostatic pressure is applied:

$$p_{hyd} = h \cdot \gamma \tag{1}$$

Therefore, at point *A*, there will be some pressure, which is given by the relationship:

$$p_A = p_{atm} + p_{hyd}$$

$$p_A = p_{atm} + h \cdot \gamma \tag{2}$$

If we define relationship (2) using the given clues, we will have:

$$p_A = p_{atm} + h \cdot \gamma = 1033.6 \frac{p}{\text{cm}^2} + 10\,\text{cm} \cdot 1 \frac{p}{\text{cm}^3}$$

$$p_A = 1033.6 \frac{p}{\text{cm}^2} + 10 \frac{p}{\text{cm}^2}$$

$$p_A = 1043.6 \frac{p}{\text{cm}^2}$$

2) Point *A* lies at depth h = 10 cm inside a fluid (Figure (2)) with specific weight $\gamma = 1 p / \text{cm}^3$. How much pressure is applied at point *A* if the pressure of the gas found in the area *X* is $p_{air} = 2067.2 p / \text{cm}^2$?

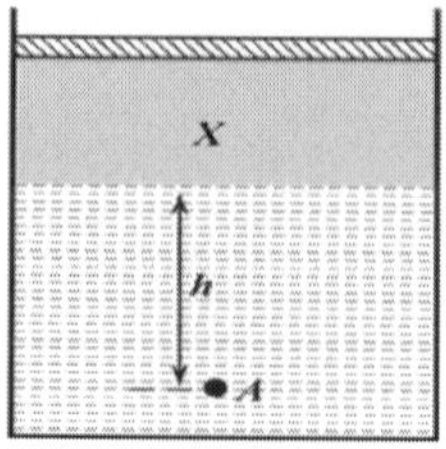

Figure (2)

Solution:

The pressure p_A found at point *A* is:

$$p_A = p_{\alpha\varepsilon\rho} + h \cdot \gamma$$

$$p_A = 2067.2 \frac{p}{\text{cm}^2} + 10\,\text{cm} \cdot 1 \frac{p}{\text{cm}^3}$$

$$p_A = 2077.2 \frac{p}{\text{cm}^3}$$

3) A man's sole, weighing $W = 80\,\text{kp}$, has a surface $S = 40\ \text{cm}^2$. Calculate the pressure *p* that is produced on the floor, when the man leans on one foot.

Solution:

The pressure *p* is exerted by the man's weight that is vertical to *S*. According to the definition of pressure, we have:

$$p = \frac{W}{S} = \frac{80\,\text{kp}}{40\,\text{cm}^2} = 2\frac{\text{kp}}{\text{cm}^2} = 2\alpha t.$$

NOTE.– When the man leans on both his feet, the pressure is 1 atm provided that the area of surface of both his feet is the same:

$$p = \frac{W}{S} = \frac{80\,\text{kp}}{80\,\text{cm}^2} = 1\frac{\text{kp}}{\text{cm}^2} = 1\alpha t.$$

4) The vessel D contains fluid with specific weight $\gamma = 13.6 p / \text{cm}^3$. How much hydrostatic pressure is exerted by the fluid at points *A*, *B* and *C* if they are located away from its free surface at distances of 5 and 10 cm, respectively (Figure (4))?

Solution:

The hydrostatic pressure *p* that the fluid with specific weight γ applies at a point that is located far away from its free surface at a distance *h* is given by the relationship:

$$p = \gamma \cdot h$$

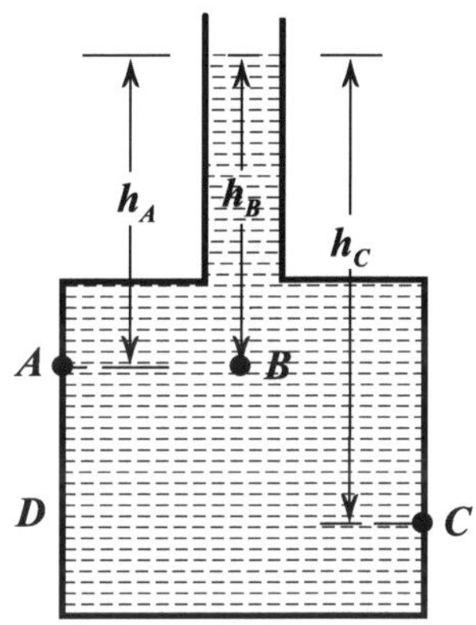

Figure (4)

So we have:

a) For point A:

$$p_A = \gamma \cdot h_A = 13.6 \frac{p}{\text{cm}^3} \cdot 5\,\text{cm} = 13.6 \times 5 \frac{p}{\text{cm}^3}$$

$$p_A = 68 \frac{p}{\text{cm}^2}$$

b) For point B:

$$p_B = \gamma \cdot h_B = 13.6 \frac{p}{\text{cm}^3} \cdot 5\,\text{cm} = 13.6 \times 5 \frac{p \cdot \text{cm}}{\text{cm}^3}$$

$$p_B = 68 \frac{p}{\text{cm}^2}$$

c) For point C:

$$p_C = \gamma \cdot h_C = 13.6 \cdot 10 = 136 \text{ N/cm}^2$$

5) How much force does water apply on a surface of area 100 cm^2 when it lies at a depth of 40 m $(\gamma = 1p/\text{cm}^3)$?

Solution:

If we denote the area of the surface by *S* and the hydrostatic pressure applied on it by *p*, then the force *F* applied on it will be:

$$F = p \cdot S \tag{1}$$

The pressure *p* is given by the relationship:

$$p = \gamma \cdot h \tag{2}$$

where γ is the specific weight of water and *h* is the depth at which the surface lies.

From relationships (1) and (2), we get:

$$F = \gamma \cdot h \cdot S \tag{3}$$

If we set the facts in relationship (3), we will have:

$$F = 1 \frac{p}{\text{cm}^3} \cdot 4{,}000 \,\text{cm} \cdot 100 \,\text{cm}^2 = 1 \cdot 4{,}000 \times 100 p$$

$$F = 4 \times 10^5 \, p = 400 kp$$

6) The vessel D (Figure (4)), contains water with specific weight $\gamma = 1p/\text{cm}^3$. How much is the difference Δp in the hydrostatic pressures which are exerted at points *A* and *B* if their vertical distance is $h = 10$ cm?

Solution:

We know the formula:

$$p_B - p_A = \Delta p = h \cdot \gamma \tag{1}$$

If we substitute the given facts in formula (1), we get:

$$\Delta p = 10\,\text{cm} \cdot 1 \cdot \frac{p}{\text{cm}^3} = 10\frac{p}{\text{cm}^2}$$

$$\Delta p = 10\frac{p}{\text{cm}}$$

7) Vessel D (Figures (7), (7a), (7b)) contains water with specific weight $\gamma = 1p/\text{cm}^3$. How much is the difference *Δp* in the hydrostatic pressures which are exerted at points C and *E* if their vertical distance is $h = 10\,\text{cm}$?

Solution:

We know the formula:

$$p_\Gamma - p_A = \Delta p = h \cdot \gamma \tag{1}$$

If we substitute the given facts in formula (1), we get:

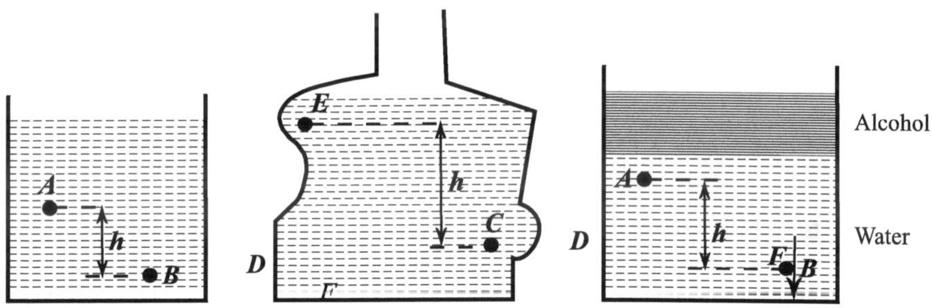

Figure (7) Figure (7a) Figure (7b)

$$\Delta p = pcm \cdot \frac{1}{\text{cm}^3} = 10\frac{p}{\text{cm}^2}$$

$$\Delta p = 10\frac{p}{\text{cm}^2}$$

8) Vessel D contains water and alcohol, as shown in Figure (7). If the specific weights of alcohol and water are $\gamma_\alpha = 0.79p/\text{cm}^3$ and $\gamma_N = 1p/\text{cm}^3$, how much is

the difference in their hydrostatic pressures which are exerted at points A and B if their vertical distance is $h = 10\,\text{cm}$ (we assume that the fluids do not mix)?

Solution:

Points A and B are points of the same fluid, and therefore, the following formula is applicable:

$$p_B - p_A = \Delta p = h \cdot \gamma_N \tag{1}$$

If we substitute the given facts in formula (1), we get:

$$\Delta p = 10\,\text{cm} \cdot 1 \frac{p}{\text{cm}^3} = 10 \frac{p}{\text{cm}^2}$$

$$\Delta p = 10 \frac{p}{\text{cm}^2}$$

9) On the side wall of a vessel (Figure (9)) that contains water, a circular hole with surface area of 1 cm^2 is opened and at a distance of 50 cm from water's free surface. Calculate the force F_1 that must be applied on a cork that shuts the hole down so that water does not come out.

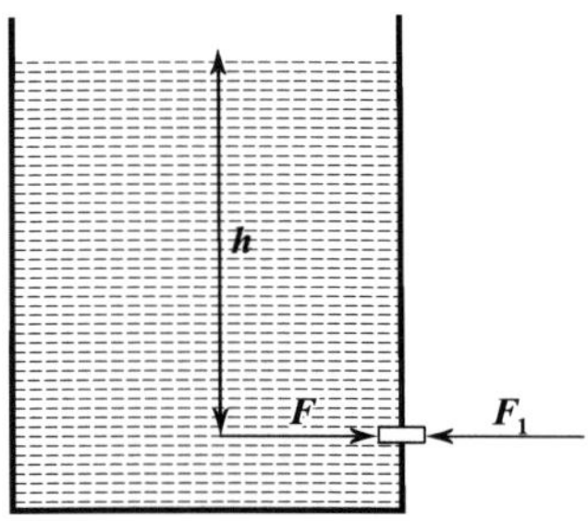

Figure (9)

Solution:

The numerical value of the force F applied by water on the cork is:

$$F = p \cdot S \tag{1}$$

where p is the hydrostatic pressure applied on the cork by water and S is the hole's surface area (of the cork).

Then, p is given by the relationship:

$$p = \gamma \cdot h \quad (2)$$

where γ is the specific weight of water (1 p/cm^3) and h is the hole's distance from the water's free surface.

From relationships (1) and (2), we get:

$$F = \gamma \cdot h \cdot S \quad (3)$$

If we substitute the given facts in relationship (3), we get:

$$F = \gamma \cdot h \cdot S = 1\frac{p}{\text{cm}^3} \cdot 50\,\text{cm} \cdot 1\,\text{cm}^2 = 1 \times 50 \times 1p$$

$$F = 50kp$$

The force F_1 that should be applied on the cork must be opposite to F, that is:

$$F_1 = -F$$

$$F_1 = F$$

Therefore, $F_1 = 50p$.

10) Inside a fluid and at different depths, we hang an iron cylinder with volume 25 cm^3, a copper sphere with volume 25 cm^3 and a nickel bar with volume 25 cm^3. How much buoyancy is applied on each of these bodies if the fluid's specific weight is $\gamma_\upsilon = 1p/\text{cm}^3$, and is it the same in all areas?

Solution:

The buoyancy A_K applied on the iron cylinder is:

$$A_K = \text{v}_\upsilon \cdot \gamma_\upsilon \quad (1)$$

where v_υ is the fluid's volume that is displaced from the cylinder. This volume is equal to the cylinder's volume, which means that $\text{v}_\upsilon = \text{v}_K = 25\,\text{cm}^3$, where γ_υ is the fluid's specific weight.

If we substitute the given facts in relationship (1), we get:

$$A_K = \mathrm{v}_\upsilon \cdot \gamma_\upsilon = \mathrm{v}_K \cdot \gamma_\upsilon = 25\,\mathrm{cm}^3 \cdot 1p/\mathrm{cm}^3$$

$$A_\sigma = 25p$$

The buoyancy A_σ applied on the copper sphere is:

$$A_\sigma = \mathrm{v}_\upsilon \cdot \gamma_\upsilon = \mathrm{v}_\sigma \cdot \gamma_\upsilon = 25\,\mathrm{cm}^3 \cdot 1p/\mathrm{cm}^3$$

$$A_K = 25p$$

The buoyancy A_ρ applied on the nickel bar is:

$$A_\rho = \mathrm{v}_\upsilon \cdot \gamma_\upsilon = \mathrm{v}_\rho \cdot \gamma_\upsilon = 25\,\mathrm{cm}^3 \cdot 1p/\mathrm{cm}^3$$

$$A_\rho = 25p$$

11) We hang an iron sphere of volume 25 cm^3 successively inside two fluids with specific weights $\gamma_1 = 1p/\mathrm{cm}^3$ and $\gamma_2 = 2p/\mathrm{cm}^3$, respectively. How much buoyancy is applied on the sphere when it is hung inside the first fluid and how much when it is hung in the second one?

Solution:

The buoyancy A_1 applied on the sphere when it is hung inside the first fluid is:

$$A_1 = \mathrm{v}_{\upsilon\gamma} \cdot \gamma_1 \tag{1}$$

where $\mathrm{v}_{\upsilon\gamma}$ is the fluid's volume that is displaced by the sphere. This volume is equal to the sphere's volume. This means that $\mathrm{v}_{\upsilon\gamma} = \mathrm{v}_\sigma$, where γ_1 is the fluid's specific weight.

If we substitute the facts given in relationship (1), we have:

$$A_1 = \mathrm{v}_{\upsilon\gamma} \cdot \gamma_1 = \mathrm{v}_\sigma \cdot \gamma_1 = 25\,\mathrm{cm}^3 \cdot 1p/\mathrm{cm}^3 = 25p$$

$$A_1 = 25p$$

The buoyancy A_2 applied on the sphere when it is hung inside the second fluid is:

$$A_2 = \mathrm{v}_{\upsilon\gamma} \cdot \gamma_2 = \mathrm{v}_{\sigma\varphi} \cdot \gamma_2 = 25\,\mathrm{cm}^3 \cdot 2p/\mathrm{cm}^3 = 50p$$

$$A_2 = 50p$$

12) A vessel contains fluid with density ρ = 1,010 kg/m^3 and is put into a car that now starts and moves with constant acceleration. The vessel's base is 8 cm and the free surface from the vessel's base is 7 cm, when it lies in a calm state. In t = 10 sec time, it has gained velocity V = 100 km/h. Calculate:

a) the highest height of the fluid and

b) the largest relative pressure that is developed inside the fluid (Figure (12)).

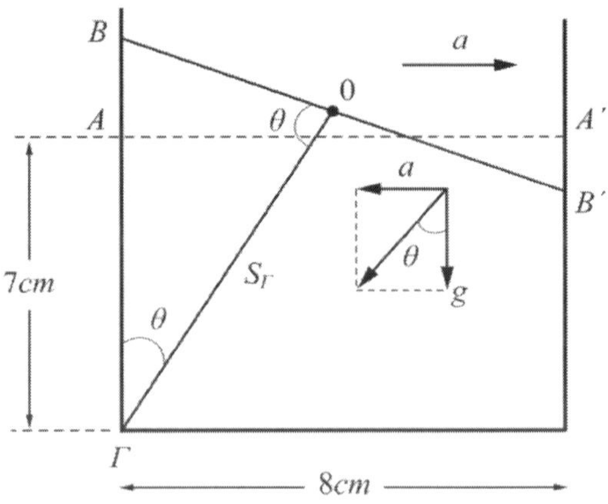

Figure (12)

Solution:

The fluid's free surface before the car's initiation was AA'. When the car started and after the application of the car's acceleration on it, BB' appeared.

a) From Figure (12), we have:

$$\tan\theta = \frac{\alpha}{g} = 0.2834 \Rightarrow \theta = 15.82°$$

$$\text{and } \tan\theta = \frac{AB}{OA} \Rightarrow AB = \tan\theta.OA \Rightarrow AB = 1.13\,\text{cm}$$

since the car's acceleration *a* from the relationship

$$\Delta V = at \text{ gives } a = \frac{\Delta V}{t} = \frac{110\,\text{km/h}}{10\,\text{sec}} \Rightarrow$$

$$\Rightarrow \alpha = 2.78\,\text{m}/\text{sec}^2$$

b) The biggest pressure inside the fluid appears at the point of its biggest removal from its free surface BB', which is the point Γ. Therefore, from the triangle $BO\Gamma$ of Figure (12), we have

$$S_\Gamma = B\Gamma\cos\theta \Rightarrow S_\Gamma = 7.82\,\text{cm}$$

and from the relationship,

$$\alpha_\Sigma = \sqrt{g^2 + \alpha^2}$$

we have

$$\alpha_\Sigma = \sqrt{9.81^2 + 2.78^2} \Rightarrow \alpha_\Sigma = 10.2\,\text{m}/\text{sec}^2$$

Therefore, from the relationship:

$$p = p_{\varepsilon\xi} + \rho\alpha_\Sigma \cdot g$$

we finally get:

$$p_{\Gamma\sigma\chi} = 0.806\,\text{KPa}$$

2.11. Problems to be solved

1) A diver lies at a depth of 300 m from the sea surface. If the specific weight of the sea water is $\gamma = 1{,}010\,kp/\text{m}^3$ and the atmospheric pressure is 1 atm, calculate the pressure applied on the diver.

2) We add mercury into a U-shaped tube, and then another fluid is added into one of its two parts. The free surfaces of the mercury and the other fluid are far away from the horizontal level $\Delta\Delta'$ that separates them at distances 20 and 40 cm, respectively. What is the specific weight of the fluid if the specific weight of mercury is 0.6 p/cm^3?

3) A body's volume is 90 cm^3 and the body is immersed within two fluids with specific weights $\gamma_1 = 2p/\text{cm}^3$ and $\gamma_2 = 1p/\text{cm}^3$, as shown in Figure (3). How much buoyancy is applied on this body if 60 cm^3 of its volume is immersed in fluid 1, while the remaining 30 cm^3 is immersed in fluid 2;

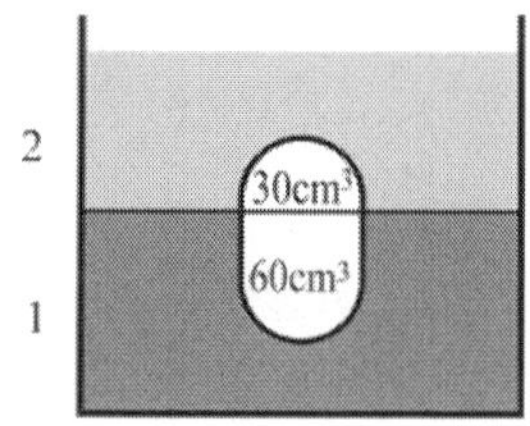

Figure (3)

4) A vessel with horizontal bottom, of surface area 100 cm^2, contains water. How much is the force applied by water at the bottom if its free surface is far away from the bottom at a distance of 30 cm?

5) The vessels A and B in Figure (5) have equal bottoms ($S_1 = S_2$) and contain fluid with specific weight $\gamma = 1p/\text{cm}^3$ at the same height ($h_1 = h_2$). How much is the force F_1 applied by the fluid at the bottom of A and how much is the force F_2 applied at the bottom of B, if $h_1 = h_2 = 10\,\text{cm}$ and $S_1 - S_2 - 4\,\text{cm}^2$?

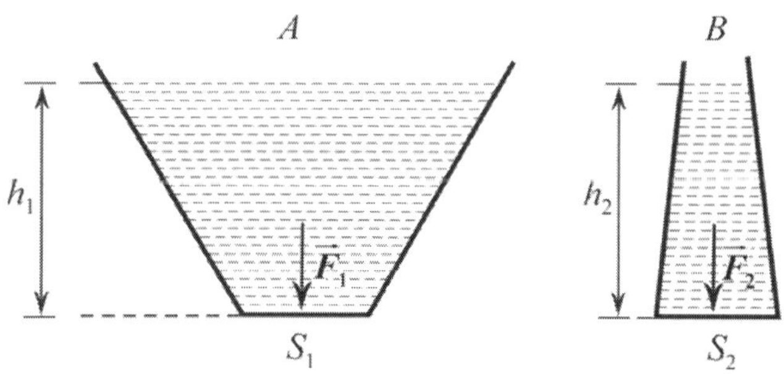

Figure (5)

6) A body's volume is 90 cm^3 and it floats in a fluid with specific weight $2\,p/\text{cm}^3$. How much buoyancy is applied on the body if the volume of the body that lies in the fluid is 60 cm^3?

7) A solid and homogeneous body weighs $500\,kgf$ in the atmosphere. When immersed in water, its bulk density is $300\,kgr$. Calculate the volume and density of the body.

3

Aerostatics

3.1. Introduction

Aerostatics is a branch of fluid mechanics that examines compressed fluids and “gases” which are in a calm state and lie under the influence of the acceleration of gravity or other celestial bodies.

The laws and principles applicable to fluids can also be applied to gases. However, as the density of gases varies with height h when its value is high, like the change in the acceleration of gravity with height, the relationships which have been used for liquids (Chapter 2) should be varied, and these variations should be taken into account to calculate the various quantities of gases accurately.

Therefore, when a gas inside a vessel is in a calm state with a small altitude range between its layers, it shows negligible differences in pressure because of its small density, and thus the quantity ρgh is small. On the contrary, for the highest values of height, the gas layers have different variations, and hence every single variation should be calculated with respect to the height for each layer.

This particularity of gases is mostly applicable in the distribution of various quantities in the earth’s atmosphere, in which the fluid mechanic and thermodynamic behavior is especially interesting for the field of meteorology and aviation. Hence, a special area of knowledge for fluid mechanics has been developed: aerodynamics.

Also, applications for the same behavior of gases are found in various technical situations where large height differences appear, such as ventilation installation in high buildings, underground mine tunnels and diffusers.

Therefore, in this chapter, we will define the basic laws of gases when they are in a calm state.

3.2. General characteristics of gases

Gases have the following general characteristics:

i) They do not have a constant volume: they occupy the entire given space.

ii) They do not form a free surface: gases that are inside a vessel do not display any free surface, but they occupy the whole space in the vessel.

iii) They do not have a constant shape: they take the shape of the vessel in which they are contained.

iv) The cohesive forces of gases, the forces by which their molecules are attracted to each other, are very small: they do not have a constant volume.

v) They show a strong tendency for expansion: this happens because their cohesive forces are very small.

vi) They are compressed.

vii) They have (perfect) volume elasticity.

viii) The motion of their molecules is continuous and their direction is without any particular order.

ix) They apply pressures on the walls of the vessel that contains them.

x) They have weight.

3.3. Pressure applied by air

A gas in a calm state applies two types of pressure forces on each point of the attached surfaces:

a) pressure caused by the motion of its molecules, which is constant and without any order;

b) pressure caused by its weight.

3.3.1. *Pressure caused by the motion of gas molecules*

The gas molecules contained in a vessel move continuously in all directions.

Due to this free motion, these molecules collide with the surfaces of the vessel walls (Figure 3.1a).

Because of this collision, a vertical force is applied on every elementary part of the vessel walls. As a result, a pressure is developed on the vessel walls.

For example, the force developed during the collision of gas molecules with the elementary surface ΔS (Figure 3.1a) gives rise to a resultant $\vec{F}$, vertical to ΔS. The $\vec{F}$ applies a pressure p on ΔS, which is given by the relationship:

$$p = \frac{F}{\Delta S} \quad [3.1]$$

It can be proved that *if a gas contained in a vessel has the same density throughout its volume, then the pressure caused by the motion of molecules has the same value at all of its points.*

3.3.2. *Pressure caused by the weight of gases*

Due to its weight, each layer of gases exerts pressure on the following layer. For example, layer A (Figure 3.1) exerts pressure due to its weight on the following layer, which is B, and layer B in turn transmits this pressure to the next layer C.

Therefore, a gas quantity exerts pressure at all of its points due to its weight. This pressure is similar to the hydrostatic pressure. If its specific weight is equal to its whole volume, then its pressure on a point X (Figure 3.1), due to its weight, is calculated by the relationship:

$$p_X = \gamma \cdot h_X \quad [3.2]$$

where γ is the specific weight of the gas and h_X is the distance of this point from its free surface.

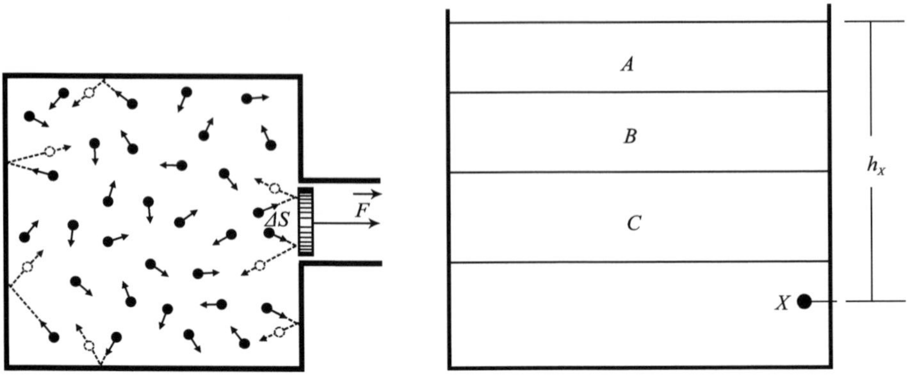

Figure 3.1. *Gas molecule movements and layers inside a vessel*

NOTES.–

The specific weight of a gas quantity that is at rest is not equal to its whole volume (lower layers have higher specific weight), and hence when we calculate the pressure that is caused by its weight, the relationship [3.1] gives us an approximate value.

Since the specific weight of the gas is small, the pressure is considered to be negligible, and therefore it is not calculated for small volumes. When we refer to the pressure of a gas, we mean the pressure that is caused by the motion of molecules, which is continuous and without any order. This means that we do not take into account the pressure due to its weight, as it is very small.

Practically, the pressure of the gas caused by the motion of its molecules has the same value in the whole volume. Therefore, the pressure of a gas inside a vessel of normal size has the same value at all its points.

3.4. Buoyancy: Archimedes' principle

When a body is immersed in a gas that is in a balance state, the gas applies a *vertical* force on every point of the body's surface. The resultant of all the forces that the balancing gas applies on a body immersed in it is called the *buoyancy of the body*. The buoyancy is vertical, acting in the upward direction, and it is applied on the *center of gravity* of the air displaced by the body. The numerical value of a body's buoyancy is equal to the numerical value of the weight of the displaced gas.

Experiment: We balance the sphere Σ (Figure 3.2a) with weights $\boldsymbol{\sigma}$, whose volume is much smaller than the volume of the sphere Σ. When the air is removed (Figure 3.2b) with a pump, we note that the horizontal balance of the phalanx is ruined and leans toward the sphere Σ. This means that the air exerts a force on the sphere Σ, resulting in a vertical resultant force that acts upward, which is the buoyancy.

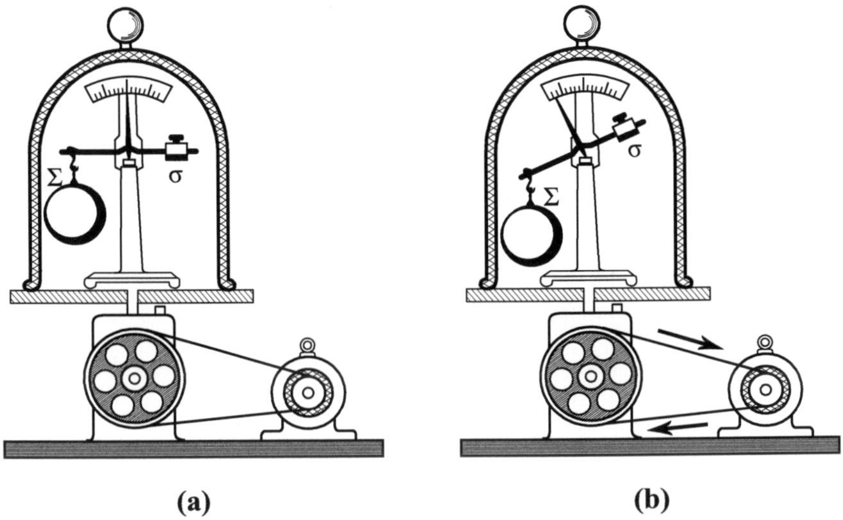

Figure 3.2. *Sphere in a balance experiment*

The air also applies buoyancy to the weights, but it is negligible, because the volume of the weights is much smaller than the volume of the sphere. The resultant of the forces applied to the body is zero, as they cancel each other out. In conclusion, the buoyancy is the resultant of the forces that the air applies to the body due to its weight. Therefore, if the air has no weight, it will not apply buoyancy to a body immersed in it.

The buoyancy A that is applied to each body which is immersed in a balancing gas is a vertical force that acts in the upward direction. Its numerical value is equal to the numerical value of the weight *W of the displaced air*, and its application point is on the center of gravity of the displaced air. This means:

$$A = W \text{ (Archimedes' principle)} \qquad [3.3]$$

If the volume of the displaced gas is v and the specific weight of the air is ε, then the following relationship applies:

$$W = \gamma \cdot \mathrm{v} \quad [3.4]$$

From relationships [3.3] and [3.4], we get:

$$A = W = \gamma \cdot \mathrm{v} \text{ and}$$

$$W = \gamma \cdot \mathrm{v} \ \textit{(Archimedes' principle)} \quad [3.5]$$

NOTES.–

1) The buoyancy that a body experiences in a gas, whose pressure is close to the atmospheric pressure value (e.g. 1 atm), is largely diminished because, in this case, its specific weight is small.

2) The buoyancy developed in light bodies (e.g. hot air balloons) is very high.

3) The buoyancy of a gas is generally smaller than that of fluids because gases have a much lower specific weight than fluids.

3.4.1. *Apparent weight of a body*

The apparent weight W' of a body that is immersed in a gas is the difference between the real (absolute) weight of the body W and its buoyancy A in gases. That is:

$$W' = W - A \quad [3.6]$$

Therefore, the *apparent weight of the body* can be found by determining the weight of a body immersed in a gas, for example, in the air.

To calculate accurately, we must consider the buoyancy caused by the gases on the bodies. The buoyancy of weights is often negligible as their volume is small.

3.5. Hot air balloons

Hot air balloons are arrangements whose total weight near the ground is smaller than the buoyancy applied by air on them, and hence they rise when they are released up in the air.

A hot air balloon (Figure 3.3) consists of an airtight bag Σ (called an envelope) full of a gas whose density is smaller than that of air.

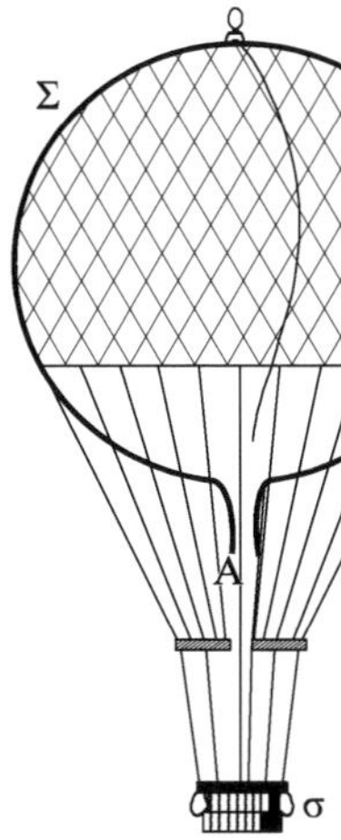

Figure 3.3. *Hot air balloon*

This envelope consists of an elastic material or cloth through which the gases cannot penetrate.

Hot air balloons are usually filled with hot air, gas, hydrogen or helium (helium has the advantage of being non-flammable). The envelope of hot air balloons is surrounded by a net made of rope whose bottom part has a ring, from which hangs a suitable vessel σ.

Aeronauts, sand bags or lead bags (ballasts), scientific instruments, etc. go on board the vessel. On the upper part of the bag, there is a hole (0) that can be shut with a valve. The aeronaut can open this hole when they want to release the gas from the bag. On the bottom part of the bag, there is an open canal A. When the hot air balloon rises, as the atmospheric pressure gets smaller, the pressure of the bag's gas becomes larger than the external pressure, and thus the gas leaves A. Therefore, in each case, the pressure of gas inside the bag is equal to the external pressure.

If the hot air balloon were totally closed, forces would be developed due to the pressure difference which is produced during its lift, and which at a certain height would break the hot air balloon.

A totally shut hot air balloon is used when the hot air balloon has no passengers and, in this case, only scientific instruments are placed inside the vessel, in order to investigate higher levels of the atmosphere. The vessel of a totally shut hot air balloon is provided with a parachute in order to fall with low velocity when its bag breaks.

3.6. Lifting force of a hot air balloon

The force F which lifts (Figure 3.4) the hot air balloon upwards is called the *lifting force (lift)* of the hot air balloon.

On the hot air balloon, the following forces are applied: its buoyancy A and its total weight W_{tot}.

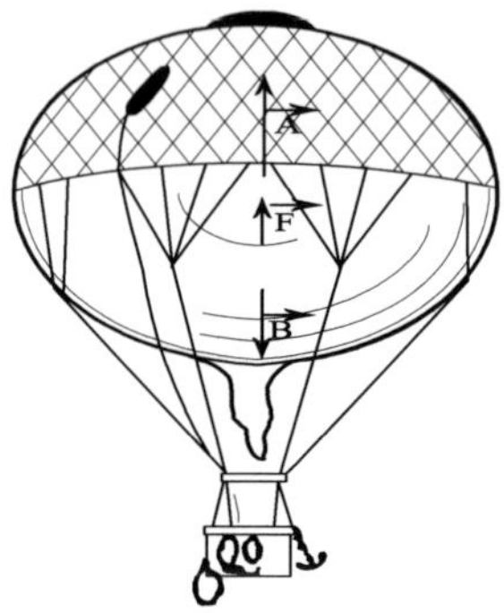

Figure 3.4. *Hot air balloon forces*

Therefore, the lifting force F of the hot air balloon will be:

$$F = A - W_{o\lambda} \text{ or}$$

$$F = A - (W_{\alpha} + W_{\Sigma}) \quad [3.7]$$

where A is the buoyancy of the hot air balloon.

W_{α} = the weight of the gas inside the bag of the hot air balloon;

W_{Σ} = the weight of the bag and various components (vessel, instruments, etc.).

If γ_A is the specific weight of air and γ_α is the specific weight of gas contained inside the hot air balloon's bag, then the following relationships are applicable:

$$A = \gamma_A \cdot \text{v} \quad [3.8]$$

$$W_0 = \gamma_\alpha \cdot \text{v} \quad [3.9]$$

where v is the volume of the hot air balloon.

From relationships [3.7], [3.8] and [3.9], we get:

$$F = A-(W_\alpha + W_\Sigma) = \gamma_A \cdot \text{v} - (\gamma_\alpha \cdot \text{v} + W_\Sigma)$$

$$F = \gamma_A \cdot \text{v} - \gamma_A \cdot \text{v} - W_\Sigma \qquad [3.10]$$

$$F = (\gamma_A - \gamma_\alpha) \cdot \text{v} = -W_\Sigma)$$

From relationship [3.10], we know that during the lift of the hot air balloon, its lifting force reduces, because the specific weight γ_A of air reduces while the hot air balloon rises.

3.7. Basic aerostatic law

We consider a gas with large volume, which is in a calm state and in balance, and assume an elementary cylinder with cross-sectional area δS and height δh filled with this gas, in such a way that gravity influences the direction of its axis. In this cylinder, as shown in Figure 3.5, a pressure p is developed on the bottom base, a pressure $p+\delta p$ on the top base and its weight W.

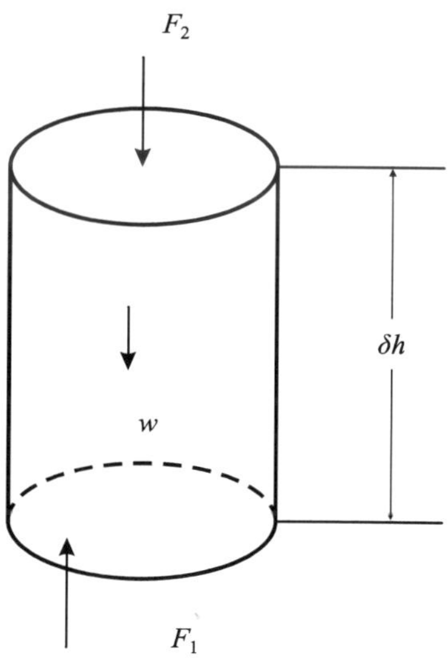

Figure 3.5. *Cylinder with gas in balance*

Also, pressures are developed on the vertical surface, but they are ignored due to the pressure balance. Therefore, as the cylinder balances, the sum of the forces applied on it will be equal to zero. Thus, we have:

$$F_2 + W - F_1 = 0$$

or $(p+\delta p)\delta S + mg - p\delta S = 0$

or $(p+\delta p)\delta S + \rho v g - p \cdot \delta S = 0$

or $(p+\delta p)\delta S + \rho \cdot \delta S \cdot \delta h \cdot g - p \cdot \delta S = 0$

or $\delta p + \rho g \ \delta h = 0$

or $\dfrac{\delta p}{\delta h} = -\rho g$ [3.11]

Considering the limit as δh tends to be zero, [3.11] becomes:

$$\frac{dp}{dh} = -\rho g \qquad [3.12]$$

This expression shows that when a fluid is at rest, its pressure varies with the height, and this constitutes the *basic aerostatic law.*

3.8. Gas pressure variations: the Boyle–Mariotte law

A certain mass of a gas can have various volumes. However, when the volume of a certain mass of the gas changes, the pressure also changes. The existing relationship between the pressure, which a gas mass obtains when it occupies a volume, and the volume is expressed by the Boyle–Mariotte law.

The Boyle–Mariotte law sets the following:

*At a constant temperature, pressure (**p**) multiplied by volume (**v**) of a certain mass (**m**) of a gas remains constant. That is*:

$p, \mathrm{v} = \text{constant}$ (Boyle–Mariotte law) [3.13]

If for instance, the pressure of a gas is p_1 when its volume is v_1 and p_2 when its volume turns to v_2, then as its temperature remains constant, the following relationship applies:

$$p_1 \cdot v_1 = p_2 \cdot v_2 = \text{constant} \qquad [3.14]$$

From the above relationships, we get:

$$p = \frac{const}{v} \qquad [3.15]$$

$$\text{or } \frac{p_1}{p_2} = \frac{v_2}{v_1} \qquad [3.16]$$

From the above relationships arises the following statement of the law of Boyle–Mariotte: the pressure of a certain mass (*m*) of a gas under constant temperature is inversely proportional to its volume.

Therefore, if the volume (*m*) of the gas doubles, its pressure halves; if its volume halves, the pressure doubles. When the volume and gas pressure vary with temperature remaining constant, it is called isothermal variation.

3.8.1. *The Boyle–Mariotte law*

The relationship $p.v = \text{constant}$ is known as the Boyle–Mariotte law, as shown in Figure 3.6, which is called the *isothermal curve*, because it represents the isothermal pressure variations of the gas and the volume.

Volume (lt)	Pressure (atm)	lt×atm
1	8	8
2	4	8
4	2	8
8	1	8

Table 3.1. *Values related to the Boyle–Mariotte graph*

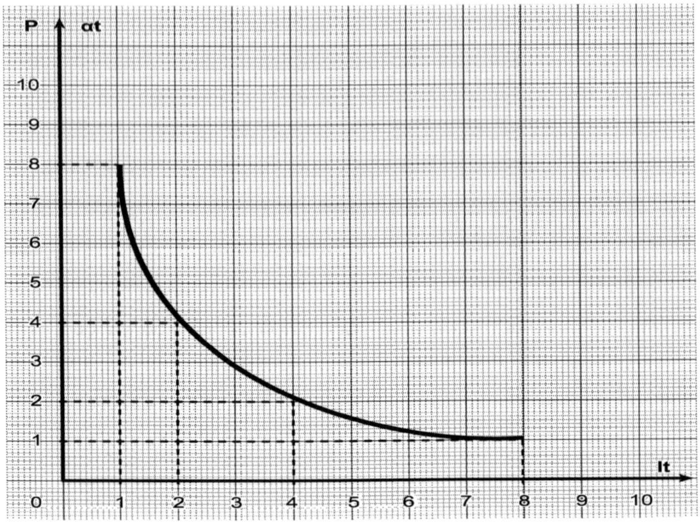

Figure 3.6. *Boyle–Mariotte graph*

3.9. Changes in gas density

If a quantity has mass m and volume v_1, then its density ρ is given by the relationship:

$$\rho_1 = \frac{m}{v_1} \quad [3.17]$$

If the volume of a quantity m of a gas turns to v_2, then its density ρ_2 is given by the relationship:

$$\rho_2 = \frac{m}{v_2} \quad [3.18]$$

By dividing relationships [3.17] and [3.18], we have:

$$\frac{\rho_1}{\rho_2} = \frac{\frac{m}{v_1}}{\frac{m}{v_2}} = \frac{v_2}{v_1} \Rightarrow$$

$$\Rightarrow \frac{\rho_1}{\rho_2} = \frac{v_2}{v_1} \qquad [3.19]$$

If p_1 is the pressure of the gas mass m with volume v_1 and p_2 with volume v_2, while its temperature remains constant, then the following relationship applies:

$$\frac{p_1}{p_2} = \frac{v_2}{v_1} \qquad [3.20]$$

From relationships [3.19] and [3.20], we have:

$$\frac{\rho_1}{\rho_2} = \frac{p_1}{p_2} \qquad [3.21]$$

Relationship [3.21] expresses that when the temperature of a gas quantity remains constant, its density is proportional to its pressure.

3.10. The atmosphere

The air that surrounds the earth is called the atmosphere. The atmospheric air is mostly a mixture of oxygen (23.1% by weight and 20.93% by volume) and nitrogen (75.5% by weight and 78.1% by volume). It also contains small quantities of other gases like hydrogen, helium and rare gases like argon, krypton, neon, etc.

It is natural that the behavior of an aerodynamic body, which is inside the moving air, will depend on its physical properties. Airplanes move in the atmosphere and therefore a detailed knowledge of the properties of air inside the atmosphere is substantial for every behavioral study of an airplane. Provided that the atmosphere can be considered to be a fluid at rest, we can use the *aerostatic theory* to calculate its macroscopic properties. The most important of these properties are temperature, pressure, density and viscosity, whose exploitation according to height variation and space is necessary.

The atmosphere can be divided into two separate areas, whose height depends on the latitude and season. The lower layer of the atmosphere until a height of 11 km is called the *troposphere*, where it has been found that the temperature reduces linearly with height. The higher area, which reaches up to 35 km, is called the stratosphere, in which the temperature remains constant with respect to height. Between these two areas there is supposed to be a limit called the *tropopause*.

The distribution of the atmospheric values is given by the basic law of aerostatics. In the real atmosphere, the pressure varies with height and the acceleration of gravity varies with height and latitude.

As the mathematical calculations for the specification of the change in atmospheric values are not accurate based on certain assumptions, defined standards of the atmosphere are used to make these calculations easier. The more complex standards are used in accurate applications, and the calculation results are given in special charts called the standard atmosphere. In general, the simpler standards have prevailed for the isothermal, isentropic and multimodal atmospheres. Today, two standard atmospheres based on the isentropic atmosphere are used, which are:

a) the Standard Atmosphere of CIRA (Cospar International Reference Atmosphere, 1972);

b) the International Standard Atmosphere (ISA, 1962).

The first one is continually improved and is used for scientific purposes, and the second one is used in aviation and spaceships following specific international agreements.

3.10.1. *International Standard Atmosphere (ISA)*

As mentioned previously, an airplane's performance depends absolutely on the physical properties of air in which it moves. Also, we mentioned that the values of those properties vary not only with the height from the sea surface but also according to space and time.

It is therefore obvious that the observations for the specification of the performances, as, for example, in the calibration of the equipment, etc., are capable of being compared with each other when they are based on similar atmospheric conditions. For this purpose, in aeronautics and meteorology, there are several standard atmospheres, national and international. One of these is the International Standard Atmosphere (ISA) by which a single value system is established for the atmospheric conditions that exist in various heights in all countries. These values have emerged as an average of the atmospheric condition values that dominate during the largest part of the year in temperate zones, such as in Europe and North America, and they are used in most countries.

Thus, the definition of the standard atmosphere is based on the following conditions:

1) The standard temperature of the air on the sea surface is:

$$T_O = 288.2°\text{K} \quad [3.22]$$

2) With an increase in height to 11 km, the temperature reduces with a constant value, which is called the *lapse rate*, denoted by λ. For the standard atmosphere, λ is:

$$\lambda = 6.50°\,\text{K/km} \qquad [3.23]$$

which is given by the relationship:

$$\lambda = -\frac{dT}{dh} \qquad [3.24]$$

Because the lapse rate is constant, we will have:

$$T = T_O - \lambda\, h \qquad [3.25]$$

where T_O is the absolute temperature on the sea surface. This area of this atmosphere is the *troposphere*.

3) In the area between 11 and 20 km, the temperature remains constant as the height increases. This area is the *stratosphere*. The point at which the temperature stops decreasing is called the *tropopause* and, at this point, the temperature is:

$$T = 216.5°\text{K} \qquad [3.26]$$

The variation of the temperature with respect to the height for the previous areas is shown in Figure 3.7.

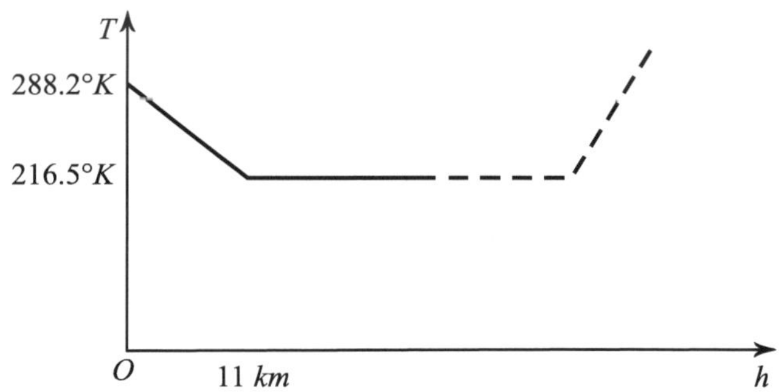

Figure 3.7. *Temperature variation due to height*

4) The standardized pressure of air at sea level is:

$$p_O = 1.013 \times 10^5\ \text{N/m}^2 \qquad [3.27]$$

The pressure caused by the weight of air reduces with increasing height. In the tropopause, the pressure decreases a little more than 1/5 of its value on the sea surface. The variation of pressure due to height is shown in Figure 3.8.

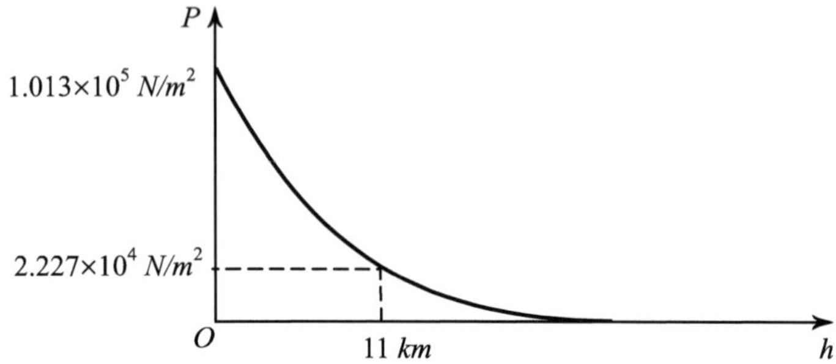

Figure 3.8. *Pressure variation due to height*

5) The standardized density on the sea level is:

$$\rho = 1.255\,\text{kg/m}^3 \qquad [3.28]$$

The density also reduces with height and, in fact, in the tropopause, its value is almost 3/10 of its value on the sea surface. The variation of density is shown in Figure 3.9.

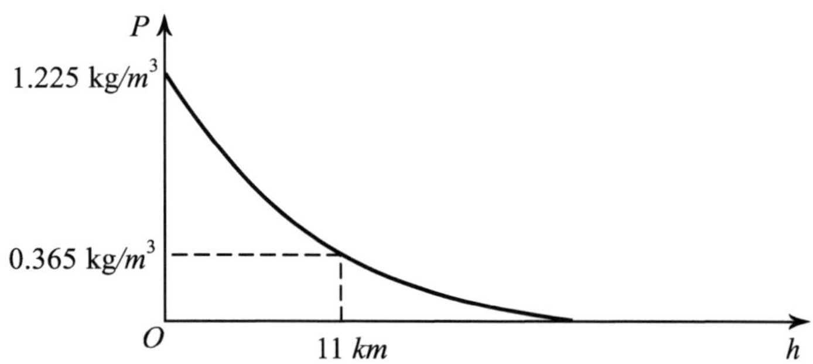

Figure 3.9. *Density variation due to height*

At this point, we note that the density ρ is associated with pressure p and the temperature T with the gas state equation, which is given by the relationship:

$$\frac{p}{\rho} = RT \qquad [3.29]$$

$$\text{or } \rho = \frac{1}{R} \cdot \frac{p}{T} \quad [3.30]$$

where R is the universal gas constant, whose numerical value is:

$$R = 287.26 \text{ J/kg°K} \quad [3.31]$$

6) The dynamic viscosity coefficient of air (μ_0) on the sea level in the standard atmosphere is:

$$\mu_O = 1.79 \times 10^{-5} = \frac{N \cdot s}{m^2} \quad [3.32]$$

The viscosity coefficient for every gas, as we have already mentioned, depends on its temperature and increases due to it. Therefore, μ reduces constantly with increasing height, until it reaches the tropopause, where its numerical value is:

$$\mu = 1.42 \times 10^{-5} \text{ kg/ms} \quad [3.33]$$

In the stratosphere, its numerical value is *constant*.

7) The kinematic viscosity v in the standard atmosphere has a numerical value on the sea level, which is given by the relationship:

$$v_O = 1.46 \times 10^{-5} \text{ m}^2/\text{s} \quad [3.34]$$

As the kinematic viscosity depends on the viscosity μ and on the density ρ ($v = \mu/\rho$), its variation according to the height will depend on the variation of those values. Therefore, the kinematic viscosity increases with height in both the troposphere and the stratosphere. This is because in the troposphere, μ and ρ reduce with height, but ρ reduces faster. In the stratosphere, ρ reduces with height, while μ remains constant.

8) Finally, we also define the value of the speed α of sound in the standard atmosphere, which on the sea level is:

$$c_O = 340 \text{ m/s} \quad [3.35]$$

The speed of sound, which will be dealt with in the next chapter, depends only on the air temperature. Therefore, in the troposphere, it reduces constantly with height, while it is constant in the stratosphere and equal to:

$$c_\sigma = 295 \text{ m/s} \quad [3.36]$$

The benefit of establishing the standard atmosphere is that using the thermodynamic relationships, which give us the variation of various values, it is possible to calculate these values for the middle heights.

The values of pressure, density, temperature, speed of sound, viscosity, kinematic viscosity and all the other relevant values, in terms of height, are given in Table 3.2 as published in the editions of the standard atmosphere.

International Standard Atmosphere						
h (km)	T (K)	c (m/s)	p (N/m^2) $\times 10^{-4}$	ρ (kg/m^3)	μ (kg/ms) $\times 10^{-5}$	ν (m^2/s) $\times 10^{-5}$
0	288.2	340.3	10.132	1.225	1.789	1.461
1.0	281.7	366.4	8.989	1.112	1.758	1.581
2.0	275.2	332.5	7.950	1.007	1.726	1.715
3.0	268.7	328.6	7.012	0.909	1.694	1.863
4.0	262.2	324.6	6.166	0.819	1.661	2.028
5.0	255.7	320.5	5.405	0.736	1.628	2.211
6.0	249.2	316.5	4.722	0.660	1.595	2.416
7.0	242.7	312.3	4.111	0.590	1.561	2.646
8.0	236.2	308.1	3.565	0.526	1.527	2.904
9.0	229.7	303.8	3.080	0.467	1.493	3.196
10.0	223.2	299.5	2.650	0.414	1.458	3.525
11.0	216.7	295.1	2.270	0.365	1.422	3.898
12.0	216.7	295.1	1.940	0.312	1.422	4.558
13.0	216.7	295.1	1.658	0.267	1.422	5.333
14.0	216.7	295.1	1.417	0.228	1.422	6.239
15.0	216.7	295.1	1.211	0.195	1.422	7.300
16.0	216.7	295.1	1.035	0.167	1.422	8.540
17.0	216.7	295.1	0.885	0.142	1.422	9.990
18.0	216.7	295.1	0.757	0.122	1.422	11.686
19.0	216.7	295.1	0.647	0.104	1.422	13.670
20.0	216.7	295.1	0.553	0.089	1.422	15.989

Table 3.2. *International standard atmosphere*

3.11. Formulae

1) Pressure *p*:

$$p = \frac{F}{\Delta S}$$

where F is the resultant force and ΔS is the elementary surface.

2) Archimedes' principle:

$$A = \gamma \mathrm{v}$$

where A is the buoyancy, γ is the specific weight of gas and v is the volume.

3) Lifting force of an air balloon (lift):

$$F = (\gamma_A - \gamma_\alpha)(\mathrm{v} - \mathrm{v}_\Sigma)$$

where F is the lifting force, γ_A is the specific weight of air, γ_α is the specific weight of gas inside the hot air balloon, v is the volume of the hot air balloon and v_Σ is the volume of the bag, vessel, instruments, etc.

4) Basic law of aerostatics:

$$\int_1^2 \delta_p = -\rho g \int_1^2 dh$$

where p is the pressure of air, ρ is the density of air, g is the acceleration of gravity and h is the height of the body.

5) Law of Boyle–Mariotte:

$$p_1 \mathrm{v}_1 = p_2 \mathrm{v}_2 = \mathrm{const}$$

or

$$\frac{p_1}{p_2} = \frac{v_2}{v_1}$$

where p_1 is the pressure of gas when its volume is v_1 and p_2 is the pressure of gas when its volume is v_2, at constant temperature.

6) Gas density changes with pressure:

$$\frac{\rho_1}{\rho_2} = \frac{p_1}{p_2}$$

where ρ_1, ρ_2 are densities of gas and p_1, p_2 are their respective pressures.

7) Lapse rate λ:

$$T = T_0 - \lambda h$$

where T is the absolute temperature in height h, T_0 is the absolute temperature on the sea surface (288.2 K), λ is the lapse rate (6.5 K/km) and h is the geometrical height from the sea surface.

8) International Standard Atmosphere (ISA) values:

$$T_0 = 288.2\,\text{K}$$
$$p_0 = 1.013\times10^{-5}\ \text{N/m}^2 = 1.013$$
$$\rho_0 = 1.255\ \text{kg/m}^3$$
$$\mu_0 = 1.79\times10^{-5}\ \text{kg}^2/\text{ms}$$
$$v_0 = 1.46\times10^{-5}\ \text{m}^2/\text{s}$$
$$c_0 = 340\,\text{m/s}$$

where T_0 is the absolute temperature, p_0 is the pressure, ρ_0 is the density, μ_0 is the viscosity, v_0 is the kinematic viscosity and c_0 is the speed of sound.

3.12. Questions

1) What are the general characteristics of a gas?

2) Does a gas have any weight? Give an experimental explanation of your answer.

3) How is the gas pressure caused by the motion of its molecules defined and how do you explain it?

4) How is the gas pressure caused by its weight defined? Give an example.

5) When Archimedes' principle is applicable for a body, what do we call buoyancy?

6) What is a hot air balloon? How can you describe one?

7) What is the lifting force of hot air balloons called? Explain it graphically and prove its relationship.

8) Which values are related to the basic law of aerostatics? How can this be proved?

9) Which law defines the pressure variation of gases to their volume? State it verbally and mathematically.

10) What is isothermal variation called? Explain it graphically, for random values of pressure and volume of a gas quantity.

11) Prove mathematically the gas density variation with pressure, when the temperature remains constant.

12) How is the earth's *atmosphere* defined?

13) How is the *temperature* varied due to height in the troposphere and the stratosphere?

14) What is International Standard Atmosphere (ISA)?

15) Why is the ISA necessary?

16) What is the *temperature* on the sea surface in the ISA? How it varies with height in the area of the *troposphere?*

17) What is the value of the temperature in the ISA in the *stratosphere?*

3.13. Problems with solutions

1) Under normal conditions ($\Theta = 0°\mathrm{C}$ and $p = 760\,\mathrm{mm\,Hg}$), what is the weight of one liter of air and one liter of water if their specific weights are

$\gamma_A = 0.001293\, p / cm^3$ and $\gamma_N = 1\, p / cm^3$, respectively? How much larger is the density of water ρ_N than the density of air ρ_A ?

Solution:

The following relationships apply:

$$\gamma_A = \frac{W_A}{v_A} \tag{1}$$

$$\gamma_N = \frac{W_N}{v_N} \tag{2}$$

From relationships (1) and (2), we get:

$$W_A = \gamma_A \cdot V_A \tag{3}$$

$$W_N = \gamma_N \cdot V_N \tag{4}$$

If we define in relationship (3) the given facts, we calculate the weight of one liter of air:

$$W_A = 0.001293 \frac{p}{cm^3} \cdot 1000\, cm^3 = 0.001293 \times 1000 \frac{p \times cm^3}{cm^3}$$

$$W_A = 1.293 \frac{p}{cm^3}$$

If we define in relationship (4) the given facts, we calculate the weight of one liter of water:

The following relationships apply:

$$\gamma_N = g \cdot \rho_N \tag{5}$$

$$\gamma_A = g \cdot \rho_A \tag{6}$$

From relationships (5) and (6), we get:

$$\frac{g \cdot \rho_N}{g \cdot \rho_A} = \frac{\gamma_N}{\gamma_A}$$

$$\frac{\rho_N}{\rho_A} = \frac{\gamma_N}{\gamma_A} = \frac{1 \cdot \frac{p}{cm^3}}{0.001293 \frac{p}{cm^3}}$$

$$\frac{\rho_N}{\rho_A} = 773 \text{ and } \rho_N = 773 \cdot \rho_A$$

2) A hot air balloon has volume $\mathrm{v} = 50\,\mathrm{m}^3$. Its casing and other equipment weigh 700 p. The hot air balloon is full of hydrogen. The external air and hydrogen have temperature 0 °C and pressure $p = 76\,\mathrm{cmHg}$. In this case, the specific weights of air and hydrogen are $\gamma_A = 1.32\, p/lt$ and $\gamma_H = 0.09\, p/lt$, respectively. What is the lifting force (F) of the hot air balloon when it takes off?

Solution:

The lifting force F of the hot air balloon is given by the relationship:

$$F = A - W_{tot} \tag{1}$$

where $W_{o\lambda}$ is the total weight of the hot air balloon.

The total weight of the hot air balloon is:

$$W_{tot} = W_1 + W_H \tag{2}$$

where W_1 is the weight of the casing and other equipment of the hot air balloon.

W_H is the weight of hydrogen contained inside the hot air balloon.

From relationships (1) and (2), we have:

$$F = A - (W_1 + W_H) \tag{3}$$

The following relationships are applicable:

$$A = \mathrm{v} \cdot \gamma_A \tag{4}$$

$$W_H = \mathrm{v} \cdot \gamma_H \tag{5}$$

where v is the volume of the hot air balloon (of course, the volumes of the displaced air and hydrogen are equal to v).

From relationships (3), (4) and (5), we get:

$$F = \mathrm{v} \cdot \gamma_A - W_1 - \mathrm{v} \cdot \gamma_H$$

$$F = \mathrm{v}(\gamma_A - \gamma_H) - W_1 \tag{6}$$

If we set the given facts in relationship (6), we get:

$$F = 50,000\,lt \cdot (1.3 - 0.09)\frac{p}{lt} - 700\,p$$

$$F = 50,000 \times 1.2\frac{lt \cdot p}{lt} - 700\,p$$

$$F = 59,800\,p = 59.8\,kp$$

3) A mass of air m has a temperature $\Theta = 20°\mathrm{C}$, a volume $\mathrm{v}_1 = 10\,\mathrm{cm}^3$ and a pressure $p_1 = 4\alpha t$. What is its pressure p_2 if its volume reaches $\mathrm{v}_2 = 20\,\mathrm{cm}^3$ while its temperature remains constant $(T = 20°\mathrm{C})$?

Solution:

The variation of air is isothermal ($T = 20° = \text{constant}$) and thus the following relationship applies:

$$p_1 \cdot \mathrm{v}_1 = p_2 \cdot \mathrm{v}_2$$

From relationship (1), we get:

$$p_2 = \frac{p_1 \cdot \mathrm{v}_1}{\mathrm{v}_2}$$

If in relationship (2) we set the given facts, we find:

$$p_2 = \frac{4\alpha t \cdot 10cm^3}{20cm^3} = \frac{4 \times 10 \cdot \alpha t \cdot cm^3}{20cm^3} = \frac{40}{20} = 2\alpha t$$

4) Vessels I and II contain a fluid of specific weight ε, which are connected to a closed manometer that contains a manometer fluid of specific weight $\gamma_m < \gamma$. The difference in the level between α and b is Δ. Calculate the pressure difference between point 1 of the first vessel and point 2 of the second vessel placed at a height Δz. The arithmetic facts are $\gamma = 960\,\text{kg/m}^3$, $\gamma_m = 860\,\text{kg/m}^3$, $\Delta z = z_1 - z_2 = 0.86\,\text{m}$, $\Delta = 0.15\,\text{m}$.

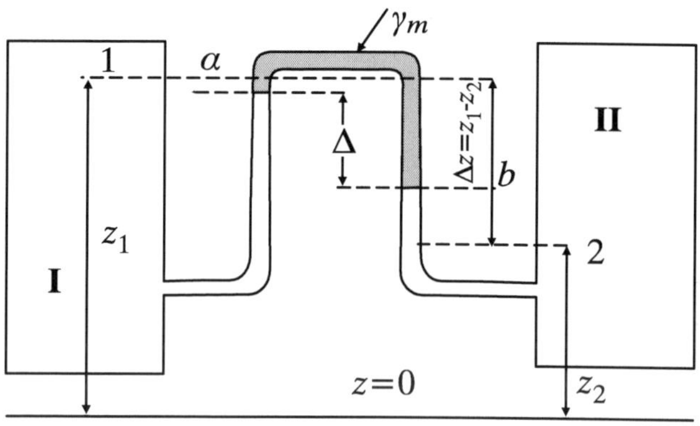

Solution:

We note that in order for the manometer shown in the figure to operate, it must be $\gamma_m < \gamma$, otherwise there will be an unstable balance state.

For the solution, we equate the pressure gauge heights between points 1 and α for the fluid of vessel I, considered to be a reference level ($z = 0$) at any horizontal level. Thus, from the altitude equation, we have:

$$\frac{dp}{dz} = -\gamma \Rightarrow dp = -\gamma dz \Rightarrow \int_{p_\alpha}^{p_1} dp = -\gamma \int_{z_\alpha}^{z_1} dp$$

$$\Rightarrow p_1 - p_\alpha = -\gamma(z_1 - z_\alpha) \Rightarrow$$

$$\Rightarrow \frac{p_1}{\gamma} - \frac{p_\alpha}{\gamma} = -z_1 + z_\alpha \Rightarrow$$

$$\Rightarrow \frac{p_1}{\gamma} + z_1 = \frac{p_\alpha}{\gamma} + z_\alpha \qquad (1)$$

Similarly, for the fluid of the tube (points α and b) and the fluid of vessel II (points b and 2), we take respectively:

$$z_\alpha + \frac{p_\alpha}{\gamma_m} = z_b + \frac{p_b}{\gamma_m} \tag{2}$$

and $z_2 + \dfrac{p_2}{\gamma} = z_b + \dfrac{p_b}{\gamma_m}$ (3)

From equation (2), we have:

$$\Delta = z_\alpha - z_b = \frac{p_b - p_\alpha}{\gamma_m} \tag{4}$$

Subtracting (3) from (1), we get:

$$z_1 - z_2 = \Delta z = \frac{p_2 - p_1}{\gamma} + (z_\alpha - z_b) + \frac{p_\alpha - p_b}{\gamma} \Rightarrow$$

$$\Rightarrow \Delta z = \frac{p_2 - p_1}{\gamma} + (z_\alpha - z_b) + \frac{p_\alpha - p_b}{\gamma}$$

$$\Rightarrow \gamma \Delta z = p_2 - p_1 + \gamma (z_\alpha - z_b) + p_\alpha - p_b$$

$$\Rightarrow p_2 - p_1 = \Delta p = \gamma \Delta z - \gamma (z_\alpha - z_b) + p_b - p_\alpha$$

$$\Rightarrow \Delta p = \gamma \Delta z - \gamma (z_\alpha - z_b) + p_b - p_\alpha .$$

Combining this with (4), we have:

$$\Delta p = \gamma \Delta z - \gamma \Delta + \gamma_m \Delta \Rightarrow \Delta p = \gamma \Delta z - \Delta (\gamma - \gamma_m) \tag{5}$$

Moreover, substituting it into the values, we have:

$$\Delta p = 960 \cdot 0.86 - 0.15(960 - 860) = 810.6 \text{ N/m}^2 . \tag{6}$$

5) Calculate the value of pressure p_2 at point P and height h_0 as shown in the following figure. $p_1 = 760 \text{ mm Hg}$, $\rho = 1000 \text{ kg/m}^3$, $h_1 = 2.2 \text{ m}$ and $h_2 = 1.6 \text{ m}$.

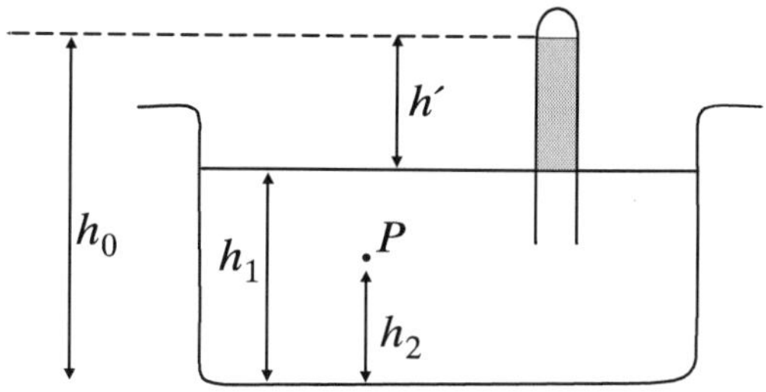

Solution:

The fundamental equation of the hydrostatics is written for this case as:

$$p_2 - p_1 = \gamma(h_1 - h_2)\,, \tag{1}$$

where p_2 and h_2 are the pressure and the height; at point *P*, p_1 and h_1 are the pressure and the height of the free surface of the fluid, respectively.

From (1), we have:

$$p_2 = p_1 + \gamma(h_1 - h_2) \tag{2}$$

and

$$p_2 = 760 \cdot 133.319 + 1000 \cdot 981(2.2 - 1.8) = 107.22\,\text{N/m}^2\,. \tag{3}$$

Using the state equation of the air statics to calculate the height h_0, we have:

$$dp = \gamma dz \Rightarrow \int_{p_1}^{p_0} dp = \gamma \int_{h_1}^{h_0} dz \Rightarrow$$

$$\Rightarrow p_0 - p_1 = -\gamma(h_0 - h_1) \Rightarrow$$

$$\Rightarrow \frac{p_0}{\gamma} - \frac{p_1}{\gamma} = -h_0 + h_1 \tag{4}$$

and because $p_0 = 0$, the above equation becomes:

$$-\frac{p_1}{\gamma} = -h_0 + h_1 \Rightarrow -\frac{p_1}{\gamma} - h_1 = -h_0 \Rightarrow h_0 = \frac{p_1}{\gamma} + h_1 \tag{5}$$

and by substituting:

$$h_0 = \frac{760 \cdot 133.319}{1000 \cdot 9.81} + 2.2 = \frac{101 \cdot 392.44}{9810} + 2.2 = 12.53\,\text{m} . \qquad (6)$$

6) If an airplane flies at an altitude, where the real values of the pressure p and the temperature T are, respectively, $p = 4.72 \times 10^4$ N/m^2 and $T = 255.7°\text{K}$, what are the altitude values of pressure, temperature and density?

Solution:

From Appendix A, we find that for $p = 4.72 \times 10^4$ N/m^2, the altitude is 6,000 m.

Likewise, for temperature, the altitude is 5,000 m.

For density, we first have to find its value from the relationship:

$$\rho = \frac{p}{RT} = \frac{4.72 \times 10^4}{287(255.7)} = 0.643\,\text{kg/m}^3 .$$

From the appendix, we find that for density $\rho = 0.643\,\text{kg/m}^3$, the altitude is approximately 6,240 m.

7) Calculate the values of temperature, pressure and density at a height of 10,200 m according to the ISA.

Solution:

From the table of the standard atmosphere, we have:

	Temperature (K)	**Pressure (P_a)**	**Density (kg/m^3)**
10,000 m	223.26	26,500	0.41351
10,500 m	220.02	24,540	0.38857
10,200 m			

We make a linear interpolation:

$$\frac{10{,}500 - 10{,}000}{10{,}200 - 10{,}000} = \frac{5}{2} = 2.5$$

$$\frac{220.02 - 223.26}{T_{10,200} - 223.26} = \frac{5}{2}$$

or $T_{10,200} = 221.964$

Likewise, for $p_{10,200}$ and $\rho_{10,200}$.

8) An airplane flying at a geometrical height of 20,000 ft has the following instrument indications: $p = 900\,\text{lb/ft}^2$ and $T = 460°\,R$. Calculate the heights of pressure, temperature and density with an accuracy of 500 ft.

Solution:

From: $p = 900\,\text{lb/ft}^2 \Rightarrow h_p = 22,000\,\text{ft}$

From: $T = 460°\,R \Leftrightarrow hr = 16,500\,\text{ft}$

It is: $\rho = \dfrac{p}{RT} = \dfrac{900}{1,716 \cdot 460} = 0.00114\,\text{slug/cuft} \Leftrightarrow h_{\tilde{n}} = 23,000\,\text{ft}$

9) Using the data from the appendix, what are the conditions of the standard atmosphere at 5,500 m and 5,700 ft?

Solution:

$T = 252.44$ K	$T = 498.38\ R$
$p = 505.34$ Pa	$p = 1,715.5$ psf
$\rho = 0.6975\,\text{kg/m}^3$	$\rho = 2005.3 \times 10^{-6}$ slug/ft^3

10) What are the conditions of the standard atmosphere at 40,000 ft?

Solution:

$T = 390\ R$

$p = 393.1\ \text{lb/ft}^2$

$\rho = 0.5873 \times 10^{-3}\ \text{slug/ft}^3$

3.14. Problems to be solved

1) It is given that the barometric pressure is 750 mm Hg. A U-shaped tube is filled with alcohol of specific weight 0.82. One edge of the tube is connected to a point of a wing model, which is placed in a wind tunnel, while the other is open to the atmosphere. The fluid in the first part comes up to 25 mm higher than the other part. Calculate the pressure difference between the model and the atmosphere and the absolute pressure at that point. The specific weight of mercury is 13.6 N/m^3 and the density of water is 1,000 kg/m^3.

2) Assuming that the pressure and the temperature on the surface of the sea are 100,500 N/m^2 and 293 K while, at an unknown height, the pressure is 71,800 N/m^2 and the temperature is 263 K, answer the following questions:

i) Is the atmosphere between the two heights constant?

ii) Calculate the height in which the second pair of measurements was taken. Can the variation of temperature and height be considered to be linear?

3) The indication of the differential manometer *A* is 0.86 atm. The differential as well as manometer *B* is exposed to the atmospheric pressure, which is 1.62 atm. If 1 atm = 760 mm Hg, calculate the absolute pressure that dominates inside vessel A:

i) in atm, ii) in N/m^2, iii) in kp/cm^2 and iv) in psi.

It is given that $\rho_{Hg} = 13,600\,\text{kg/m}^3$.

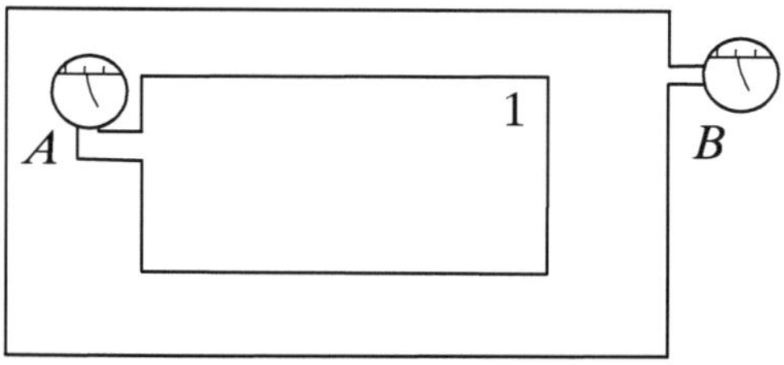

4) A manual hydraulic presser, shown in the figure, is used for the formation of a metal leaf in a specific shape. To do so, if the presser must apply a force 3t, calculate the force F that is required to achieve this formation from the operator. It is given that the diameters of the plunger are, respectively, $D_1 = 5\,\text{cm}$ and $D_2 = 5\,\text{cm}$, and the fluid can be considered to be incompressible.

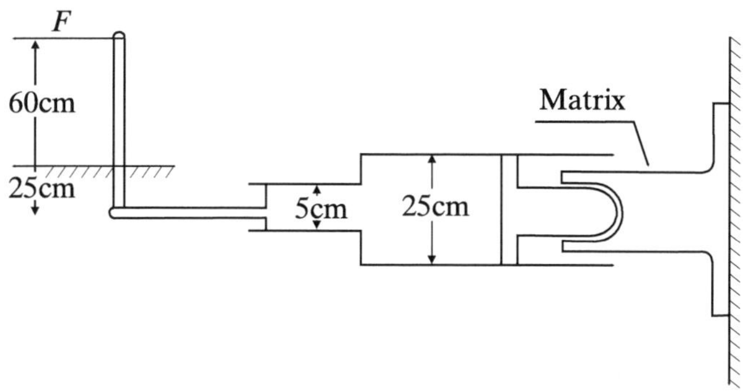

5) A cylinder vessel of diameter D = 10 m filled with water has a cap on the vertical tube of diameter 0.01 m that communicates with the internal part of the vessel. Ignoring the compressibility of water, calculate the mass m of water in kg that must be added to the vertical tube, so that the total force applied by the water on the cap due to the hydrostatic pressure is equal to 400 N.

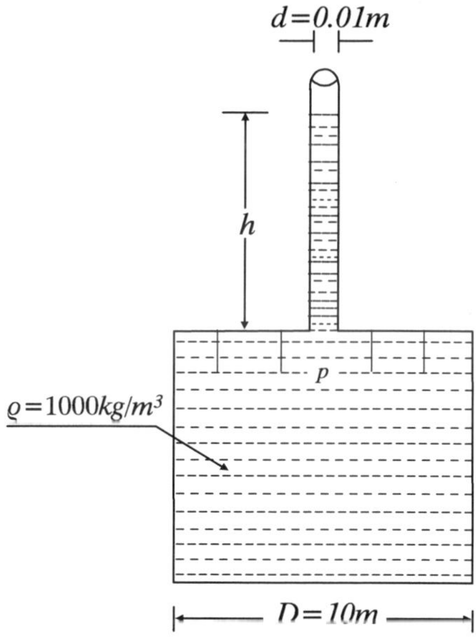

6) For the following figure where $p_A = 40\,\text{psi}$ and $p_B = 20\,\text{psi}$, calculate the differential height h of mercury columns.

It is given that $\rho_{water} = 100\,\text{kg/m}^3$ $\rho_{Hg} = 13{,}600\,\text{kg/m}^3$

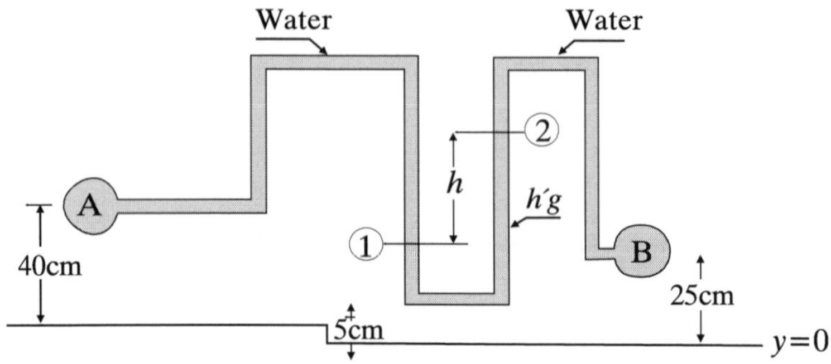

7) In the standard atmosphere for an altitude of 12 km, the pressure, density and temperature are $1.9399\times10^{4}\ \text{N/m}^{2}$, $3.1194\times10^{-1}\ \text{kg/m}^{3}$ and 216.66 K, respectively. Using these values, calculate the values of pressure, density and temperature in the standard atmosphere at an altitude of 18 km and check their accuracy using the tables in the appendix.

8) Consider an airplane which flies at a certain altitude, where the external temperature and pressure are 220 K and $2.65\times10^{4}\ \text{N/m}^{2}$, respectively. What are the heights of the pressure and density.

9) Consider an airplane that flies at a pressure height of 33,500 ft and a density height of 32,000 ft. Calculate the external temperature of air.

10) A passenger jet flies in the standard atmosphere at a height of 30,000 ft with a velocity of 500 mi/h. What is the Mach number of the flow?

11) A bubble of air, with volume $v_1 = 0.03\,\text{cm}^3$, is stuck to the wall of a vessel containing water. The bubble lies 12 cm below the free surface of the water. The atmospheric pressure is $p_A = 74\,\text{cm Hg}$. What is the volume of the bubble if the atmospheric pressure reaches $p_A = 76\,\text{cm Hg}$ while its temperature remains constant?

12) Under normal conditions ($\Theta = 0$ °C and $p_0 = 1\,\text{atm}$), the specific weight γ of a gas is $\gamma_0 = 1.293\,\text{p/lt}$. What is the specific weight γ at temperature 0 °C and pressure 50 atm?

4

Fluid Flow

4.1. Introduction

The study of kinematic and dynamic characteristics of fluid flow is based on certain principles or universal natural laws, which have been stated as a conclusion of natural observations and experience. These laws have axiomatic statements and constitute the principles on which fluid mechanics is based, and they are indirectly proved with comparison of the results of mathematical analysis and experimental facts.

These principles are:

a) mass conservation;

b) momentum conservation;

c) energy conservation.

These three principles constitute the bases on which fluid flow is defined, and they provide the laws that rule it.

4.2. Flow field

When a fluid moves in a certain direction, we say that it flows. This motion along a direction is called *flow*. The area in which it flows is called *flow field*.

A flow field is absolutely defined when the velocity of the fluid in all points of the field is known. This means that the characteristic value of the flow field is *the velocity* V *of the fluid's molecules in each point of the field.*

The flow fields are distinguished into steady or laminar flow fields and unsteady or turbulent flow fields. A *steady* or *laminar flow field* is the one in which the fluid's velocity does not change with time at all points of the field. *Unsteady* or *turbulent flow field* is the one in which the fluid's velocity changes with time.

4.3. Fluid velocity

We consider that in a certain time t, an elementary part of the fluid is in a position $P(r)$ of the area, and in another time $t+\delta t$, it lies in the position $Q(r+\delta r)$, as shown in Figure 4.1.

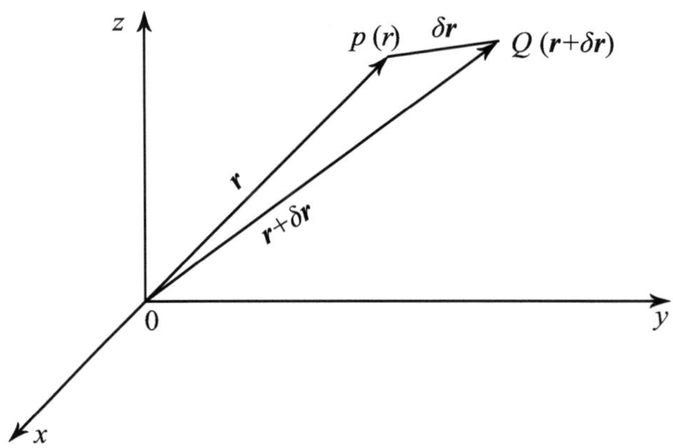

Figure 4.1. *Fluid particle position*

We define the velocity V of the fluid on point $P(r)$ by the relation:

$$V = \lim_{\delta t \to 0} \frac{(r + \delta r - r)}{\delta t} \quad [4.1]$$

$$\text{or } V = \lim_{\delta t \to 0} \frac{(\delta r)}{\delta t} \quad [4.2]$$

$$\text{or finally } V = \frac{\delta r}{\delta t} \quad [4.3]$$

In fact, in this definition, we suppose that the limit exists and it is only one. This assumption is correct when the fluid is in a continuous flow.

It arises from definition [4.3] that the velocity V of the fluid depends on the vector of the position r and on time t, that is:

$$V = V(r,t) \qquad [4.4]$$

Also, it is known that each vector is expressed from an arranged trinity of real numbers. Therefore, provided that the vector r is expressed by the relation:

$$r = r(x,y,z) \qquad [4.5]$$

we can write velocity V as:

$$V = V(x,y,z,t). \qquad [4.6]$$

Finally, if we define that the velocity is expressed as:

$$V = ui + \upsilon j + wk \qquad [4.7]$$

where u, υ, w are its Cartesian coordinates, then we will have:

$$V = ui + \upsilon j + wk = \frac{dx}{dt}i + \frac{dy}{dt}j + \frac{dz}{dt}k \qquad [4.8]$$

$$\text{or } u = \frac{dx}{dt}, \quad \upsilon = \frac{dy}{dt}, \quad w = \frac{dz}{dt}. \qquad [4.9]$$

4.4. Fluid's acceleration

In order to clarify the meaning of acceleration as well as to calculate it using the Euler method, we must take into account the following: first, the variation in the components of velocity, u, υ, w, can be achieved on a random steady point of the area $P(x,y,z)$ due to the variation of time; second, the elementary part of the fluid, which passes from the random point $P(x, y, z)$, can, during its movement, maintain or alter its velocity.

As we mentioned in the previous paragraph, the component of velocity in the x-direction is given by the relation:

$$u = u(x,y,z,t), \qquad [4.10]$$

In order to achieve the calculations for a flow particle, we assume its velocity as it moves from a position $P(x, y, z)$ in time $t + dt$ to a nearby position $Q(x + dx, y + dy, z + dz)$ in time $t + dt$.

Therefore, for the velocity component u from relation [4.10], we will have its total variation:

$$du = \frac{gu}{gt} dt + \frac{gu}{gx} dx + \frac{gu}{gy} dy + \frac{gz}{gz} dz . \qquad [4.11]$$

Also, the distance components dr along which the particle moved is:

$$dx = udt , \quad dy = \upsilon dt , \quad dz = wdt . \qquad [4.12]$$

Substituting [4.12] into [4.11], we have:

$$du = \frac{gu}{gt} dt + \frac{gu}{gx} udt + \frac{gu}{gy} \upsilon dt + \frac{gu}{gz} wdt \qquad [4.13]$$

$$\text{or } \frac{du}{dt} = \frac{\partial u}{\partial t} + u \frac{\partial u}{\partial x} + \upsilon \frac{\partial u}{\partial y} + w \frac{\partial u}{\partial z} . \qquad [4.14]$$

This expression shows the fluid's acceleration based on the Euler's theory, which is called the *total derivative*. It consists of two variations: $\partial u / \partial t$, which is called the *local derivative* because it is combined with the variation of velocity with time at this point of the area, and $u\partial u / \partial x + \upsilon\partial u / \partial y + w\partial u / \partial z$, which is called the *commutative acceleration*.

The total derivative denoted by D / Dt is given by Stokes. So:

$$\frac{Du}{Dt} = \frac{\partial u}{\partial t} + u \frac{\partial u}{\partial x} + \upsilon \frac{\partial u}{\partial y} + w \frac{\partial u}{\partial z} , \qquad [4.15]$$

$$\frac{D\upsilon}{Dt} = \frac{\partial \upsilon}{\partial t} + u \frac{\partial \upsilon}{\partial x} + \upsilon \frac{\partial \upsilon}{\partial y} + w \frac{\partial \upsilon}{\partial z} , \qquad [4.16]$$

$$\frac{Dw}{Dt} = \frac{\partial w}{\partial t} + u \frac{\partial w}{\partial x} + \upsilon \frac{\partial w}{\partial y} + w \frac{\partial w}{\partial z} , \qquad [4.17]$$

are the expressions of the total derivative and the three velocity components.

4.4.1. *Steady and unsteady flows*

Now, if all the components of the local acceleration are equal to zero, that is, if

$$\frac{\partial u}{\partial t} = \frac{\partial \upsilon}{\partial t} = \frac{\partial w}{\partial t} = 0 \text{ or } \frac{\vec{V}}{\partial t} = 0 \qquad [4.18]$$

then the flow is called *steady*. In this case, the velocity may vary from one point of the area to another, but it remains steady with time at each point of the area. This means that in the steady flow the velocity is only a function of space and not time. When the components of the local acceleration are not zero, that is,

$$\frac{\partial u}{\partial t} = \frac{\partial \upsilon}{\partial t} = \frac{\partial w}{\partial t} \neq 0 \text{ or } \frac{V}{\partial t} \neq 0 \qquad [4.19]$$

the flow is called *unsteady*.

4.4.2. *Compressible and incompressible flows*

In Chapter 1, we mentioned that fluids are categorized into two big groups: liquids and gases. Gases are considered to be compressible and their density varies with temperature and pressure. Therefore, in problems involving gas flows, their compressibility cannot be ignored and the flow is considered to be "*compressible*". On the contrary, liquids are practically incompressible, and their flow is generally considered to be "*incompressible*", thereby presenting mathematical simplifications in their motion study.

4.4.3. *Subsonic and supersonic flows*

When the velocity V of a fluid (e.g. velocity of the atmospheric air around the fuselage and the wings of the airplane) is lower than the velocity $\boldsymbol{\alpha}$ of the sound on this fluid, the flow is considered to be "*subsonic*". However, when $\mathbf{V} > \boldsymbol{\alpha}$, it is considered to be "*supersonic*". A criterion to characterize a flow as subsonic or supersonic is the Mach number, which is defined by the relationship:

$$M = \frac{V}{a} \quad (\alpha \simeq 1080 \text{ ft/sec}) \qquad [4.20]$$

where V is the velocity of the fluid and α is the local speed of the sound.

When $M < 1$ ($V < \alpha$), the flow is considered to be subsonic, when if $M > 1$, it is considered to be supersonic. When $M < 0.4$, the flow is considered to be incompressible.

It is possible for the Mach number in a fluid's flow around a solid limit to be < 1 in some areas of the flow and > 1 in others, so in some areas, it will be $M = 1$. In these areas, the flow is considered to be "*transitional*" from subsonic to supersonic.

There are also other types of flow: *laminar, turbulent, external, internal, turbulent* or *non-turbulent.* These definitions can be presented before other definitions are established. These types of flow are addressed in the following sections.

4.5. Streamlines

The velocity field is known when the velocity is known at every point of the flow field and for every single moment. If at a time $t = t_0$ inside a certain flow (flow field), we draw curves where in every curve's point, the vector of the fluid velocity is tangential, then these lines are called *streamlines*. From the definition of these streamlines arises the fact that the streamlines are tangent curves of the velocity vector of the fluid particles that lie on them at the moment $t = t_0$ (Figure 4.2).

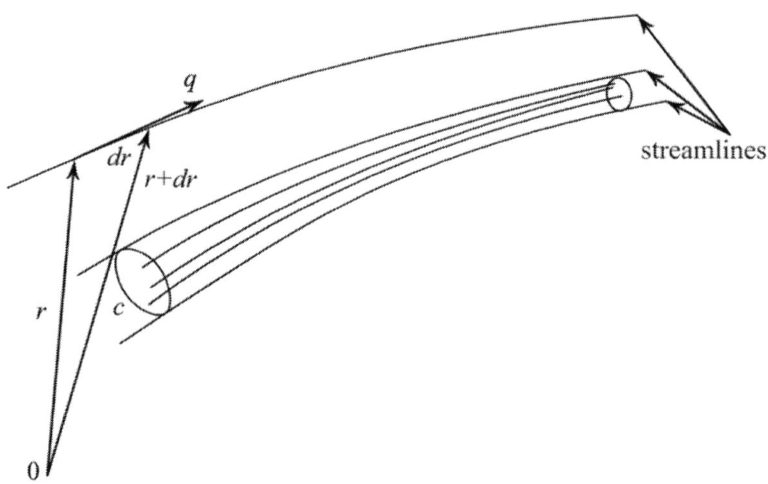

Figure 4.2. *Streamlines in a flow field*

Because on each point of the streamlines the differential element dr is collinear to the velocity V at this point, we conclude that:

$$dr \times V = 0 \quad [4.21]$$

The relationship [4.22] is a differential vector equation of the streamlines. In a system of Cartesian coordinates, we have:

$$\frac{dx}{u} = \frac{dy}{\upsilon} = \frac{dz}{w}, \quad [4.22]$$

where $dr = dxi + dyj + dzk$ and $V = ui + \upsilon j + wk$.

Because the velocity of the fluid in area B (Figure 4.3) is higher than that in areas A and C, the density of the streamlines in area B is higher than that in areas A and C.

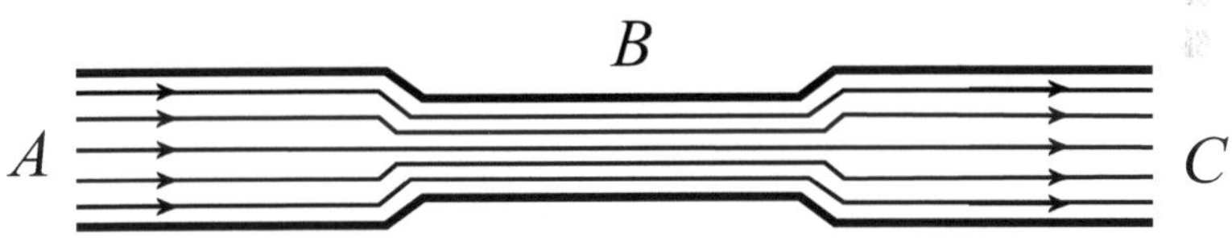

Figure 4.3. *Streamlines in a stenosed tube*

4.6. Mass conservation (continuity equation)

The mass, as it is known, is the quantity of the material. It can be neither created nor destroyed due to the principles of classical mechanics. During a fluid's motion, the mass transfers from one position to the other, but remains constant. The law of conservation of mass is expressed by the relation:

$$\frac{D}{Dt}\int dm = 0, \quad [4.23]$$

where $\int dm$ denotes a point mass and $\frac{D}{Dt}$ denotes the total derivative per time.

Therefore, when a fluid is in motion and there are no flow inlets or leakages, the law of conservation of mass due to the fluid's motion is expressed by an equation called the *continuity equation*. For the expression of the continuity equation, we consider a three-dimensional Cartesian coordinate reference system, and an infinitesimal volume of the moving fluid as shown in Figure 4.4.

If we call M its total mass, then we have:

$$M = \rho dxdydz \quad [4.24]$$

where ρ is the density of the fluid and $dxdydz = v$, which is the volume of the prism.

However, the mass M of the infinitesimal prism, because it is in motion, will be a function of space and time, which is:

$$M = f(x, y, z, t) \quad [4.25]$$

We take its differential and we have:

$$dM = \frac{\partial M}{\partial x}dx + \frac{\partial M}{\partial y}dy + \frac{\partial M}{\partial z}dz + \frac{\partial M}{\partial t}dt \quad [4.26]$$

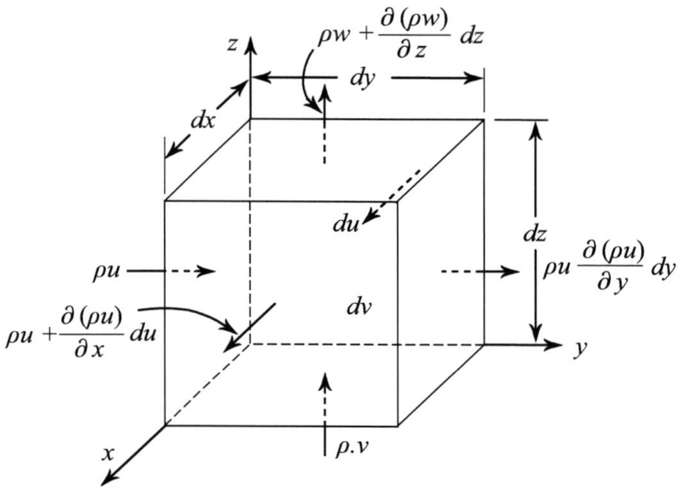

Figure 4.4. *Volume of moving fluid*

However, if u, υ, w are the components of velocity V, then we will have:

$$u = \frac{dx}{dt}, \ \upsilon = \frac{dy}{dt}, \ w = \frac{dz}{dt} \quad [4.27]$$

or $dx = udt$, $dy = \upsilon dt$, $dz = wdt$, [4.28]

substituting [4.28] into [4.26]:

$$dM = \frac{\partial M}{\partial x} udt + \frac{\partial M}{\partial y} \upsilon dt + \frac{\partial M}{\partial z} wdt + \frac{\partial M}{\partial t} dt, \qquad [4.29]$$

and because of [4.24], we have:

$$d(\rho dxdydz) = \frac{\partial}{\partial x}(\rho dydzudt)dx +$$

$$+\frac{\partial}{\partial y}(\rho dxdz\upsilon dt)dy +$$

$$+\frac{\partial}{\partial z}(\rho dxdywdt)dz +$$

$$+\frac{\partial}{\partial t}(\rho dxdydzdt). \qquad [4.30]$$

However, because the mass M does not change with time, its differential is zero, which means that:

$$d(\rho dxdydz) = dM = 0, \qquad [4.31]$$

Thus, [4.30] becomes:

$$\frac{\partial}{\partial x}(\rho dydzudt)dx + \ldots + \ldots - 0 \qquad [4.32]$$

$$\text{or } \left(\frac{\partial(\rho u)}{\partial x} + \frac{\partial(\rho \upsilon)}{\partial y} + \frac{\partial(\rho w)}{\partial z}\right)dxdydzdt +$$

$$+\frac{\partial \rho}{\partial t} dxdydzdt = 0 \qquad [4.33]$$

$$\text{or } \frac{\partial \rho}{\partial t} + \frac{\partial(\rho u)}{\partial x} + \frac{\partial(\rho \upsilon)}{\partial y} + \frac{\partial(\rho w)}{\partial z} = 0. \qquad [4.34]$$

For the case of a steady incompressible flow (ρ = steady), [4.34] becomes:

$$\frac{\partial u}{\partial x}+\frac{\partial \upsilon}{\partial y}+\frac{\partial w}{\partial z}=0 \qquad [4.35]$$

The relationship [4.35] gives the mathematical expression of the law of conservation of mass, and it is a differential equation with partial derivatives, which is satisfied from the velocity components at each point of the flow field.

4.7. Continuity equation for flow in pipes

In order to provide a mathematical expression of the law of conservation of mass for the case of fluid flow inside *pipes*, we define *the flow rate* Q of a pipe as the volume dv of the fluid that goes through the cross-section of the pipe in time dt. That is:

$$Q=\frac{dv}{dt} \qquad [4.36]$$

Provided that:

$$dv = S \cdot ds \qquad [4.37]$$

[4.36] becomes:

$$Q = S \cdot V \qquad [4.38]$$

where S is the surface area of the cross-section and V is the flow velocity.

We now consider a pipe with a non-constant cross-sectional area S, as shown in Figure 4.5, through which a fluid flows with velocity V and density ρ. Because there are no flow inlets or leakages on the pipe walls, the law of conservation of mass enforces that, at a certain time, the fluid with the same volume flows through both cross-sections. Thus, from equation [4.38], we have:

$$Q = \rho_1 S_1 V_1 = \rho_2 S_2 V_2 \,. \qquad [4.39]$$

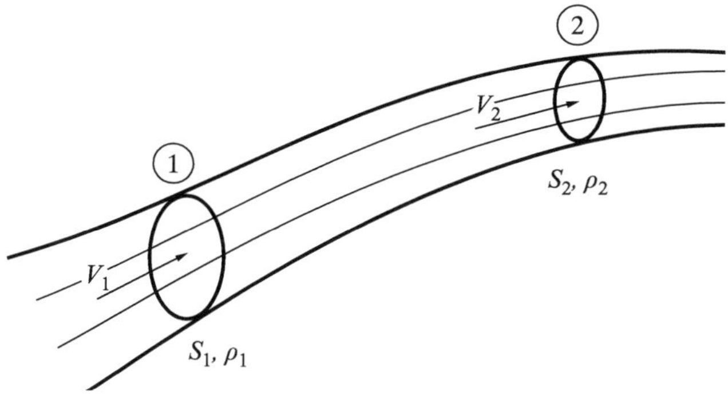

Figure 4.5. *Flow inside a pipe of non-constant diameter*

This relation gives the *continuity equation for the fluid flow inside a pipe.*

When there is no variation of density between cross-sections (1) and (2), that is, when *the flow is incompressible*, equation [4.39] becomes:

$$S_1V_1 = S_2V_2 \qquad [4.40]$$

$$\text{or } \frac{S_1}{S_2} = \frac{V_2}{V_1} \qquad [4.41]$$

From these equations, we note that, in the case of an incompressible flow, the cross-section of the pipe decreases with increasing the flow velocity. In this case, there is also a densification of flow lines. Later, we will see that the opposite occurs in a supersonic flow.

4.8. Energy conservation for incompressible flows (Bernoulli equation)

The Bernoulli equation is a basic relation that relates the pressure and flow velocity of a fluid; however, it is applicable only for the points of one and only streamline.

For deriving this equation, we consider an elementary part of the fluid, a cylinder with surface area of the base *S* and height *ds*, which moves along a streamline, as shown in Figure 4.6. Then, in cross-section (1), a pressure force will be applied, which will be given by the relation:

$$F_{P_1} = pS \qquad [4.42]$$

On the contrary, in the cross-section area (2), another pressure force will be applied, which is given by the relationship:

$$F_{P_2} = (p + \frac{\partial p}{\partial s} ds)S \qquad [4.43]$$

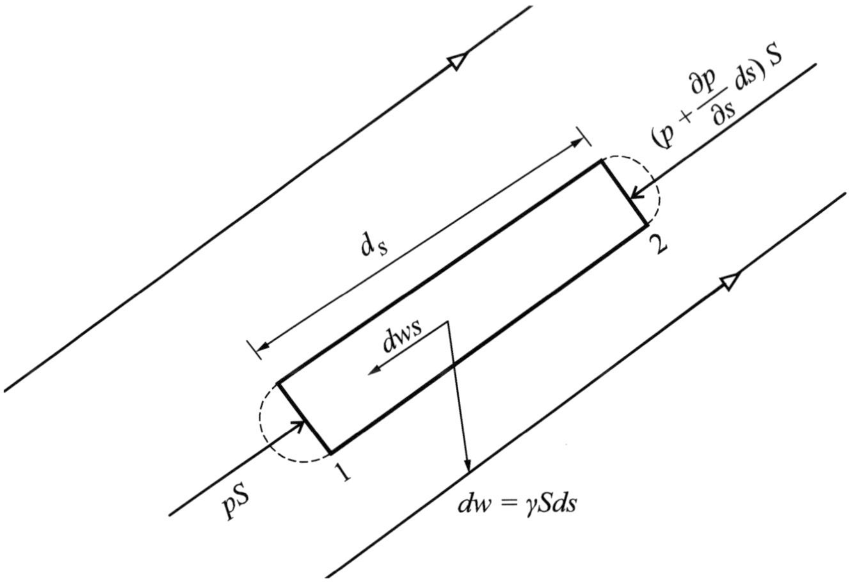

Figure 4.6. *Moving fluid element*

The resultant pressure force along the direction of motion will be equal to the difference in the forces applied on the two bases. This means that:

$$Fp = Fp_1 - Fp_2 =$$

$$= pS - (p + \frac{\partial p}{\partial s} ds)S \qquad [4.44]$$

$$\text{or } Fp = -S \frac{dp}{ds} ds, \qquad [4.45]$$

because pressure $p = p(s)$.

Also, a component of the element's weight applied along a streamline, which is shown in Figure 4.3, is given by the relation:

$$dW_s = -dW \cdot \frac{dh}{ds} \qquad [4.46]$$

or $dW_s = -\gamma S \cdot dh$ [4.47]

So, the resultant force $Ftot$ applied on the elementary cylinder part of the fluid along a streamline is:

$$Ftot = -\gamma Sdh - S\frac{dp}{ds}ds \qquad [4.48]$$

Now, if V is the total velocity of the elementary cylinder, then the acceleration along a streamline is:

$$\frac{DV}{Dt} = \frac{\partial V}{\partial t} + V\frac{\partial V}{\partial s} \qquad [4.49]$$

Hence, using Newton's second law, we have:

$$Ftot = dm\frac{DV}{Dt}, \qquad [4.50]$$

From the values of expressions [4.48] and [4.49], we have:

$$-\gamma Sdh - S\frac{dp}{ds}ds = dm(\frac{\partial V}{\partial t} + V\frac{\partial V}{\partial s}) \qquad [4.51]$$

or $\rho Sds(\frac{\partial V}{\partial t} + V\frac{\partial V}{\partial s}) - \gamma Sdh - S\frac{dp}{ds}ds$. [4.52]

If the flow is steady, that is:

$$\frac{\partial V}{\partial t} = 0, \qquad [4.53]$$

then: $\frac{\partial V}{\partial t} + V\frac{\partial V}{\partial s} = V\frac{dV}{ds} = \frac{d}{ds}(\frac{V^2}{2})$. [4.54]

Thus, [4.52] becomes:

$$\rho ds\left[\frac{d}{ds}(\frac{V^2}{2})\right]=-\gamma dh-dp \qquad [4.55]$$

$$\text{or } \frac{d}{ds}(\frac{V^2}{2})=-g\frac{dh}{ds}-\frac{1}{\rho}\frac{dp}{ds} \qquad [4.56]$$

$$\text{or } \frac{1}{\rho}\frac{dp}{ds}+g\frac{dh}{ds}+\frac{d}{ds}(\frac{V^2}{2})=0 \qquad [4.57]$$

$$\text{since: } \gamma=\rho\cdot g \qquad [4.58]$$

In the case of an incompressible flow, ρ = constant, and integrating relation [4.57], we find:

$$\frac{p}{\rho}+gh+\frac{V^2}{2}=\text{const} \qquad [4.59]$$

$$\text{or } p+\gamma h+\rho\frac{V^2}{2}=\text{const} \qquad [4.60]$$

Relation [4.60] expresses the *Bernoulli theorem* for the case of a steady incompressible flow and is applicable for the points of one streamline.

Bernoulli's law is the basis of fluid mechanics as it connects the *static pressure P* with the *dynamic pressure* $\frac{1}{2}\rho V^2$, and it provides us with the knowledge that the sum of those two pressures remains constant. This means that:

$$p+\frac{1}{2}\rho V^2 = \text{const} \qquad [4.61]$$

First, this leads to an increase in the velocity of the fluid accompanied by a decrease in the static pressure and vice versa, and second, it is applicable to flows in which there is no absorption or attribution of energy. Therefore, in general, we can say that the Bernoulli law is not applicable for the two sides of a propeller, fan or windmill, as these arrangements add or subtract energy from the air current. Also, it is not applicable in the areas of the boundary layer as well.

Therefore, in the case of air, if the internal frictions are not considered to be negligible then the principle of conservation of mechanical energy is not applicable, because there are energy losses due to friction, and the same is applicable for the Bernoulli equation. However, as in low velocities of air, which is in the interest of subsonic aerodynamics, the extent of the boundary layer is small enough to be ignored compared to the rest of the flow; therefore, the Bernoulli equation can be used everywhere in the *same* flow. This means that, for example in the case of Figure 4.5, *we can apply it between any two points inside the pipe, but not between a point inside it and another point which lies outside of the pipe.*

4.9. Applications of the Bernoulli law

4.9.1. *Venturi tube*

The equation that was previously analyzed offers as a direct practical application the possibility of measuring the velocity of a fluid.

One of the first devices used for the measurement of the velocity of fluids was the tube invented by Venturi (1746–1822). This tube is connected to either a manometer or an instrument of differential pressure, as shown in Figure 4.7. The Venturi tube is used only in small velocities and therefore in the case of air, the flow is considered to be incompressible (ρ = steady). Therefore, if point 1 represents the conditions of the free flow stream and point 2 represents the conditions during the stenosis of the Venturi tube, then according to the Bernoulli equation, we have:

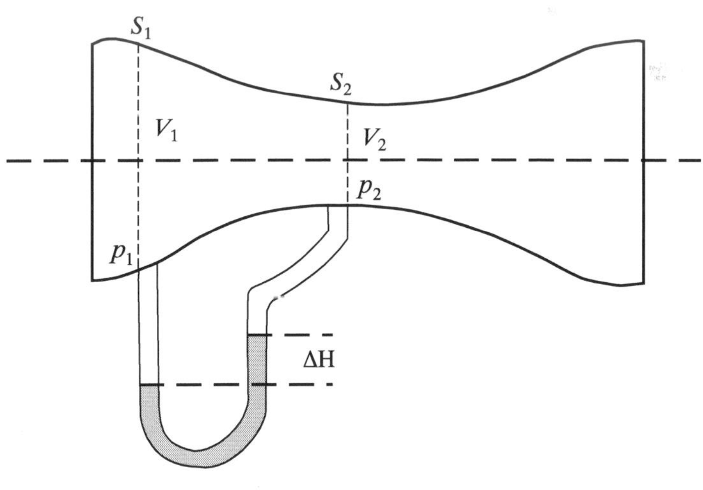

Figure 4.7. *Venturi tube*

$$p_1 + \frac{\rho V_1^2}{2} = p_2 + \frac{\rho V_2^2}{2} \qquad [4.62]$$

Also, the continuity equation gives:

$$S_1 V_1 = S_2 V_2 \qquad [4.63]$$

By combining these relations and by solving for V_1, we find:

$$V_1 = \sqrt{\frac{2(p_1 - p_2)}{\rho(\frac{S_1^2}{S_2^2 - 1})}}. \qquad [4.64]$$

The pressure difference between points 1 and 2, which is $p_1 - p_2$, is a function of ΔH, given by the equation:

$$p_1 - p_2 = \rho g \Delta H \qquad [4.65]$$

and expression [4.64] is finally formed:

$$V_1 = \sqrt{\frac{2g\Delta H}{(\frac{S_1^2}{S_2^2 - 1})}} \qquad [4.66]$$

The value of V_1 is given if we attach a suitably calibrated instrument. Because the Venturi tube:

a) creates a high drag,

b) internally oxidizes and

c) is prone to ice formation,

it generally does not provide us with the required accurate measurement of air velocity, and its use is limited to slow-moving airplanes and sailplanes.

4.9.2. *Pitot tube*

Due to all of the above disadvantages of the Venturi tube, all modern airplanes are equipped with the Pitot tube, invented in 1732, in order to measure their velocity. We must note here that, in order to measure the velocity of air, which is the flight velocity of the airplane, we must know the dynamic pressure, which depends on the density and velocity of air. Let us assume that the density is known. We know that the dynamic pressure is the difference between the total and static flow pressures. The total pressure is constant on all the flow points, but the measurement of the static pressure needs considerable attention in order to ensure that we measure the static pressure of the free flow. The total pressure is measured using a Pitot tube (Figure 4.8).

The Pitot tube, as shown in Figure 4.8, is open at one end and connected to a differential manometer at the other end. Inside the tube, the air is motionless, such that the total pressure dominates.

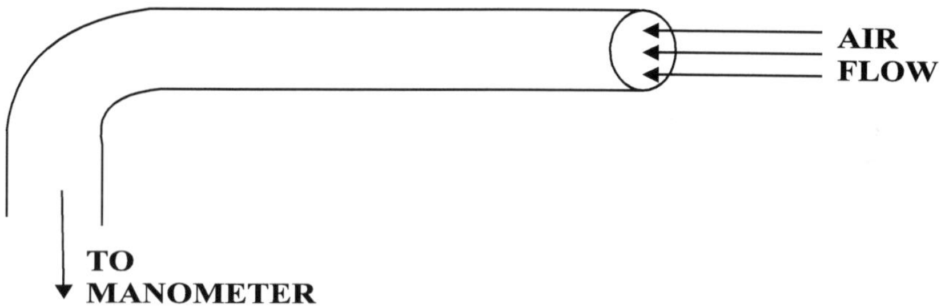

Figure 4.8. *Pitot tube*

The tube that measures the static pressure is closed on one of its sides and has holes on its walls, while its other side ends at the differential manometer of the Pitot tube (Figure 4.9).

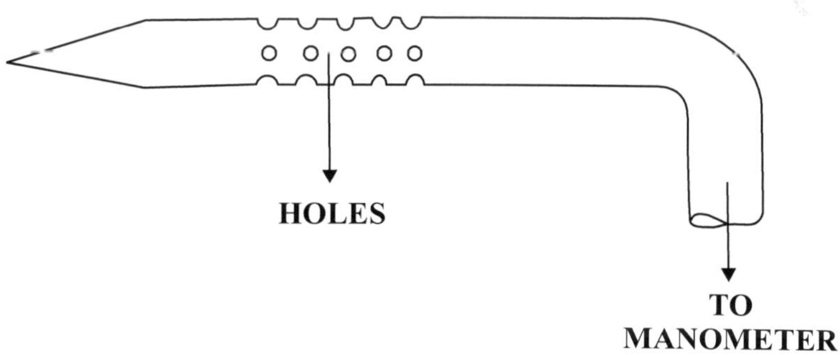

Figure 4.9. *Static pressure measurement tube (Proult)*

When pressure balance occurs, the pressure inside the tube is the same as the static pressure in the areas of the holes. Hence, the tube must be shaped in such a way to cause the smallest possible disorder around it, and thus the pressure measurement will be accurate enough.

In practice, in airplanes, these two tubes are usually together, one inside the other as shown in Figure 4.10.

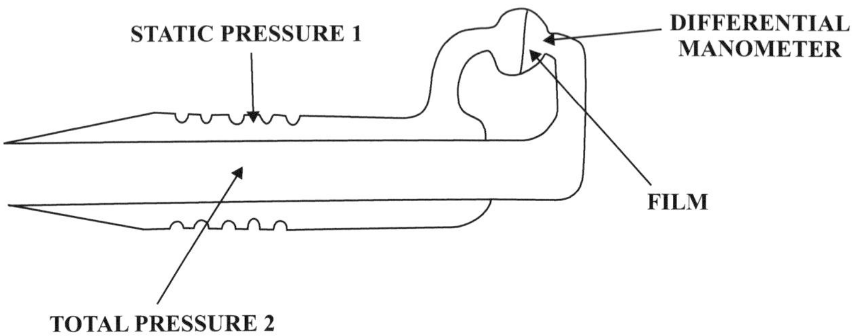

Figure 4.10. *Pitot and Proult tubes*

The differential manometer contains a flexible membrane that distorts depending on the pressure difference, which is the dynamic pressure. The manometer's bar is classified in such a way that a needle moving from the membrane directly shows the velocities. The velocity from the Pitot tube is given by the relation:

$$V_2 = \sqrt{\frac{2(p_2 - p_1)}{\rho}} \qquad [4.67]$$

where $p_2 - p_1$ is the difference between the total pressure and the static pressure obtained from the differential manometer and ρ is the *density of air that depends on the flight height.*

4.10. Euler equations

Euler equations are the general motion equations of the fluids when we ignore the friction forces (inviscid flows), and they are applicable for both compressible and incompressible flows. These equations are a result of the conservation of momentum, and they are extracted from the basic law of mechanics (Newton's law):

$$dF = \gamma \cdot dm , \quad [4.68]$$

considering all the forces that act on an elementary parallelepiped of the fluid, besides the friction forces. Working in the same way as we worked to prove the Bernoulli equation, but for all three flow directions now, we obtain the Euler motion equation, which is given in vector form by the relation:

$$\rho \frac{DV}{Dt} = F - grad\ p \quad [4.69]$$

Euler motion equations for the three Cartesian coordinates x, y and z are written as:

$$\frac{\partial u}{\partial t} + u\frac{\partial u}{\partial x} + \upsilon\frac{\partial u}{\partial y} + w\frac{\partial u}{\partial z} = F_x - \frac{1}{\rho}\frac{\partial p}{\partial x}, \quad [4.70]$$

$$\frac{\partial \upsilon}{\partial t} + u\frac{\partial \upsilon}{\partial x} + \upsilon\frac{\partial \upsilon}{\partial y} + w\frac{\partial \upsilon}{\partial z} = F_y - \frac{1}{\rho}\frac{\partial p}{\partial y}, \quad [4.71]$$

$$\frac{\partial w}{\partial t} + u\frac{\partial w}{\partial x} + \upsilon\frac{\partial w}{\partial y} + w\frac{\partial w}{\partial z} = F_z - \frac{1}{\rho}\frac{\partial p}{\partial z}, \quad [4.72]$$

where all the symbols are described as above. The motion equations of [4.70]–[4.72] with the continuity equation:

$$\frac{\partial \rho}{\partial t} + \frac{\partial}{\partial x}(\rho u) + \frac{\partial}{\partial y}(\rho \upsilon) + \frac{\partial}{\partial z}(\rho w) = 0 \quad [4.73]$$

are applicable to both compressible and incompressible inviscid fluids and constitute a system of four differential equations with four unknown parameters, u, υ, w and pressure p. For compressible fluids, density ρ is an unknown variable as well, and the above equations are not sufficient to specify the unknown functions.

The fifth equation is a relation of the form $\rho = f(p)$ which gives density as a function of pressure, and in this way, all five equations fully describe the motion of inviscid fluids.

4.11. Navier–Stokes equations

As is known, Newton's second law states that the total of the external forces dF_{tot}, acting on an element of mass dm, is equal to the mass of the element multiplied by acceleration:

$$dF_{tot} = \gamma \cdot dm \qquad [4.74]$$

$$\text{or } F_{tot} = \int \gamma \cdot dm \qquad [4.75]$$

As the acceleration γ is equal to the total variation of velocity $\mathbf{V}$ of an elementary particle per unit time, we have:

$$\gamma = \frac{D\mathbf{V}}{Dt} \qquad [4.76]$$

and [4.74] becomes:

$$F_{tot} = \frac{D}{Dt}\int \mathbf{V}dm \qquad [4.77]$$

The application of the Newton's law on fluids leads to the discovery of the motion equations, which are known as Navier–Stokes equations. To conclude, we consider the following force categories, which act upon an elementary fluid mass:

a) gravitational forces, which are forces that act on the whole extent of the fluid mass and

b) pressure and friction forces, which are forces that act on the solid bound.

If $\mathbf{F} = \rho g$ is the gravity force per unit volume and P is the force that acts on the solid bound per unit volume, then the motion equations from relation [4.77] in vector form are given by the expression:

$$\rho \frac{D\mathbf{V}}{Dt} = \mathbf{F} + \mathbf{P} \qquad [4.78]$$

with

$$\mathbf{F} = X\mathbf{i} + Y\mathbf{j} + Z\mathbf{k} \text{ (body force)} \qquad [4.79]$$

$$\mathbf{P} = P_x\mathbf{i} + P_y\mathbf{j} + P_z\mathbf{k} \text{ (surface force)} \qquad [4.80]$$

$$\mathbf{V} = u\mathbf{i} + \upsilon\mathbf{j} + w\mathbf{k} \text{ (velocity vector)} \quad [4.81]$$

and

$$\frac{D}{Dt} = \frac{\partial}{\partial t} + u\frac{\partial}{\partial x} + \upsilon\frac{\partial}{\partial y} + w\frac{\partial}{\partial z} \quad [4.82]$$

If we assume that the fluid is incompressible, which means that the shear stresses are negligible, i.e.:

$$\tau_{xy} = \tau_{yz} = \tau_{xz} = 0 \quad [4.83]$$

then Navier–Stokes equations are:

$$\begin{aligned}
\rho\left(\frac{\partial u}{\partial t} + u\frac{\partial u}{\partial x} + \upsilon\frac{\partial u}{\partial y} + w\frac{\partial u}{\partial z}\right) &= X - \frac{\partial p}{\partial x} + \mu\left(\frac{\partial^2 u}{\partial x^2} + \frac{\partial^2 u}{\partial y^2} + \frac{\partial^2 u}{\partial z^2}\right) \\
\rho\left(\frac{\partial \upsilon}{\partial t} + u\frac{\partial \upsilon}{\partial x} + \upsilon\frac{\partial \upsilon}{\partial y} + w\frac{\partial \upsilon}{\partial z}\right) &= Y - \frac{\partial p}{\partial y} + \mu\left(\frac{\partial^2 \upsilon}{\partial x^2} + \frac{\partial^2 \upsilon}{\partial y^2} + \frac{\partial^2 \upsilon}{\partial z^2}\right) \\
\rho\left(\frac{\partial w}{\partial t} + u\frac{\partial w}{\partial x} + \upsilon\frac{\partial w}{\partial y} + w\frac{\partial w}{\partial z}\right) &= Z - \frac{\partial p}{\partial z} + \mu\left(\frac{\partial^2 w}{\partial x^2} + \frac{\partial^2 w}{\partial y^2} + \frac{\partial^2 w}{\partial z^2}\right)
\end{aligned} \quad [4.84]$$

where all the symbols are defined as above.

We note here that partial cases of Navier–Stokes equations are Euler equations. As shown in relations [4.84], Euler equations are different, only with respect to the term of coherence. Euler and Navier–Stokes equations are non-linear differential equations with partial derivatives of second order, the linearity of which makes their solution difficult.

We also recall that a flow field is generally considered known, if we know the pressure p at every point of the flow and their velocity field. This occurs with the solution of the system of the continuity equation and the Navier–Stokes equations, which constitute a system of four differential equations with partial derivatives. The unknown variables that have to be defined are also four, which are the three components of velocity, u, υ, w, and pressure p.

4.12. Formulae

1) Air velocity V to Euler:

$$V = \frac{dr}{dt}$$

where r is the position vector ($r = xi + yj + zk$) and t is the time.

2) Air acceleration γ to Euler:

$$\gamma = \frac{DV}{Dt}$$

where $\frac{D}{Dt} = \frac{\partial}{\partial t} + u\frac{\partial}{\partial x} + \upsilon\frac{\partial}{\partial y} + w\frac{\partial}{\partial z}$ = Stokes operator, V is the air velocity $(V = ui + \upsilon j + wk)$ and t is the time.

3) Differential equations of flow lines:

$$\frac{dx}{u} = \frac{dy}{\upsilon} = \frac{dz}{w}$$

where u, υ, w are components of the air velocity V at the three axes x, y, z, respectively ($V = ui + \upsilon j + wk$), and x, y, z are components of the position vector r ($r = xi + yj + wk$).

4) Continuity equation:

$$\frac{\partial \rho}{\partial t} + \frac{\partial(\rho u)}{\partial x} + \frac{\partial(\rho \upsilon)}{\partial y} + \frac{\partial(\rho w)}{\partial z} = 0$$

where ρ is the fluid density, t is the time, u, υ, w are velocity components V and x,y,z are components of the position vector.

5) Continuity equations for a steady incompressible flow:

$$\frac{\partial u}{\partial x}+\frac{\partial \upsilon}{\partial y}+\frac{\partial w}{\partial z}=0$$

because ρ = density = const.

6) Pipe flow rate Q:

$$Q=\frac{dv}{dt}S\cdot V$$

where dv is the volume of the fluid, dt is the time, S is the cross-sectional area and V is the velocity of the fluid.

7) Continuity equation for flow in pipes:

$$\rho_1 S_1 V_1 = \rho_2 S_2 V_2$$

and for an incompressible flow, ρ = const.

$$S_1 V_1 = S_2 V_2$$

where ρ_1, ρ_2 are fluid densities, S_1, S_2 are surface areas of the cross-sections and V_1, V_2 are fluid velocities at points (1) and (2), respectively.

8) Bernoulli equation:

$$p_1+\frac{1}{2}\rho V_1^2 = p_2+\frac{1}{2}\rho V_2^2$$

where ρ is the fluid density, p_1, p_2 are fluid pressures and V_1, V_2 are fluid velocities on the two cross-sectional areas, respectively.

9) Fluid velocity, with a Venturi tube for an incompressible flow:

$$V_1 = \sqrt{\frac{2(p_1 - p_2)}{\rho\left(\frac{1-S_2^2}{S_1^2}\right)}}$$

and as $p_1 - p_2 = \rho g \Delta H$ (aerostatic equation), we also have:

$$V_1 = \sqrt{\frac{2\rho g \Delta H}{\left(\frac{1-S_2^2}{S_1^2}\right)}}$$

where $p_1 - p_2$ is the pressure difference between points (1) and (2), S_1, S_2 = surface areas of the cross-sections, ρ is the fluid density, $g = 9.81$ m/sec^2 and ΔH is the range altitude of the manometer's fluid.

10) Pressure equation:

$$p + \frac{1}{2}\rho V^2 = p_0$$

where p is the static pressure of the fluid, $\frac{1}{2}\rho V^2 = q_\infty$ is the dynamic pressure of the fluid and p_0 is the total pressure of the fluid. This relation is only applicable for an incompressible flow.

11) Velocity measurement of air with a Pitot tube:

$$V = \sqrt{\frac{2(p_0 - p)}{\rho}}$$

where $p_0 - p$ is the pressure difference between the total and static pressures measured by the Pitot tube and ρ is the fluid density.

12) Euler equations:

$$\rho\left(\frac{\partial u}{\partial t}+u\frac{\partial u}{\partial x}+\upsilon\frac{\partial u}{\partial y}+w\frac{\partial u}{\partial z}\right)=X-\frac{\partial p}{\partial x}$$

$$\rho\left(\frac{\partial \upsilon}{\partial t}+u\frac{\partial \upsilon}{\partial x}+\upsilon\frac{\partial \upsilon}{\partial y}+w\frac{\partial \upsilon}{\partial z}\right)=Y-\frac{\partial p}{\partial y}$$

$$\rho\left(\frac{\partial w}{\partial t}+u\frac{\partial w}{\partial x}+\upsilon\frac{\partial w}{\partial y}+w\frac{\partial w}{\partial z}\right)=Z-\frac{\partial p}{\partial z}$$

where ρ is the fluid density; u, υ, w are components of the flow velocity; X, Y, Z are components of the total force F_{tot}; p is the pressure of the fluid; t is the time and x,y,z are components of the unit vector at position r.

13) Navier–Stokes equations:

$$\rho\left(\frac{\partial u}{\partial t}+u\frac{\partial u}{\partial x}+\upsilon\frac{\partial u}{\partial y}+w\frac{\partial u}{\partial z}\right)=X-\frac{\partial p}{\partial x}+\mu\left(\frac{\partial^2 u}{\partial x^2}+\frac{\partial^2 u}{\partial y^2}+\frac{\partial^2 u}{\partial z^2}\right)$$

$$\rho\left(\frac{\partial \upsilon}{\partial t}+u\frac{\partial \upsilon}{\partial x}+\upsilon\frac{\partial \upsilon}{\partial y}+w\frac{\partial \upsilon}{\partial z}\right)=Y-\frac{\partial p}{\partial y}+\mu\left(\frac{\partial^2 \upsilon}{\partial x^2}+\frac{\partial^2 \upsilon}{\partial y^2}+\frac{\partial^2 \upsilon}{\partial z^2}\right)$$

$$\rho\left(\frac{\partial w}{\partial t}+u\frac{\partial w}{\partial x}+\upsilon\frac{\partial w}{\partial y}+w\frac{\partial w}{\partial z}\right)=Z-\frac{\partial p}{\partial z}+\mu\left(\frac{\partial^2 w}{\partial x^2}+\frac{\partial^2 w}{\partial y^2}+\frac{\partial^2 w}{\partial z^2}\right)$$

where ρ is the density, u, υ, z are velocity components, X, Y, Z are components of the body force, μ is the viscosity of the fluid, p is the pressure of the fluid, t is the time and x, y, z are system Cartesian coordinates.

4.13. Questions

1) What is flow field of a fluid and how many kinds of flow fields are there? Define each of them.

2) Define the fluid velocity. What is its expression in Cartesian coordinates?

3) Define steady and unsteady flows. Under what conditions do they occur?

4) When is a flow considered to be compressible and incompressible?

5) When is a flow considered to be subsonic and supersonic?

6) Define Mach number. How can it characterize the flows according to its value?

7) Define streamlines in a flow field. What are their equations?

8) Which are the main principles of fluid mechanics?

9) Which principle results in the *continuity equation* and how is it expressed in Cartesian coordinates?

10) Define the *mass flow rate* and the volume flow rate in a pipe.

11) Derive the relation that expresses the *continuity equation* in pipes.

12) What is the importance of the *Bernoulli principle* in fluid mechanics and what is the principle to which it relates?

13) Prove the *Bernoulli* theorem. When is it applicable?

14) What is a *Venturi tube*? Calculate the velocity of the fluid that flows through it.

15) Which are the disadvantages of the *Venturi tube* and where is it commonly used today?

16) What is a *Pitot tube*, and how is it described?

17) Which *Euler equations* are represented in Cartesian coordinates, and on which fluids are they used?

18) Which Navier–Stokes equations are represented in Cartesian coordinates for incompressible fluids?

4.14. Problems with solutions

1) In a flow field, velocity is measured in m/s and is given by the relation:

$$V = xe^{-1}i + 3yj$$

Here, x and y are measured in m and time t in sec. Calculate the streamlines going through point (1,1) and the path of the fluid particle that lies in position (1,1) when $t = 0$.

Solution:

From the equation of velocity, we have:

$$u = xe^{-1} \quad \upsilon = 3y \text{ and } w = 0 \tag{1}$$

The differential equations of the flow lines for the two-dimensional flow are given by the relation:

$$\frac{dx}{u} = \frac{dy}{\upsilon} \tag{2}$$

Substituting (2) into (1), we have:

$$\frac{dx}{xe^{-1}} = \frac{dy}{3y} \tag{3}$$

Integrating (3) for point (1,1) gives:

$$\int_1^x \frac{dx}{xe^{-1}} = \int_1^y \frac{dx}{3y} \tag{4}$$

$$\text{or } e^t \int_1^x \frac{dx}{x} = \frac{1}{3}\int_1^y \frac{dy}{y} \tag{5}$$

$$\text{or } e^t \left[\ln x\right]_1^x = \frac{1}{3}\left[\ln y\right]_1^y \tag{6}$$

$$\text{or } e^t \ln x = \frac{1}{3}\ln y \tag{7}$$

$$\text{or } \ln y = \ln x^{3e^t} \tag{8}$$

$$\text{or } y = x^{3e^t} \tag{9}$$

The orbit is calculated from the relation:

$$\frac{dx}{u} = \frac{dy}{\upsilon} = t \tag{10}$$

$$\text{or } \frac{dx}{xe^{-t}} = \frac{dy}{3y} = dt \tag{11}$$

$$\text{or } \frac{dx}{xe^{-t}} = dt \text{ and}$$

$$\frac{dy}{3y} = dt \tag{12}$$

Integrating these, we have:

$$e^{t}\int_{1}^{x}\frac{dx}{x} = \int_{0}^{t} dt \tag{13}$$

$$\text{or } e^{t}\left[\ln x\right]_{1}^{x} = \left[t\right]_{0}^{t} \tag{14}$$

$$\text{or } e^{t}\ln x = t \;\; \ln x = te^{-t} \tag{15}$$

$$\text{or } x = e^{te^{-t}} \tag{16}$$

$$\text{Also, } \frac{1}{3}\int_{1}^{y}\frac{dy}{y} = \int_{0}^{t} dt \tag{17}$$

$$\text{or } \frac{1}{3}\ln y = t \;\; \ln y = 3t \tag{18}$$

$$\text{or } y = e^{3t} \tag{19}$$

By deleting t between (16) and (19), we finally have:

$$y = \frac{1}{(1-\ln x)^{3}} \tag{20}$$

2) In a constant means, the velocity on each point (x, y, z) of the flow field and for every single moment t is given by the relation:

$$V = ui + uj + wk = \frac{x}{1+t}i + \frac{2y}{1+t}j + \frac{3z}{1+t}k$$

Calculate the acceleration to Lagrange and Euler.

Solution:

We have $u = u(x, y, z, t) = \dfrac{x}{1+t}$, $\upsilon = \upsilon(x, y, z, t) = \dfrac{2y}{1+t}$ and

$$w = w(x, y, z, t) = \frac{3z}{1+t}, \tag{1}$$

which means that the components of velocity are the functions of position and time. In this case, the components to Euler will be:

$$a_x = \frac{Du}{Dt} = \frac{\partial u}{\partial t} + u\frac{\partial u}{\partial x} + \upsilon\frac{\partial u}{\partial y} + w\frac{\partial u}{\partial z} =$$

$$= \frac{-x}{(1+t)^2} + \frac{x}{1+t}\cdot\frac{1}{1+t} = 0, \tag{2}$$

$$a_y = \frac{D\upsilon}{Dt} = \frac{\partial \upsilon}{\partial t} + u\frac{\partial \upsilon}{\partial x} + \upsilon\frac{\partial \upsilon}{\partial y} + w\frac{\partial \upsilon}{\partial z} =$$

$$= \frac{-2y}{(1+t)^2} + \frac{2y}{1+t}\cdot\frac{2}{(1+t)} = \frac{2y}{(1+t)^2}, \tag{3}$$

and $a_z = \dfrac{Dw}{Dt} = \dfrac{\partial w}{\partial t} + u\dfrac{\partial w}{\partial t} + \upsilon\dfrac{\partial w}{\partial t} + w\dfrac{\partial w}{\partial z} =$

$$= -\frac{3z}{(1+t)^2} + \frac{3z}{1+t}\cdot\frac{3}{(1+t)} = \frac{6z}{(1+t)^2}, \tag{4}$$

respectively.

But these components to Lagrange are calculated after calculating the velocity to Lagrange. We have:

$$u = \frac{dx}{dt} \text{ or } \frac{x}{1+t} = \frac{dx}{dt} \to \frac{dx}{x} = \frac{dt}{1+t}$$

or $\ln x = \ln(1+t) + \ln c_1$ or finally $x = c_1(1+t)$.

Similarly, we have:

$$\upsilon = \frac{dy}{dt} \text{ or } \frac{2y}{1+t} = \frac{dy}{dt} \to \frac{dy}{2y} = \frac{dt}{1+t}$$

or $\ln y^{\frac{1}{2}} = \ln(1+t) + \ln c_2$ or $y = c_3(1+t)^2$ $(c_3 = c_2^2)$.

Finally, $z = c_4(1+t)^3$

For $t = 0$, the above relations give:

$x = x_0 = c_1$, $y = y_0 = c_3$ and $z = z_0 = c_4$. So

$$x = x_0(1+t),\ y = y_0(1+t)^2,\ z = z_0(1+t)^3. \tag{5}$$

Lagrange's velocity is:

$$u(x_0,t) = \frac{dx}{dt} = x_0,\ \upsilon(y_0,t) = 2y_0(1+t) \text{ and } z(z_0,t) = 3z_0(1+t)^2$$

while the acceleration is:

$$a_x = \frac{d^2x}{dt} = 0$$

$$a_y = 2y_0 \text{ and } a_z = 6z_0(1+t)$$

We note that by replacing y_0, z_0 in these expressions with relation (5), we obtain acceleration expressions to Euler.

3) A flow in a nozzle has a velocity of 100 ft/sec in the inlet of a cross-sectional area 6 ft^2. If the density of air is 0.002378 slug/ft^3, what is the velocity and the dynamic pressure at the exit of a cross-section of 4 ft^2?

Solution:

From the continuity equation for an incompressible flow, we have:

$$S_1 \cdot V_1 = S_2 \cdot V_2$$

So, the velocity at the exit is:

$$V_2 = 600/4$$

or $V_2 = 150\,ft/\sec$

The dynamic pressure at the exit is:

$$q = \frac{1}{2}\rho V^2$$

or $q = 26.75 lb/ft^2$

4) We consider an incompressible flow on a converged nozzle with inlet cross-section $S_1 = 5m^2$. The air at the inlet of the nozzle has a velocity $V_1 = 10m/\sec$ and comes out with a velocity $V_2 = 30m/\sec$. If the pressure and temperature at the inlet are $p_1 = 1,2 \cdot 10^5\,Pa$ and $T_1 = 330K$, respectively, calculate the surface area of the cross section and the pressure at the exit.

Solution:

From the continuity equation for an incompressible flow:

$$S_1 \cdot V_1 = S_2 \cdot V_2 \Leftrightarrow S_2 = \frac{S_1 \cdot V_1}{V_2} \Leftrightarrow S_2 = 1.67m^2$$

From the state equation:

$$\rho = \frac{p}{R \cdot T}$$

or $\rho = 1,27kgr / m^3$

From the Bernoulli equation:

$$p_1 + \rho\frac{V_1^2}{2} = p_2 + \rho\frac{V_2^2}{2}$$

or $p_2 = 1,195 \cdot 10^5 \, Pa$

5) Calculate the velocity at certain point A of a wing, with $p_A = 2070lb / ft^2$, if the free current of air is $p_1 = 2116lb / ft^2$, $V_1 = 100m / h$ and $\rho = 0,002377 slug / ft^3$.

Solution:

From the Bernoulli equation:

$$p_1 + \rho\frac{V_1^2}{2} = p_A + \rho\frac{V_A^2}{2} \Leftrightarrow V_A = 245,4 ft / \sec$$

6) A U-shaped tube is given. One edge of the tube is connected to a pipe, that is, inside an air current, with a velocity $V = 120km / h$. Calculate up to what height Δh the water level will rise, which initially is in a calm state in the next tube. The given facts are $\rho_{air} = 1,225kg / m^3$ and $\rho_{water} = 1.000kg / m^3$.

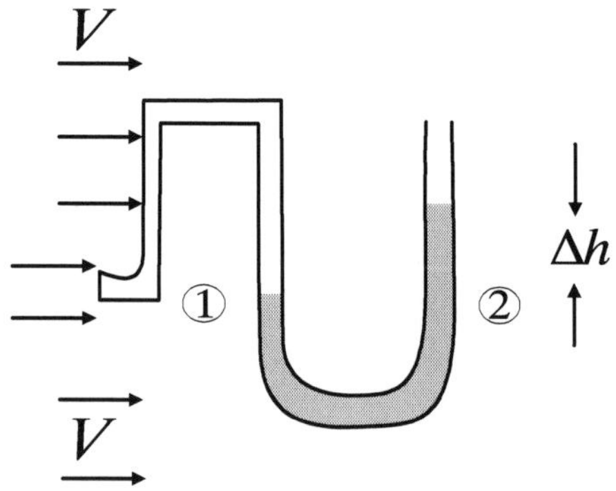

Solution:

By applying the Bernoulli equation for points (1) and (2), it is applicable:

$$p_0 + \frac{1}{2}\rho_{air} \cdot V^2 = \rho_{water} \cdot g \cdot \Delta h + p_0 \Rightarrow \Delta h = \frac{1}{2} \cdot \frac{\rho_{air}}{\rho_{water}} \cdot \frac{V^2}{g} \Rightarrow$$

$$\Rightarrow \Delta h = \frac{1}{2} \cdot \frac{1,251}{1.000} \cdot \frac{(33,33)^2}{9,81} \cdot \frac{kg/m^3}{kg/m^3} \cdot \frac{m^2/\sec^2}{m/\sec^2}$$

$$\Rightarrow \Delta h = 0,071m \Rightarrow \Delta h = 71mm$$

7) In the pipe of the following figure, which is leaked by a fluid of density (ρ), a manometer has been placed, whose indication is Δh. If $D_1 = 2D_2$, calculate:

a) The dynamic pressure in position (2).

b) The pipe's flow rate, for $D_1 = 20cm$

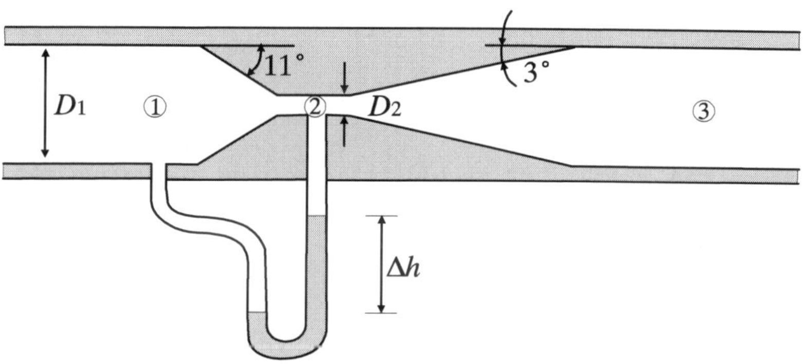

Solution:

a) By applying the Bernoulli equation from point 1 to point 2, it is applicable:

$$\frac{1}{2}\rho V_1^2 + p_1 = \frac{1}{2}\rho V_2^2 + p_2 \Rightarrow p_1 - p_2 = \frac{1}{2}\rho V_2^2 - \frac{1}{2}\rho V_1^2 \quad (1)$$

It is also applicable:

$$\Delta p = p_1 - p_2 = \rho \cdot g \cdot \Delta h \quad (2)$$

We apply the continuity equation, wherever applicable:

$$S_1V_1 = S_2V_2 \Rightarrow \frac{\pi D_1^{\,2}}{4}V_1 = \frac{\pi D_2^{\,2}}{4}V_2$$

So, $V_1 = \frac{V_2}{4}$

From equations (1), (2) and (3), we finally have:

$$\Delta p = \frac{1}{2}\rho V_2^{\,2} - \frac{1}{2}\rho V_1^{2} \Rightarrow \frac{1}{2}\rho V_2^{\,2} = \frac{16}{15}g\Delta h \qquad (3)$$

b) The pipe's flow rate is calculated as shown below:

$$Q = S_1 \cdot V_1 = \frac{\pi D_1^{\,4}}{4} \cdot \sqrt{\frac{2\frac{16}{15}g \cdot \Delta h}{\rho}} \Rightarrow Q = \frac{0.4\pi}{4} \cdot \sqrt{\frac{\frac{32}{15} \cdot g \cdot \Delta h}{\rho}}$$

$$\Rightarrow Q = 0.1\pi \cdot \sqrt{\frac{32g \cdot \Delta h}{15 \cdot \rho}}$$

8) We consider that the instrument given in the above figure (Pitot tube with a closed mercury manometer) is put in a pipe, through which water passes. If the indication of the manometer $\Delta = 20cm$ is known, calculate the water velocity along the axis of the tube.

$$\gamma = 100kg / m^3 \quad g = 9{,}81m / \sec^2 \quad \varepsilon_m = 136kg / m^3$$

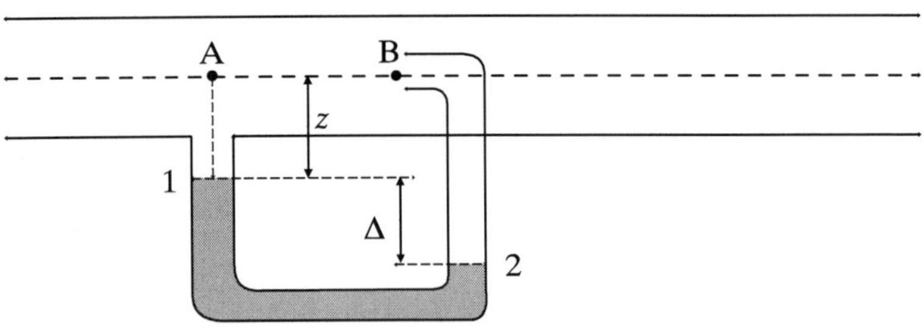

Solution:

We take points A and B on the axis of the tube and intersections 1 and 2 for the separation of water mercury.

Therefore, the basic equation of the aerostatics for points 1 and A gives:

$$dp = -\gamma dz \Rightarrow \int_1^A dp = -\gamma \int_z dz \Rightarrow p_A - p_i = -\gamma z \Rightarrow p_1 = p_A + \gamma z \quad (1)$$

Similarly, for points 2 and B, it gives:

$$p_2 = p_B + \gamma(z + \Delta) \quad (2)$$

And for the points 2 and 1:

$$p_2 = p_1 + \gamma_m \Delta \quad (3)$$

Subtracting (1) from (2), we have:

$$\begin{aligned} & p_2 - p_1 = p_B - p_A + \gamma(z+\Delta) - \gamma z \Rightarrow \\ & p_B - p_A = p_2 - p_1 - \gamma(z+\Delta) + \gamma z \Rightarrow \\ & p_B - p_A = p_2 - p_1 - \gamma z - \gamma\Delta + \gamma z \Rightarrow \\ & p_B - p_A = p_2 - p_1 - \gamma\Delta \end{aligned} \quad (4)$$

Substituting (3) into (4), we have:

$$p_B - p_A = \gamma_m - \gamma\Delta \Rightarrow p_B - p_A = \Delta(\gamma_m - \gamma) \quad (5)$$

Now, by applying the Bernoulli theorem for points A and B, which lie on this horizontal level, we have:

$$p_A + \frac{1}{2}\rho V_A^2 = p_B + \frac{1}{2}\rho V_B^2$$

And because $V_B = 0$ due to the presence of the Pitot tube, the above becomes:

$$p_a + \frac{1}{2}\rho V_B^2 = p_B \Rightarrow p_A + \frac{1}{2}\frac{\gamma}{g} V_B^2 = p_B \Rightarrow$$

$$\Rightarrow p_A + \frac{V_B^2}{2g} = \frac{p_B}{\gamma} \Rightarrow \frac{p_B - p_A}{\gamma} = \frac{V_B^2}{2g}$$

And from (5), we have:

$$\frac{V_B^2}{2g} = \frac{\Delta(\gamma_m - \gamma)}{\gamma} \Rightarrow$$

$$\gamma V_B^2 = 2g\Delta(\gamma_m - \gamma) \Rightarrow$$

$$V_B = \sqrt{\frac{2g\Delta(\gamma_m - \gamma)}{\gamma}} \Rightarrow$$

$$V_B = \sqrt{\frac{19,6 \cdot 0,2(13.600 - 1000)}{1000}} = 7,02 m / \sec$$

9) In the subsonic wind tunnel shown in the figure, one edge of a mercury manometer is put at the entrance of the wind tunnel with the fan and the other edge is put in the test chamber. The ratio of the cross-sections of the nozzles is $S_2 / S_1 = 1/15$.

The pressure and temperature of the tank are $p_1 = 1,1 atm$ and $T_1 = 300 K$, respectively. When the tunnel operates, the range altitude between the two mercury columns is $10 cm$. The density of fluid mercury is $\rho_H = 1.36 \times 10^4 kg / m^3$. Calculate the flow velocity of air V_2 in the area of the measurements.

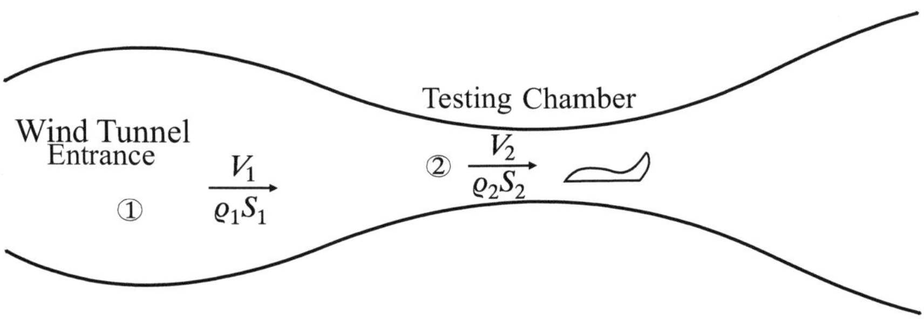

Solution:

The altitude range Δh between the two mercury columns is:

$$\Delta h = 10 cm = 0,1 m$$

For points (1) and (2), the Bernoulli equation and the continuity equation give us, respectively:

$$p_1 + \frac{\rho V_1^2}{2} = p_2 + \frac{\rho V_2^2}{2} \tag{1}$$

$$\text{and } S_1 V_1 = S_2 V_2 \tag{2}$$

By combining these, we have:

$$V_2 = \sqrt{\frac{2(p_1 - p_2)}{\rho\left[1-(S_2/S_1)^2\right]}} \tag{3}$$

Therefore, velocity V_2 of the air flow in the test chamber will be calculated from (3), provided density ρ and pressure difference $p_1 - p_2$ .are known. Therefore, from the state air equation, we have:

$$\rho = \frac{p_1}{RT_1} = \frac{1,1\left(1.01\times10^5\right)}{287(300)} = 1,29 kg / m^3$$

which is:

$$\rho = 1,29 kg / m^3 \tag{4}$$

We recall that:

$$R = 287 J / (Kg)(K)$$

$$1 At = 1,01\times10^5 N / m^3$$

The pressure difference $p_1 - p_2$ is found from the relation:

$$\int_1^2 dp = -\int_1^1 \rho g dh \Rightarrow p_1 - p_2 = \rho_H \bullet g \bullet \Delta h \Rightarrow$$

$$\Rightarrow p_1 - p_2 = \left(1,36\times10^4\ kg/m^3\right)\left(9.8\, m/S^2\right)(0,1) \Rightarrow$$

$$p_1 - p_2 = 1,33\times10^4\ N/m^2 \tag{5}$$

Therefore, substituting (4) and (5) into (3) gives:

$$V_2 = \sqrt{\frac{2\left(1.33\times10^4\right)}{1,29\left[1-\left(\frac{1}{15}\right)^2\right]}} = 144m/s$$

or otherwise $V_2 = 0,4M \left(M = \frac{V}{a}\right)$

10) A rectangular tank of large dimensions contains fluid, above which there is air overpressure $p_u = 0,07$ atm. At a depth of $1,2m$ from the fluid's surface, there is a hole, from which the fluid flows. What is the flow velocity?

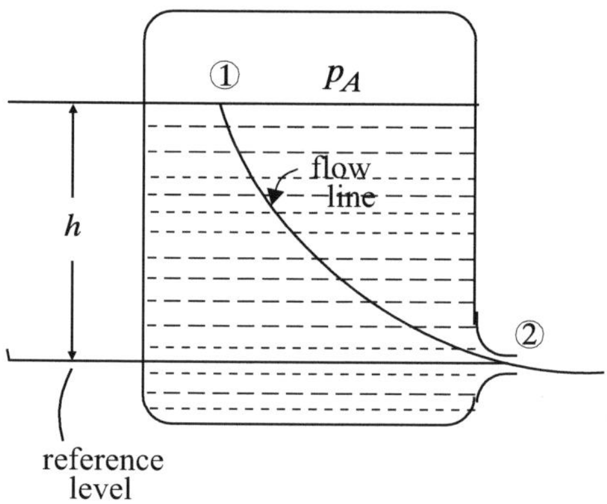

Solution:

For an ideal flow, we can use, for the flow line $1\to2$, the Bernoulli equation in the form:

$$p_1 + \rho\frac{u_1^{\,2}}{2} + \rho g h_1 = p_2 + \rho\frac{u_2^{\,2}}{2} + \rho g h_2 \tag{1}$$

If we assume that the reference level is as it is shown above, along point (2), we have:

$h_2 = 0$ and $h_1 = h = 1,2m$

As the tank has large dimensions, the fluid's velocity at point (1) is $u_1 = 0$ (the level decreases very slowly) and the pressure p_1 is considered constant at $1,07 atm$. Thus, (1) becomes:

$$p_1 + pgh = p_2 + \rho \frac{u_2^2}{2}$$

$$u_2 = \sqrt{\frac{2}{\rho}(p_1 - p_2 + \rho gh)}$$

$$u_2 = 6,1 m/s$$

11) The illustrated tube is leaking a gas of constant density. The pressure on the surface S_2 is equal to the external pressure. Calculate the force K on the formed diffuser, as a function of S_1, S_2, W_2 from the momentum equation and the integration of the pressure on the walls.

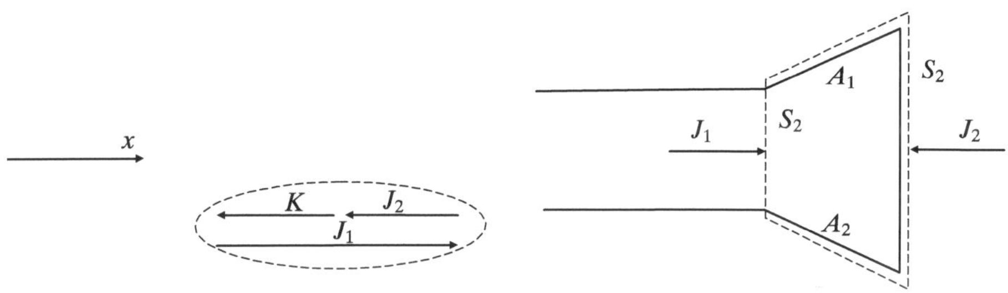

Solution:

a) With application of the momentum theorem to the x axis according to which:

$$\rho \int_{(K)} V dQ = \vec{F} + \vec{\sigma} + \vec{w} \tag{1}$$

we have that:

$$K = S_1\left(p_1 + \rho_1 V_1^2\right) - F_2\left(p_2 + \rho V_2^2\right) + \left(A_2 - A_1\right) p_0 \tag{2}$$

where p_i, V_i, ρ_i are the static pressure, density and velocity for the cross-section i(i=1,2).

Because of the pressure equation on the side surfaces A_1 and A_2, $(A_2 - A_1)\,p_0 = 0$. The continuity equation gives:

$$V_1 S_1 = V_2 S_2 \Rightarrow V_1 = \frac{V_2 S_2}{S_1} \tag{3}$$

By applying the Bernoulli equation in cross-sections 1 and 2, respectively:

$$p_1 + \left(\frac{\rho_1}{2}\right) V_1^2 = p_0 + \left(\frac{\rho}{2}\right) V_2^2 \Rightarrow p_1 = p_0 + \left(\frac{\rho}{2}\right)\left(V_2^2 - V_1^2\right) \Rightarrow$$
$$\overset{(2)}{\Rightarrow} p_1 = p_0 + \left(\frac{\rho}{2}\right) V_2^2 \left(1 - \frac{S_2^2}{S_1^2}\right) \tag{4}$$

Substituting (4) into (2), we have:

$$K = S_1\left[\rho/2\left(V_2^2 - V_2^2\frac{S_2^2}{S_1^2}\right) + \rho V_2^2 \frac{S_2^2}{S_1^2}\right] = S_2 \rho_2 V_2^2 =$$
$$= (\rho/2) V_2^2 \left(S_1 - S_1\frac{S_2^2}{S_1^2} + 2S_1\frac{S_2^2}{S_1^2} - 2S_2\right) = (\rho/2)\left[S_1\left(1 + \frac{S_2^2}{S_1^2}\right) - 2S\right]$$

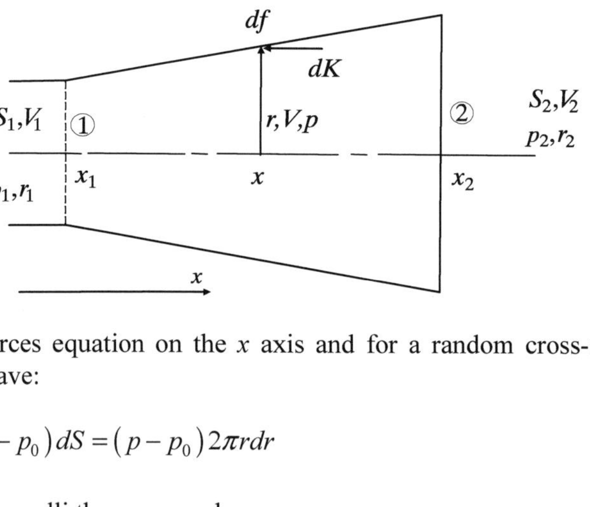

From the forces equation on the *x* axis and for a random cross-section in the position *x*, we have:

$$dk = (p - p_0)\,dS = (p - p_0)\,2\pi r dr \tag{5}$$

From the Bernoulli theorem, we have:

$$p + (\rho/2)V^2 = p_0 + (\rho/2)V_2^2 \tag{6}$$

The continuity principle gives us the equation:

(applied between x and x_2 cross-sections):

$$VS = V_2 S_2 \Rightarrow V = V_2 \frac{S_2}{S} = V_2 \frac{S_2}{\pi r^2} \tag{7}$$

From (6), we have:

$$p = p_0 + (\rho/2)V_2^2 - (\rho/2)V_2^2\left(S_2^2/\pi^2 r^4\right)$$

So $K = \int_{r_1}^{r_2} dK = \int_{r_1}^{r_2} (\rho/2)\left(V_2^2 - V_2^2 S_2^2/\pi^2 r^4\right)2\pi r dr =$

$$= \pi\rho V_2^2 \int_{r_1}^{r_2}\left(r - \frac{S_2^2}{\pi^2 r^3}\right)dr = \pi\rho V_2^2\left[\frac{r^2}{2} - \frac{S_2^2}{\pi^2}\frac{1}{2r^2}\right]_{r_1}^{r_2} \Rightarrow$$

$$\Rightarrow K = \frac{\rho V_2^2}{2}\left[2S_2 - S_1\left(\frac{S_2^2}{S_1^2} + 1\right)\right]$$

The force K is reverse and hence:

$$K = \left(\frac{\rho V_2^2}{2}\right)\left[S_1\left(\frac{S_2^2}{S_1^2} + 1\right) - 2S_2\right]$$

12) The illustrated body is given, which is surrounded by a parallel dynamic flow. What is the resistance of the body to the flow?

Solution:

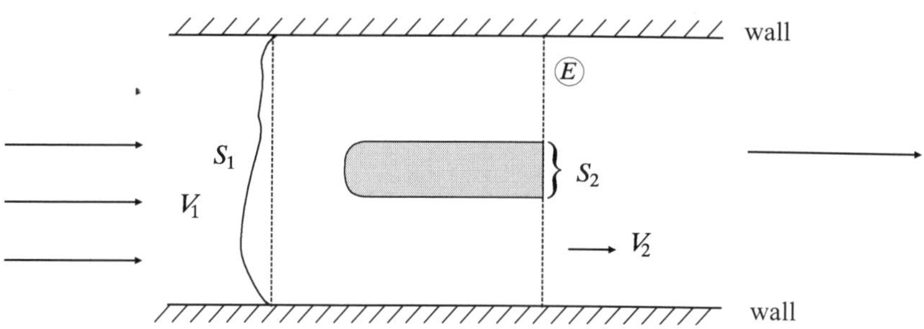

The application of the momentum theorem to the x axis gives:

$$K = p_2 S_1 - p_2 S_2 + \rho V_1^2 S_1 - p_2 (S_1 - S_2) - \rho V_2^2 (S_1 - S_2) \tag{1}$$

We take E, shown in the figure, as the surface.

Relation (1) gives:

$$\begin{aligned} K &= S_1 (p_1 - p_2) + \rho S_1 V_1^2 - \rho V_2^2 (S_1 - S_2) = \\ &= S_1 (p_1 - p_2) + \rho V_1^2 S_1 - \rho S_1 V_1 V_2 \end{aligned}$$

From the continuity principle, this relationship applies:

$$S_1 V_1 = (S_1 - S_2) V_2 \Rightarrow (S_1 - S_2) = \frac{S_1 V_1}{V_2} \tag{2}$$

$$\text{Finally, } K = S_1 (p_1 - p_2) + \rho S_1 V_1 (V_1 - V_2) \tag{3}$$

From the Bernoulli theorem, we have:

$$p_1 + (\rho_1/2) V_1^2 = p_2 + (\rho_2/2) V_2^2 \Rightarrow (p_1 - p_2) = (\rho/2)\left(V_2^2 - V_1^2\right) \tag{4}$$

Substituting (4) into (3), we have:

$$\begin{aligned} K &= S_1 (\rho/2)\left(V^2{}_1 - V_2^2\right) + \rho F_1 V_1 (V_1 - V_2) = \\ &= (\rho/2) S_1 \left(V_2^2 - V_1^2 + 2V_1^2 - 2V_1 V_2\right) = \\ &= (\rho/2) S_1 \left(V_2^2 - V_1^2\right)^2 = (\rho/2) S_1 V_2^2 \left(1 - \frac{V_1}{V_2}\right)^2 = \\ &= (\rho/2) S_1 V_2^2 \left[1 - \frac{(S_1 - S_2)}{S_1}\right] \Rightarrow K = (\rho/2) \frac{S_2^2}{S_1^2} V_2^2 \end{aligned}$$

For a parallel infinite flow, we have:

$S_1 \to \infty$ and S_2 has a finite value, which is:

$$K = 0$$

13) In the middle of the conical part (2) in the shape of a rectilinear pipe (1)-(2)-(3), which has a circular cross-section with diameters D_3=2D_1=1000 mm, there is a mill M of negligible thickness and negligible axis diameter, hub, bearings, etc., which rotates by the air energy (density ρ = 0.1 kpm^{-4}•sec^2) resulting from the flow rate Q=20 m^3/sec. Velocities V_1 and V_2 on parts (1) and (3) are axial and steadily distributed, and static pressures on points A and B of the manometer are equal to the external pressure. The following are asked:

a) Are the static pressures on parts (1) and (3) steadily distributed as well, and why?

b) Does the mill M absorb or deliver power, and what is its value?

c) How much force in *kp* should be applied on the walls of cone (2)?

d) If, for the same dimensions and the same flow rate Q, we assume that the mill rotates freely (without any energy exchange), which value would change from the given facts of the problem and how much would the change be?

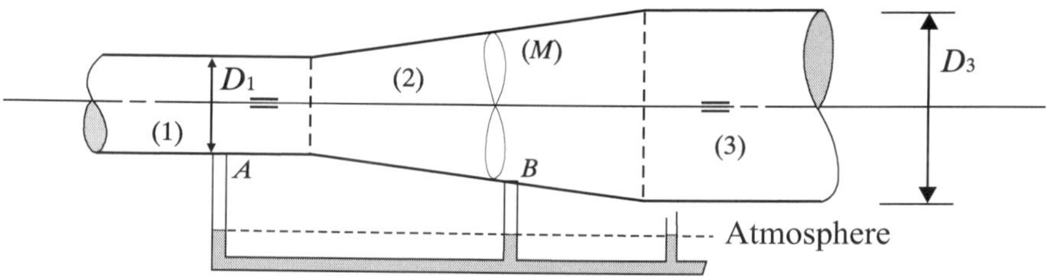

Solution:

a) From the given facts, we have:

$$V_1\text{=const and } V_2\text{=const} \tag{1}$$

and in the axial direction. From the Bernoulli theorem, we have:

$$p_{01} = const = p_1 + \rho\left(V_1^2/2\right) \text{ and } p_3 + \rho\left(V_3^2/2\right) = p_{03} const. \tag{2}$$

Therefore, p_1 = const and p_2 = const.

This means that we also have a constant static pressure.

b) From the application of the Bernoulli theorem, between cross-sections (1) and (3), we have:

$$\left.\begin{array}{ll} p_{01} = p_1 + (\rho/2)V_1^2 & [1] \\ p_{03} = p_3 + (\rho/2)V_3^2 & [2] \end{array}\right\} \Rightarrow$$

$$\Rightarrow p_{01} - p_{03} = (\rho/2)\left(V_1^2 - V_3^2\right) \quad (3)$$

because $\Rightarrow p_1 = p_3 = p_{atm}$

Moreover, the flow rate is given by the relations:

$$Q = V_1 \frac{\pi}{4} D_1^2 = V_3 \frac{\pi}{4} D_3^2 = const \quad (4)$$

It is given that $D_3 > D_1$. Therefore, (4) gives:

$$V_1 > V_3 \Rightarrow (\rho/2)\left(V_1^2 - V_3^2\right) > 0 \Rightarrow p_{01} > p_{03} \quad (5)$$

This means that the mill M absorbs power. This has a value:

$$N_m = Q \cdot (p_{01} - p_{03}) = 20(m3/5) \cdot \left(0.1kpm^4 S^2\right) \cdot \left(V_1^2 - V_3^2\right).$$

Relation (4) gives value to V_1 and V_4, which is:

$$V_1 = 101.86(m/\sec) \text{ and } V_3 = 25.46(m/\sec).$$

Finally, $N_m = 19454.50(kpm/\sec)$.

c) We consider the cone of the pipe with the mill as a reference surface. We apply the momentum theorem, according to which the total force from the fluid on the walls Kr on the mill km is equal to the momentum flow inside the reference surface (from the theory). This means that:

$$K_x = K_{tot} = -\rho Q(V_3 - V_2) = -m(V_3 - V_2) \quad (6)$$

is the momentum flow inside the reference surface.

The minus (–) was used in relation (6) because in the theory the calculating force is opposite, *which is the one that the fluid takes*. The force km on the wings is equal to:

$$Km = (\delta p)S, \tag{7}$$

where S is the mill's surface and (δp) is the pressure difference before and after the mill:

$$Km = \frac{N_m}{Q} \cdot S = \frac{N_m}{Q} \cdot \frac{\pi}{4}\left[(D_1 + D_3)/2\right]^2 = 214{,}9(kp).$$

And $K_{tot} = m(V_1 - V_3) = 152{,}8(kp)$.

So, $kr = -62{,}1(kp)$ with opposite direction from the one that we supposed.

d) If $p_{01} = p_{03}$ (the mill does not absorb power),

then $p_1 - p_3 = (\rho/2)(V_3^2 - V_1^2) = 486{,}4(kp/m^2)$.

14) The rectangular branch shown in the figure overleaf with tube diameters $D_1 = D_3 = 2D_2 = 0.5m$ leaks air, which is considered to be an incompressible inviscid fluid of density $\rho = 0.1kpm^{-4}\sec^2$. If the velocities and the static pressures on the straight parts of the tube are steadily distributed and $V = 50(m/\sec)$, find:

a) The flow rate form sectors (2) and (3) in $(m^3/\sec)$.

b) The boost that the system kp receives and the discovery of its direction.

c) The head $h(mm)$ of the column, in which the static pressure of branch (3) is expressed, in connection with the other branches. The obstructive fluid of the multiple manometer is water with $\gamma = 1000\,kp/m^3$.

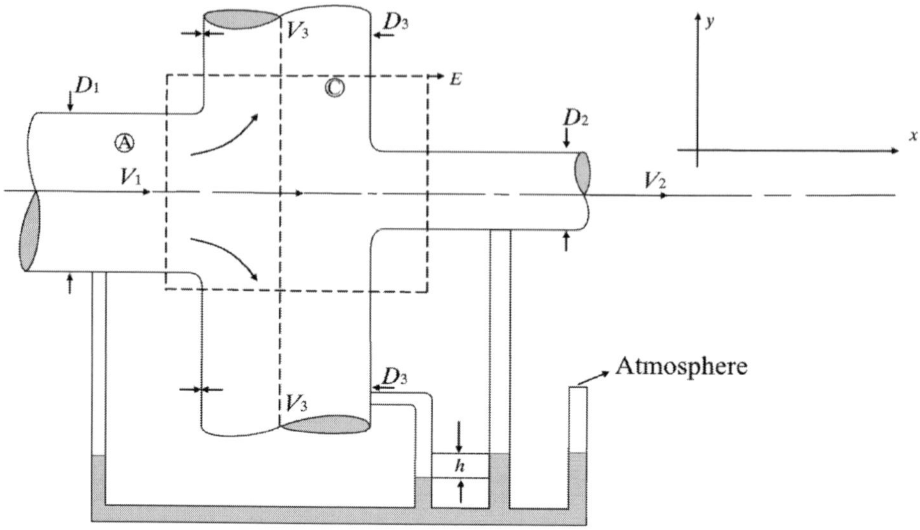

Solution:

a) Flow rate from branch 1 is:

$$Q_1 = \left(\frac{\pi}{4}\right) D_1^{\ 2} V_1 = \frac{\pi}{4} \cdot 0.5^2 \left(m^2\right) \cdot 50\left(m/\sec\right) \approx 9.89\left(m^3/\sec\right).$$

As shown, $p_1 = p_2$ and $V_1 = V_2$ from the Bernoulli theorem:

$$\text{So, } Q_2 = \left(\frac{\pi}{4}\right) D_2^{\ 2} V_2 = 2,45\left(m^3/\sec\right)$$

$$\text{and } Q_3 = \left(Q_1 - Q_2\right)/2 = 7,37/2 = 3,68\left(m^3/\sec\right).$$

b) The boost that the system receives comes from the momentum variation (initial momentum minus final momentum).

$$K_x = \left(\frac{\pi}{4}\right)\rho\left[D_1^{\ 2} V_1^{\ 2} - D_2^{\ 2} V_2^{\ 2}\right] = \left(\frac{\pi}{4} D_1^{\ 2}\right)\cdot \rho \cdot V_1 V_1 - \left(\frac{\pi}{4} D_2^{\ 2}\right)\cdot \rho \cdot V_2 \cdot V_2 =$$

$$= \left(S_1 \cdot \rho \cdot V_1\right) = V_1 = \left(S_2 \cdot \rho \cdot V_2\right)\cdot V_2 = m_1 \cdot V_1 - m_2 \cdot V_2 = J_1 - J_2,$$

Due to symmetry, we do not have any force in the y axis.

Finally: $K_x = \frac{\pi}{4}\rho V_1^2\left[D_1^2 - D_2^2\right]$ because $V_1 = V_2 = V$, as we previously proved.

So: $K_x = 36,8kp$

c) From the Bernoulli equation for points A and C, we have:

$$p_1 + (\rho/2)V_1^2 = p_3 + (\rho/2)V_3^2.$$

So, $p_3 = p_1 + (\rho/2)\left[V_1^2 - V_3^2\right] \Rightarrow p_3 - p_1 = (\rho/2)\left[V_1^2 - V_3^2\right]$,

where $V_1 = 50(m/\sec)$,

and $Q_3 = (\pi/4)\cdot D_3^2 \cdot V_3 \Rightarrow V_3 = \dfrac{4Q_3}{\pi D^2} \Rightarrow V_3 = 18,74(m/\sec)$.

And from the relation:

$$p_3 - p_1 - \gamma h \Rightarrow h = \frac{p_3 - p_1}{\gamma} = 107.4mm \Rightarrow h = 107,4(mm)$$

15) In a training airplane, the altitude indicator shows 5,000 ft. An independent external thermometer shows $505°R$ and a Pitot measures a pressure of $1818lb/ft^2$. Calculate the true and indicating velocities of the airplane.

Solution:

From the table of standard atmosphere, we have the following:

For pressure height $h_p = 5,000\ ft$, we have static pressure:

$$p = 1761lb/ft^2.$$

So, the dynamic pressure of the air is:

$$q = 1818 - 1761 = 57 lb / ft^2.$$

From the state equation:

$$p = \rho RT \Leftrightarrow \rho = \frac{p}{R \cdot T}$$

or $\rho = 0,002032 slug / ft^3$

The true velocity of the airplane is:

$$V_\delta = \sqrt{\frac{2q}{\rho_0}}$$

From the table of the standard atmosphere, we have $\rho = 0.002377 slug / ft^3$. Therefore, finally, $V_\delta = 219 ft / \sec$.

4.15. Problems to be solved

1) The vector of the velocity V of a flow field is given as:

$$V = (6 + 2xy + 3t^2)i + (2xy - 5t)j - xy^2 tk$$

a) Calculate the velocity for $t = 0, x = 1, y = 0$

b) Calculate the acceleration on the same position, but in two different times $t_1 = 0$ and $t_2 = 1$

2) Calculate the streamlines when:

a) $u = ax$, $\upsilon = ay$, $w = -2az$

b) $u = ax$, $\upsilon = -ay$, $w = C$

c) $u = \dfrac{C(y^2 + z^2 - 2x^2)}{r^5}$, $\upsilon = \dfrac{-3cxy}{r^5}$, $w = \dfrac{-3cxz}{r^5}$,

where a and c are constant and $r^2 = x^2 + y^2 + z^2$.

3) If the components of velocity in a flow field of an incompressible fluid are given by the relations:

$$u = -c^2 \frac{y}{r^2}, \quad \upsilon = c^2 \frac{x}{r^2}, \quad w = 0$$

where r is the distance from the z axis, solve the differential equation of the flow lines.

4) If a flow field has velocity:

$$V = (2x^2 - y)i + (3xy + x^2)j$$

calculate:

a) Is the flow one-dimensional, two-dimensional or three-dimensional?

b) Is the field steady or unsteady?

c) How much is the acceleration?

5) The velocity vector is given by:

$$V = 2kxi - kx^2 j$$

Calculate the equation of the laminar flow that goes through the point $(1, -1)$.

6) A wind tunnel has the following characteristics:

At the entrance:

$$p_1 = 101000 P\alpha$$

$$S_1 = 1m^2$$

$$T_1 = 288K$$

$$V_1 = 200m/\sec$$

At the test chamber:

$$S_2 = 0,25m^2$$

$$V_2 = 900m/\sec$$

Calculate:

a) The mass flow in the wind tunnel.

b) The density of air in the test chamber.

7) Consider an airfoil in a flow of air, where the pressure p, velocity V and density ρ are 10,331.2 kg/m^2, 44.4 m/sec and 1,225 kg/m^3, respectively, ahead of the airplane. At a certain point A of the wing, pressure p_A is $10.106,6 kg/m^2$. What is the velocity at point A?

8) Suppose a converging air flow, where $S_1 = 5m^2$ and $S_2 = 1,67m^2$. If $V_1 = 10\,m/\sec$, $V_2 = 30\,m/\sec$, and the pressure and temperature at the inlet are $p_1 = 1,2\times10^5\,N/m^2$ and $T_1 = 330\,K$, respectively, then what is the pressure at the exit?

9) Calculate the velocity V of an airplane, whose Venturi tube has a ratio of the cross-sections' surface area $S_2/S_1 = 1/4$ and a pressure difference between the entering and narrowing points of the tubes $p_1 - p_2 = 80lb/ft^2$. It is given that $\rho = 1,255kg/m^3$.

10) Inside a vertical tube, $5lt$ of water flows per second without any friction from top to bottom. The inlet diameter of the tube is $12cm$, while at one of its positions that lies $12cm$ below, the diameter becomes $8cm$. Calculate the pressure difference between these two points.

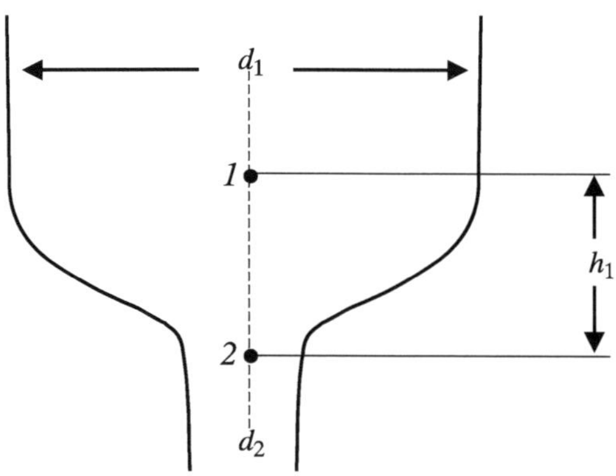

11) A slow-moving airplane flies with a velocity of $90m/\sec$ at sea level. The velocity is measured with a Venturi tube with surface area ratio 13/1. Inside the cockpit, there is a velocity indicator, which is connected to a pressure counter, which measures the pressure difference $p_1 - p_2$ (calibrated to show velocity). What is the

maximum differential pressure $p_1 - p_2$, that will be measured by the instrument (ρ = 1,255 kg/m^3).

12) An airplane flies with a velocity of $60m/\sec$ at a height of standard velocity of $3km$. At a point of the wing, the flow velocity is $70m/\sec$. Calculate the pressure on this point. We assume that the flow is incompressible.

13) We assume water flow in a diffuser with surface area 10 ft^2, and velocity at the inlet is 5 ft/sec. If the surface area of the exit is four times that of the entrance, then:

a) Calculate the exit velocity of the water flow.

b) Calculate the pressure difference between the inlet and the outlet of the diffuser ($\rho_{water} = 62,4lb/ft^3$).

14) A training airplane flies at 2,100 m, where the static pressure and the density of air are $p_{static} = 78.520N/m^2$ and $\rho = 9,9649kg/m^3$, respectively. Calculate the true and indicated velocities of the airplane if a Pitot tube put at the wing tip measures the pressure $p_{tot} = 9,132N/m^2$ ($\rho_0 = 1,2250kg/m^3$).

15) A Pitot tube is adjusted at the test chamber of a subsonic wind tunnel of low velocity. The flow at the test chamber has velocity, pressure and temperature of 150 m/h, 1 atm and $70°F$, respectively. Calculate the pressure that is measured from the Pitot tube.

16) The altimeter of an airplane shows 8,000 ft. The Pitot tube measures 1,650 lb/ft^2. If the external temperature is $500R$, what is the real air velocity of the airplane?

17) During 1910–1930, in order to measure the velocity of slow-moving airplanes, the Venturi tube was used. This mechanism, shown in the following figure, was put in a special place in the airplane, where the inlet velocity of air was almost the same as that of the free flow. If we know that $S_1/S_2 = 4$ and $p_1 - p_2 = 4.000\,N/m^2$, calculate the velocity of the airplane at the sea level.

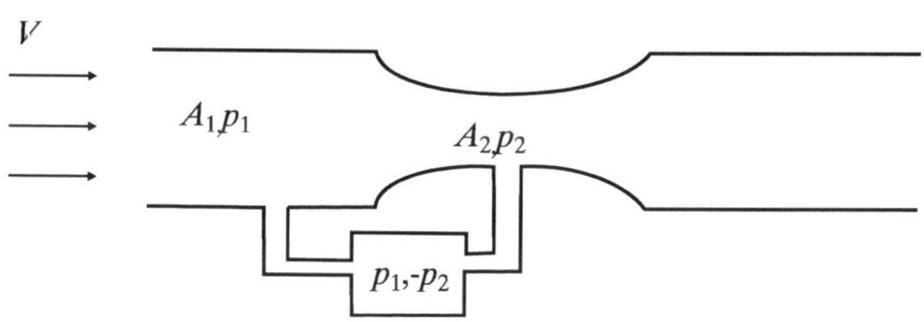

18) A tube manometer U is used for the measurement of the pressure of an incompressible fluid in a cylinder at sea level. The fluid in the tube U is water with density 1,94 $slug/ft^3$, with an altitude range h =5 ft. Calculate the pressure inside the cylinder.

19) A pilot C-130 wants to fly at a low level, for a navigation mission, at 4 NM/min (240 knots ground velocity). The winds are north (they blow to the south) at 20 knots, the pressure height is 10,000 ft and the temperature is $80°F$. What is the indicative velocity at which the pilot must fly?

20) In a wind tunnel, the following flow principles are applicable at the inlet:

$$p = 101.000\, N/m^2,\, S_1 = 1m^2,\, T = 228\, K,\, V_1 = 200\, m/\sec.$$

In the test chamber, the following are applicable: $S_2 = 0,25m^2$ and $V_2 = 900m/\sec$.

a) What is the mass flow rate in the wind tunnel?

b) What is the density in the test chamber?

21) The flow in the entrance of a low-velocity jet has the following values:

$$p = 14,5\, psi,\, S_1 = 2,65\, ft^2,\, V_1 = 300\, ft/\sec \text{ and } \rho = 0,0024\, slug/ft^3.$$

At the end of the entrance, there is a compression surface, where $p_2 = 15\, psi$. Assume that the flow is steady, incompressible, one-dimensional and inviscid.

a) What is the mass flow rate?

b) Calculate the area surface of the compression surface.

22) We consider an airplane that flies with a velocity $V_1 = 60m/\text{s}$ at height $h = 2000m$ under ISA conditions. Using the Bernoulli equation, calculate the static pressure at point A of the airplane, if the velocity at that point is $V_2 = 80m/s$.

Afterwards, if you know that the velocity at another point B is $V_3 = 63\,m/s$, calculate the pressure difference between those two points.

5

Flow in Pipes

5.1. Introduction

In the previous chapters, we dealt with the kinematics of real fluids in its general form, stating its basic concepts, categorizing it as laminar and turbulent and mentioning the factors that influence it as well as the laws and the principles that define it.

In this chapter, we will deal with the flow of real fluids inside closed pipes, meaning that the fluid takes up the entire cross-sectional area of the pipe during its motion. Furthermore, the fluids are considered incompressible, which means their density is constant (ρ = constant). Also, for gases, the Mach number is considered not to be greater than 0.4 ($M < 0.4$), which means that the flow rate remains constant in any cross-sectional area of the pipe and the gas flow can be considered incompressible.

Finally, we will have to mention that the flows of this category are of great importance in everyday life, not only nowadays, but since ancient times, given the fact that their use has been lost over time, as nature itself uses them on both plant and animal organizations. The only difference in the evolution of humans constitutes the material with which the pipes are constructed. Pipes were made of clay initially (Mycenaean Civilization, Roman Empire), and then metal in the industrial era and plastic in modern times. With the improvement of materials, losses associated with the flow of fluids during their osculation with pipe walls were reduced, thus allowing us to use them in all productive procedures, like water flow rate, cold stores and the transportation of fuels including gases and steam, as well as in various scientific fields like medicine, which has developed a separate field of science, bio-fluid mechanics, significantly helping researchers in the construction of artificial veins, thereby improving people's lives.

5.2. Physical quantities

In the study of flow of an incompressible real fluid, the physical quantities influencing it are the quantities mentioned in the geometric characteristics of the *pipe* (length (L), diameter $(\mathbf{d})$ and roughness $(\boldsymbol{\varepsilon})$); quantities of the *fluid* (density $(\boldsymbol{\rho})$, weight $(\mathbf{w})$, pressure $(\mathbf{p})$, kinematic viscosity $(\boldsymbol{\nu})$); *flow* quantities (flow rate $(\mathbf{Q})$ and velocity $(\mathbf{V})$) and *energy quantities* (head $(\mathbf{h})$, pressure $(\mathbf{p})$, velocity $(\mathbf{V})$, *head losses* $(\Sigma\mathbf{y}, \mathbf{y}_\mathbf{f}, \mathbf{y}_\mathbf{k})$, *pump head* $(\mathbf{y}_\mathbf{p})$ and *turbulent* $(\mathbf{y}_\mathbf{t})$).

5.3. Laminar and turbulent flows in pipes

5.3.1. *Reynolds number in pipes*

A basic factor in the study of these flows is the Reynolds number, which is given by the relation:

$$\mathrm{Re} = \frac{\rho V L}{\mu} = \frac{VL}{\nu} = \text{ constant} \quad [5.1]$$

where ρ is the fluid density, V is the flow velocity, L is the characteristic flow length, μ is the fluid viscosity and ν is the kinematic viscosity $\left(\nu = \dfrac{\mu}{\rho}\right)$.

Now, in flow inside cylindrical pipes, we consider the internal diameter of the pipe as the characteristic flow length. Therefore, we have:

$$\mathrm{Re} = \frac{\rho V d}{\mu} = \frac{Vd}{\nu} \quad [5.2]$$

where d is the internal diameter of the cylindrical pipe

When the pipe *is not a cylinder*, we consider four times the *hydraulic radius*, $\mathbf{r_h}$, as the characteristic length, which is given by the relation:

$$r_h = \frac{S}{P} \quad [5.3]$$

where S is the surface area of the pipe's cross sectional area and P is its perimeter. Therefore, the Reynolds number for the pipes is expressed by the equation:

$$\mathrm{Re} = \frac{4r_h \rho V}{\mu} = \frac{4r_h V}{\nu} \qquad [5.4]$$

When the Reynolds number is smaller than 2,100 (Re < 2,100), the flow is laminar. A critical area follows, in which the flow gradually turns turbulent (2,100 < Re < 4,000 approximately). For values larger than Re 4,000 (Re > 4,000), the flow is turbulent.

Relation [5.2] shows that the *laminar flow is favored in pipes of small diameter, with small flow velocities and in fluids of high viscosity (thick)*. When we study the flow of a certain fluid in a pipe of diameter d, we will have laminar flow for low velocities V (Re < 2,100). By increasing the velocity, the laminar flow is disturbed, as sporadic turbulences are created (transitional area, 2,100 < Re < 4,000). With a further increase in velocity, the flow will become turbulent (Re > 4,000), with the formation of turbulences in the whole space of the flow, except for a very thin layer that lies in osculation with the walls of the pipe (*viscous sub-layer*). Respective results arise if the velocity does not vary, but the diameter of the pipe or the kinematic viscosity of the fluid varies.

5.3.2. *Average velocity and velocity distribution*

As the flow develops inside the pipe, velocity varies from one point to the other. Therefore, the velocity increases with the distance from the wall and becomes maximum at the center of the pipe.

If we denote the local velocity by u and the distance from the center of the circular cross-sectional area by r, we have:

– For $r = R$ (contact with the wall): $u = 0$.

– For $r = 0$ (center of the tube): $u = u_{\max}$.

The distribution of velocities in the cross-sectional area of a cylindrical pipe is only calculated accurately in the case of laminar flow. Particularly, in the *laminar flow*, the value of the local velocity is equal to:

$$V_{md} = V_{\max} \cdot \left(1 - \frac{r^2}{R^2}\right) \qquad [5.5]$$

In turbulent flow, the local velocity increases at a faster pace away from the wall. For velocity distribution, the empirical relation is applicable:

$$V_{md} \approx V_{\max} \cdot \left(1 - \frac{r}{R}\right)^n \qquad [5.6]$$

where the exponent n varies between 1/9 and 1/5: $1/9 \le n \le 1/5$. However, for the calculations, we use the *average flow velocity* ***V***, which is defined by the continuity equation:

$$V_{nd} = \frac{m}{\rho \cdot S} \qquad [5.7]$$

The relationship between the average velocity and the local velocity is:

$$V_{md} = \frac{1}{S} \cdot \int_S u \cdot dS \qquad [5.8]$$

For laminar flow, we have $V_{md} = 0,5V_{\max}$, and for turbulent flow, we have $0,5V_{\max} < V \le 0,8V_{\max}$. In problems, we generally use this average velocity.

5.3.3. *Shear stress in a horizontal cylindrical pipe*

In the horizontal cylindrical pipe shown in Figure 5.1, we consider a small cylindrical volume element of fluid of length δx and radius r, co-axial to the main pipe, whose radius is R, and the velocity of the fluid is in conformity with the whole length with symmetrical distribution to the main axis of the pipe. The forces applied on the elementary cylinder are exclusively pressure forces caused by the pressure difference between the two vertical surfaces of the element to the axis and forces resulting from the shear stress $\boldsymbol{\tau}$ that acts on the cylinder surface of the element.

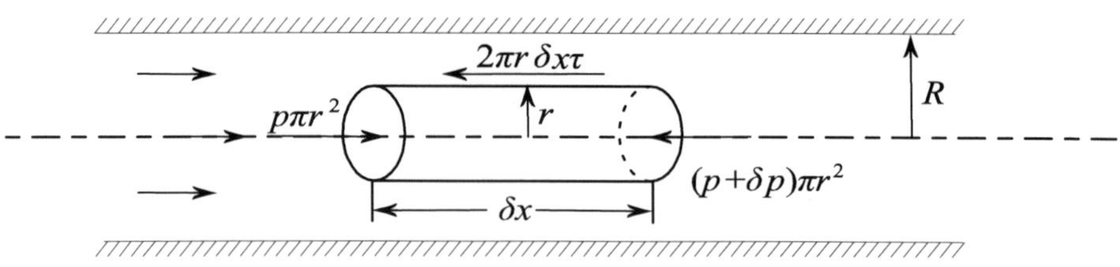

Figure 5.1. *Cylindrical volume element of fluid*

Because the flow is considered uniform and constant, the sum of these forces will be zero. Therefore:

$$F_1 - F_2 - F_3 = 0 \qquad [5.9]$$

And by replacing these with the parameters in Figure 5.1, we get:

$$p\pi r^2 - (p+\delta p)\pi r^2 - \tau 2\pi r \delta x = 0 \qquad [5.10]$$

Performing the sums and solving for r, we have for the shear class:

$$\tau = -\underset{\delta x \to \infty}{\frac{\delta p}{\delta x}} \cdot \frac{r}{2} \Rightarrow \qquad [5.11]$$

$$\tau = -\frac{1}{2} r \frac{dp}{dx} \qquad [5.12]$$

However, pressure is independent of r and because velocity is considered to be uniform along the pipe and has a symmetrical distribution to the main axis of the pipe, we conclude that it does not vary along the pipe; also, the shear stress τ is independent of x and only a function of r. Therefore:

$$\frac{dp}{dx} = \text{const} \qquad [5.13]$$

Thus, by integrating relation [5.12] between two points along the pipe, we have:

$$\tau \int_{\ell_1}^{\ell_2} dx = -\frac{r}{2} \int_{p_1}^{p_2} dp \Rightarrow \qquad [5.14]$$

$$\Rightarrow \tau(\ell_2 - \ell_1) = r \frac{(p_2 - p_1)}{2} \qquad [5.15]$$

and setting $\ell = \ell_2 - \ell_1$, [5.15] becomes:

$$\tau = \frac{r}{2\ell}(p_1 - p_2). \qquad [5.16]$$

Relation [5.15] is a linear function of the radiant distance r, so its value at the pipe walls is:

$$\tau_0 = \tau(R) \tag{5.17}$$

and substituting [5.16] into [5.17]:

$$\tau_0 = \frac{R}{2\ell}(p_1 - p_2). \tag{5.18}$$

Relation [5.18] gives the highest value that the shear stress can take inside the pipe.

5.3.4. *Pressure drop in a horizontal cylindrical pipe*

If we solve relation [5.18] to $p_1 - p_2$ and if we use diameter d of the pipe instead of radius R, then we have:

$$p_1 - p_2 = \tau_0 \frac{2\ell}{R} \Rightarrow \tag{5.19}$$

$$p_1 - p_2 = \tau_0 \frac{4\ell}{d} \Rightarrow \tag{5.20}$$

$$\text{because } R = \frac{d}{2} \tag{5.21}$$

$$\text{and } \Delta p = p_1 - p_2 = 4\frac{\ell}{d} \cdot \tau_0 \tag{5.22}$$

Relation [5.22] provides us with the pressure drop in a length ℓ of the pipe when the value of the shear stress τ on its walls is known.

5.3.5. *Pressure drop in a horizontal non-cylindrical pipe*

When the closed pipe does not have a cylindrical cross-sectional area, we use the *equivalent or hydraulic diameter*, ***dh***, instead of diameter *d*, as:

$$d_h = 4r_h \tag{5.23}$$

where r_h is the *hydraulic radius*, defined from relation [5.3] as:

$$\tau_h = S / P, \tag{5.24}$$

substituting [5.24] into [5.23]:

$$d_h = 4\frac{S}{P} \tag{5.25}$$

and using [5.25] in [5.22] gives:

$$\Delta_p = p_1 - p_2 = 4\frac{\ell}{4\frac{S}{P}} \cdot \tau_0 \Rightarrow \Delta_p = \frac{\ell p}{S} \cdot \tau_0 \tag{5.26}$$

5.3.6. *Shear stress τ_0 and friction coefficient f*

From the dimensional analysis by using the basic dimensions *M*, *L* and *T* of the values F_D, *r*, *V*, ρ and μ for a long cylinder of radius *r*, with a vertical axis in a uniform flow of real fluids of density ρ and viscosity μ, which moves with velocity *V*, the pressure drop $\Delta p = p_1 - p_2$ between two positions (1) and (2) according to the Buckingham method is given by the relation:

$$\Delta p = \frac{f}{2} \cdot \frac{\ell}{d} \cdot \rho V^2 \tag{5.27}$$

where *f* is the friction coefficient of the flow. Combining relationships [5.22] and [5.27], we find that the friction coefficient *f* is given by the relation:

$$f = \frac{8\tau_0}{\rho V^2}. \tag{5.28}$$

Therefore, if one of the two terms (*f* or τ_0) is calculated from experimental measurements, we can calculate the other one from relation [5.28].

5.3.7. *Pressure drop and friction coefficient relationship for a horizontal pipe*

1) *Cylindrical pipe*

Combining [5.22] with [5.28], for the pressure drop in a cylindrical horizontal pipe, we have:

$$\Delta p = 4\frac{\ell}{d} \cdot \frac{f\rho V^2}{8} \Rightarrow$$

$$\Delta p\left(p_1 - p_2\right) = \frac{f}{2}\frac{\ell\rho V^2}{d} \quad [5.29]$$

2) *Non-cylindrical pipe*

Combining [5.26] with [5.29], for the pressure drop in a horizontal non-cylindrical pipe, we get (Figure 5.2):

$$\Delta p = 4\frac{\ell P}{S} \cdot \frac{f\rho V^2}{8} \Rightarrow$$

$$\Delta p\left(p_1 - p_2\right) = \frac{f}{8} \cdot \frac{P\ell}{S}\rho V^2 \quad [5.30]$$

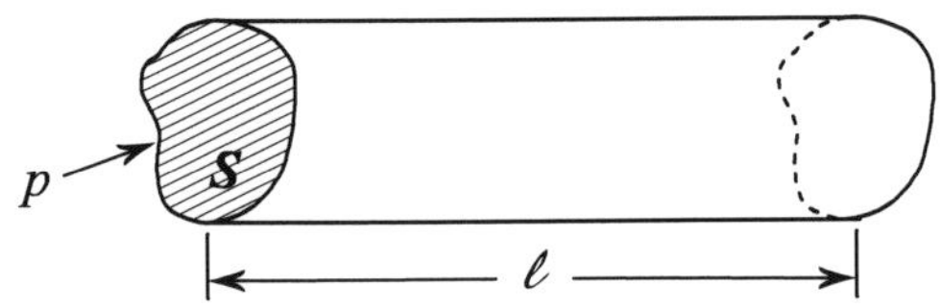

Figure 5.2. *Non-cylindrical pipe*

5.4. Basic equations

The steady flow of an incompressible fluid in pipes is defined by two laws: the law of conservation of mass (*continuity equation*) and the principle of energy conservation (*Bernoulli equation*). We have studied these laws in a previous chapter and, therefore, here we will give an expression that defines them for the steady and incompressible flow in cylindrical pipes.

5.4.1. *Continuity equation*

In a steady flow of an incompressible fluid in a pipe, the volume flow rate Q is constant in any cross-sectional area of the pipe:

$$Q = S \cdot V = \text{const} \tag{5.31}$$

where Q is the volume flow rate, V is the fluid velocity and S is the cross-sectional area at a point of the pipe.

For cylindrical pipes, it is $S = \pi \cdot d^2 / 4$; therefore:

$$Q = \frac{\pi \cdot d^2}{4} \cdot V = \text{const} \tag{5.32}$$

Solving [5.32] for the velocity of the fluid, we have:

$$V = \frac{4 \cdot Q}{\pi \cdot d^2} . \tag{5.33}$$

This means that *the velocity is proportional to the flow rate and inversely proportional to the square of the diameter*. For a certain flow rate Q, the increase in diameter leads to a significant decrease in velocity and vice versa. If, for example, the diameter is doubled, then the velocity will be decreased four times. Under steady flow, when the diameter is not varied, the velocity remains constant at the whole length of the pipe.

5.4.2. *Energy equation (Bernoulli equation)*

The Bernoulli equation, as we have proved, is expressed by the relationship:

$$p + \frac{1}{2}\rho V^2 = \text{const} \tag{5.34}$$

In order to give an expression for this flow, we consider that in a cross-sectional area of the pipe, the fluid has energy per weight unit as follows:

$$\frac{E}{mg} = \frac{E}{\gamma \cdot Q} = h + \frac{V^2}{2 \cdot g} + \frac{p}{\gamma} \tag{5.35}$$

Each term in equation [5.35] is expressed in length dimensions as follows:

$$\left(M \cdot L^2 \cdot T^3\right) / \left(M \cdot T^{-1} \cdot L \cdot T^{-2}\right) = L$$

This means that, in the SI system, each term has meter as its unit:

$$\left(J/N = N \cdot \frac{m}{N} = m \right)$$

Hence, the above terms are called *energy heads*:

h : head of dynamic energy

$\frac{V^2}{2g}$: head of kinetic energy

$\frac{p}{\gamma}$: head of pressure energy

Between two cross-sectional areas (1) and (2) of the pipe, there are energy losses per weight unit of the fluid because of the frictions $\boldsymbol{\Sigma h}$, which are called head losses. Therefore, when the fluid flows from the cross-sectional area (1) to (2), the total head losses of energy decreases to Σh:

$$\left(\frac{p_1}{\gamma} + \frac{V_1^2}{2 \cdot g} + h_1 \right) - \left(\frac{p_2}{\gamma} + \frac{V_2^2}{2 \cdot g} + h_2 \right) = \Sigma h$$

$$\frac{p_1 - p_2}{\gamma} + \frac{V_1^2 - V_2^2}{2 \cdot g} + (h_1 - h_2) = \Sigma h \qquad [5.36]$$

We note that the energy equation, also known as the Bernoulli, does not have any special form for the cylindrical pipes. Generalizing, if between the points (1) and (2) there is a pump which adds energy to the fluid or turbulence, which subtracts energy, then the energy equation becomes:

$$\frac{p_1 - p_2}{\gamma} + \frac{V_1^2 - V_2^2}{2g} + (h_1 - h_2) = \Sigma h + h_t - h_p \qquad [5.37]$$

where h_t is the head of turbulent energy and h_p is the energy head of the pump.

During the study of steady incompressible flow in pipes, if there is no remarkable difference between two points that we choose, then [5.33] differs proportionally. Therefore:

1) if $p_1 - p_2 \approx 0$, it means that there is no remarkable variation in pressure: for example, when two open tanks are connected to pipes that have an altitude range *H*, considering the cross-sectional areas (1) and (2) as their free surfaces, then we will have:

$$p_1 = p_2 = p_{atm}$$

and $V_1 = V_2 \approx 0$

Relation [5.36] becomes:

$$h_1 - h_2 = \Sigma h \Rightarrow H = \Sigma h \qquad [5.38]$$

2) if $V_1^2 = V_2^2 \approx 0$, which means that we do not have a notable variation of the velocity, like in a pipe of constant cross-sectional area and steady flow, then we will have:

$$V_1 = V_2 = \frac{4Q}{\pi d^2},$$

and [5.36] will be:

$$\frac{p_1 - p_2}{\gamma} + (h_1 - h_2) = \Sigma h \qquad [5.39]$$

and if it is horizontal, then every $h_1 = h_2$ and [5.36] will be:

$$\frac{p_1 - p_2}{\gamma} = \Sigma h \qquad [5.40]$$

Finally, we have to mention that proportionally to the quantity to which the energy values are expressed, the Bernoulli equation takes various forms. Therefore, we have another expression for energy per time unit, another per mass volume unit, etc.

5.5. Friction coefficient of a laminar flow of real fluid in a horizontal cylindrical pipe

We consider the horizontal pipe of radius *R*, as shown in Figure 5.3

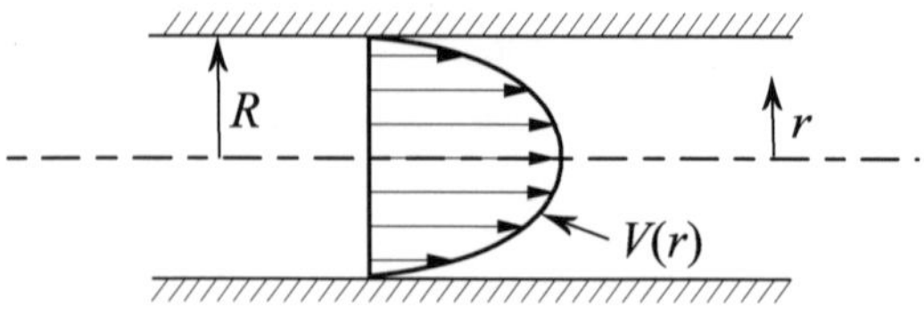

Figure 5.3. *Laminar flow in horizontal pipe*

As we know, the shear stress for a laminar flow is exclusively caused by the fluid viscosity and is given by the relationship:

$$\tau = -\mu \frac{dV}{dr}, \text{ where } V = V(r) \tag{5.41}$$

The minus (−) has been included because the shear stress increases close to the pipe walls, while the velocity on the contrary decreases.

However, the shear stress on a horizontal cylindrical pipe is also given by the relationship:

$$\tau = \frac{r}{2\ell}(p_1 - p_2) \tag{5.42}$$

Therefore, from [5.41] and [5.42], we get:

$$\frac{r}{2\ell}(p_1 - p_2) = -\mu \frac{dV}{dr} \tag{5.43}$$

With $V = 0$ on the walls, as well as $r = R$

If in [5.43] we separate the variables, and by integrating, we get:

$$\frac{p_1 - p_2}{2\ell} \int_{\tau}^{R} r dr = -\mu \int_{V}^{0} dV \Rightarrow$$

$$V(r) = \frac{(p_1 - p_2) R^2}{4\mu\ell} \cdot \left(1 - \frac{r^2}{R^2}\right). \tag{5.44}$$

Therefore, [5.44] shows us that the velocity distribution in a cross-sectional area of the pipe is a *parabola*. In order to find the *maximum velocity* V_{max} of the fluid inside the pipe, we calculate the derivative of $V(r)$:

$$\frac{dV}{dr} = 0 = -\frac{2r}{4\mu\ell}(p_1 - p_2), \tag{5.45}$$

so the velocity reaches its maximum for $r = 0$, which is along the axis of the pipe. Therefore, the maximum velocity is given by the relationship:

$$V_{\max} = \frac{R^2}{4\mu\ell}(p_1 - p_2). \tag{5.46}$$

In order to calculate the *average velocity* V_{md} of the fluid inside the horizontal cylinder, we use the relationship:

$$V_{md} = \frac{1}{S}\int VdS = \frac{R^2}{\pi R^2}\int_D^R \frac{(p_1 - p_2)}{4\mu\ell}\left(1 - \frac{r^2}{R^1}\right)\cdot 2\pi r dr =$$

$$= \frac{p_1 - p_2}{2\mu\ell}\left\{\int_0^R r dr - \frac{1}{R^2}\int_0^R r^3 dr\right\} =$$

$$= \frac{p_1 - p_2}{2\mu\ell}\left(\frac{1}{2}R^2 - \frac{1}{4}R^2\right). \tag{5.47}$$

And by performing the sums, we find for V_{md} the expression:

$$V_{md} = \frac{R^2}{8\mu\ell}(p_1 - p_2) = \frac{V_{\max}}{2} \tag{5.48}$$

And $V_{\max}$ is given by relationship [5.46].

Therefore, the average velocity of the fluid that is moving inside a cylindrical horizontal pipe is the half of its maximum velocity.

If relationship [5.48] is solved to $p_1 - p_2$, we have:

$$p_1 - p_2 = \Delta_p = \frac{8\mu\ell}{R^2}V_{md}. \tag{5.49}$$

However, from relationship [5.25], we have found that the pressure drop is given by the expression:

$$p_1 - p_2 = \Delta p = \frac{f}{2} \cdot \frac{\ell}{d} \rho V^2 . \qquad [5.50]$$

Equating [5.49] and [5.50], we have:

$$\frac{8\mu\ell}{R^2} V_{md} = \frac{f}{2} \cdot \frac{\ell}{d} \rho V_{md}^2 \qquad [5.51]$$

And because $R = \frac{d}{2}$, [5.51] becomes:

$$\frac{8\mu\ell}{\frac{d^2}{4}} = \frac{f}{2} \cdot \frac{\ell}{d} \rho V_{md}^2 \Rightarrow$$

$$f = \frac{64\mu}{\rho d V_{md}} \qquad [5.52]$$

and because

$$\mathrm{Re} = \frac{\rho d V_{md}}{\mu}, \qquad [5.53]$$

The relationship [5.52] using [5.53] gives us *for the friction coefficient* ***f*** of a laminar flow the relationship:

$$f = \frac{64}{\mathrm{Re}} \qquad [5.54]$$

Therefore, if we know the geometry of the pipe and the average velocity of the fluid, then the pressure drop is calculated from the relationship:

$$\Delta p = p_1 - p_2 = \frac{f}{2} \frac{\ell}{d} \rho V_{md}^2 = \frac{32}{\mathrm{Re}} \frac{\ell}{d} \rho V_{md}^2 \Rightarrow$$

$$\Rightarrow \Delta p = 32 \frac{V_{md}}{d^2} \cdot \mu \cdot \ell . \qquad [5.55]$$

From [5.55], we conclude that in the laminar flow, the pressure drop is proportional to the average velocity V_{md} of the fluid and independent of its density ρ. Finally, from [5.54], we have that for the average velocity V_{md}, the following expression arises:

$$V_{md} = (p_1 - p_2)\frac{d^2}{32\mu\ell} \qquad [5.56]$$

Relation [5.56] is known as the equation of *Hagen* (1839) and *Poiseuille* (1840), which each one proved it experimentally while its analytical expression took place thanks to *Wiedemann*, who stated it in 1856.

5.5.1. *Inlet conditions*

The previous relationships are applicable provided the fluid is uniform. This means that they are not applicable for the part of the pipe near the entrance. For example, if the fluid that is transferred from a pipe begins from a big tank and goes through the pipe from a rounded mouth, then the distribution of the velocity at the entrance is uniform. However, as the fluid starts feeling the presence of the pipe walls, because of the shear stresses caused by the zeroing of the velocity on the walls, the velocity must increase at the center of the pipe and decrease at the edges, but must always satisfy the continuity relationship:

$$\int VdS = \text{const} \qquad [5.57]$$

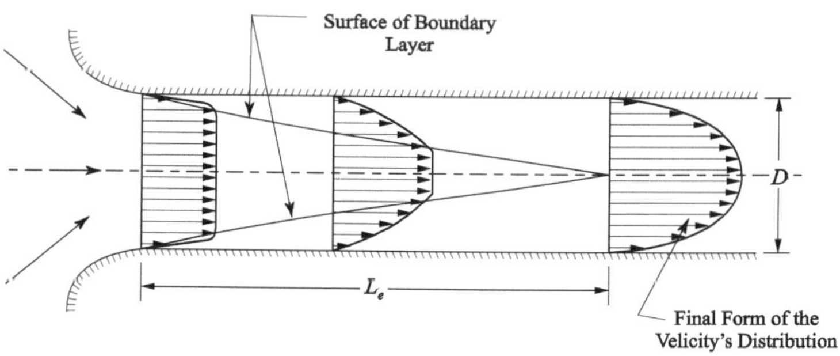

Figure 5.4. *Velocity distribution along a pipe*

Boussinesq was the first to calculate the entrance length ℓ_c, where:

$$\frac{\ell_c}{d} = 0,03\text{Re} \qquad [5.58]$$

Also, *Langhaar* found the theoretical relationship:

$$\frac{\ell}{d} = 0,058\text{Re}, \qquad [5.59]$$

that is verified by experimental measurements.

Both relationships give the entrance length that the flow needs to gain parabolic velocity distribution, concerning *laminar flow*. In the case *where the flow is turbulent, there is no relationship that gives the entrance length.* However, generally, we can say that the *entrance length in the turbulent flow* is sometimes *longer* and sometimes *shorter* than the corresponding length in the laminar flow, which usually depends on local and geometrical conditions.

During the transportation of the fluids in pipes, the entrance length is often shorter than other distances, and so the losses are calculated by considering the whole length of the pipe under laminar or turbulent flow conditions. Therefore, from relationship [5.55]:

$$\Delta p = \frac{f}{2}\frac{\ell}{d}\rho V^2, \qquad [5.60]$$

if we divide both sides by ρg and the expression $\Delta p = h_1 \rho g$, with h_1 being the *head losses*, then:

$$h_1 = \frac{\Delta p}{\rho g} = f\frac{\ell}{d}\frac{V_{md}^2}{2g} \qquad [5.61]$$

This means that the *head losses* in a part of the pipe, h_l, is:

$$h_1 = f\frac{\ell V_{md}}{d2g}, \qquad [5.62]$$

where ℓ is the total length of the pipe, d is the pipe's diameter, V_{md} is the average fluid velocity, g is the acceleration of gravity and f is the coefficient of friction.

This relationship is called the equation of *Darcy* and *Weisbach* or the equation of *Fanning*, who reached this relationship by looking for the loss in pressure during the motion of the fluid in a horizontal pipe.

5.6. Turbulent flow in pipes

In general, turbulence is created from the shearing forces that are developed because of the viscosity of the fluid and from the motions of the neighboring layers of the fluid with different velocities. Therefore, we can say that in the turbulent flow, the total shear stress is caused by both tiny (molecular) motion and macroscopic motion (of the fluid's elements), while in the laminar flow, the shear stress is a negligible phenomenon.

In general, the velocity in a turbulent flow can be defined as the sum of two velocities:

$$V = V_{md} + u' \qquad [5.63]$$

where u' is the function of velocity variation from the average value V_{md}, and V_{md} is the average value of the velocity, which is given by the relationship:

$$V_{md} = \frac{1}{\Delta t}\int_0^{\Delta t} V dt \,. \qquad [5.64]$$

Because of these variances u', the distribution of the velocity in a cross-sectional area of a cylindrical pipe is flatter than the one that we would have in the case where the flow would be laminar. Therefore, comparatively, we have the two distributions given in Figure 5.5.

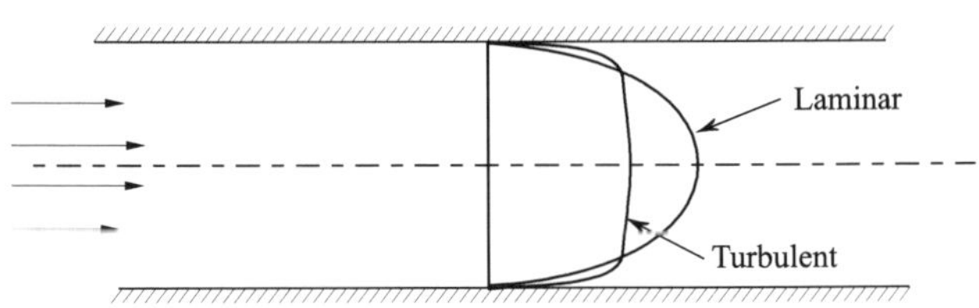

Figure 5.5. *Laminar and turbulent velocity distribution along a pipe*

However, although there is an analytical relationship that expresses the velocity distribution for the case of the laminar flow, which is given by the relationship:

$$V_{(r)} = \frac{(p_1 - p_2)d^2}{16m\ell}\left(1 - \frac{r^2}{R^2}\right), \qquad [5.65]$$

when the flow is turbulent, there is no particular expression for this distribution. Of course, for smooth pipes, there are analytical relationships based on various theories, but from the practical side, they only contribute a little, because most of the practical applications refer to pipes that are far from smooth. In these cases, we make use of empirical equations and diagrams.

5.6.1. *Turbulent flow in smooth pipes*

Smooth pipes are made of glass or metal with a shiny surface without any welding or stitches.

Blasius in 1911, one of the pioneers in the section of hydrodynamics, after experimental measurements, proposed a relationship that gives the friction coefficient f for a turbulent flow in smooth pipes and for values of the Reynolds number up to 10^5:

$$f = \frac{0,316}{\mathrm{Re}^{0,25}}, \text{ where } \mathrm{Re} < 10^5 \qquad [5.66]$$

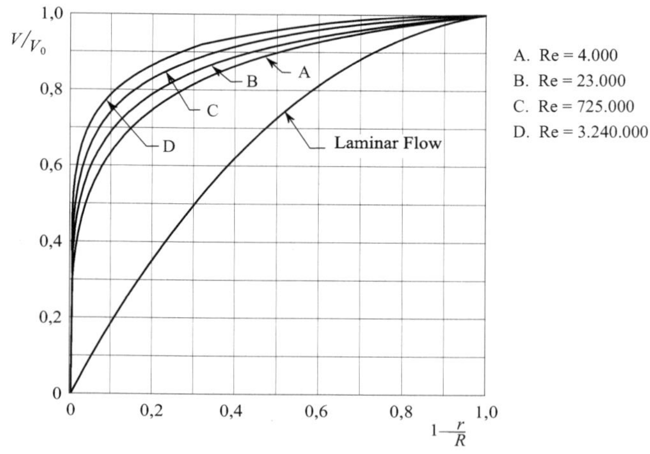

Figure 5.6. *Nikuradse measurements*

Nikuradse, one of the classical researchers in the turbulent flow in closed pipes, after measurements, found that the distribution of the velocity becomes flatter as the

Reynolds number increases. Some distributions from his measurements are shown in Figure 5.6, where V_{max} is the maximum velocity and r is the distance from the center of the pipe.

Based on the measurements of Nikuradse, an empirical relationship for the expression of velocity distribution in smooth pipes (it is not applicable for positions that lie very close to the walls) is as follows:

$$\frac{V_{md}}{V_{max}} = \left(1 - \frac{r}{R}\right)^{1/\eta} \quad [5.67]$$

The exponent η depends on the value of the Reynolds number. Table 5.1 gives the η values according to the Reynolds number.

η	Re
6	4×10^3
7	$1,1 \times 10^5$
10	$3,2 \times 10^6$

Table 5.1. *Exponent constant values, Nikuradse*

A combination of the Nikuradse equation and Prandtl theory that is attributed to a semi-empirical equation results in a relevantly simple expression for the friction coefficient in *smooth* pipes:

$$\frac{1}{\sqrt{f}} = 0,87\ln\left\{\mathrm{Re}\sqrt{f}\right\} - 0,8 \quad [5.68]$$

where f is the friction coefficient and Re is the Reynolds number.

Relationship [5.68] completely agrees with the experimental measurements that have been taken from smooth pipes.

5.6.2. *Turbulent flow in pipes with roughness*

Examining turbulent flow in pipes whose walls are not considered to be smooth is of great importance because it has practical applications: for example, pipes with roughness are often used in hydraulic constructions. This *roughness* is expressed by the average height ε of the various small prominences that extend from the surface. In fact, it is the non-dimensional ratio $\frac{\varepsilon}{d}$ that is responsible for the behavior of the flow. This ratio is called *relative roughness*. Nikuradse proved with his experiments the practical importance of the relative roughness $\frac{\varepsilon}{d}$ using smooth pipes, on the walls of which he had glued sand grains. By varying the size of the diameter, he managed to form the representative curves of the flows for various values of $\frac{\varepsilon}{d}$. As the Reynolds number increases, the curves are detached from the line (for smooth pipes), and by making a concave curve, afterward they are paralleled with the axis of the Reynolds values. It must be noted that initially the curves with the highest $\frac{\varepsilon}{d}$ are detached as shown in Figure 5.7.

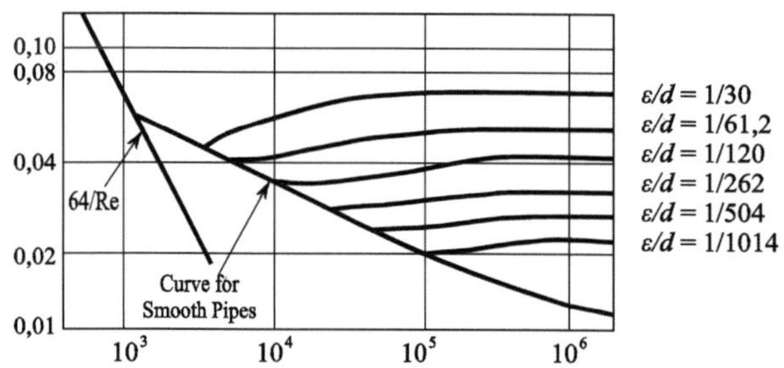

Figure 5.7. *Relative roughness*

Material	ε (in)	ε (ft)	ε (cm)	ε (m)
commercial steel or iron	0.0018	0.00015	0.0046	0.000046
asphalted cast iron	0.0048	0.0004	0.012	0.00012
galvanized iron	0.006	0.0005	0.0152	0.000152
cast iron	0.0102	0.00085	0.0259	0.000259
smooth wood	0.0072–0.036	0.0006–0.003	0.0183–0.091	0.000183–0.00091
steel with riveting	0.036–0.36	0.003–0.03	0.091–0.91	0.00091–0.0091
concrete	0.012–0.12	0.001–0.01	0.03–0.3	0.0003–0.003
drawn tubing	0.00006	0.000005	0.00015	0.0000015

Table 5.2. *Relative roughness values*

Nikuradse was forced to use artificial roughness because the roughness of the natural pipes caused some problems in the calculations (Figure 5.7). Later, in 1944, *Moody* constructed a diagram which allows us to define the friction coefficient f for common pipes. Today, the Moody diagram is the most common diagram used by all researchers in hydraulic constructions. This also relates the friction coefficient f to the Reynolds number for various values of $\frac{\varepsilon}{d}$. Table 5.2 gives the average value of the roughness for various materials.

5.6.3. *The Moody diagram*

Nikuradse's experimental measurements were taken in pipes with artificially developed roughness. Moody, by using empirical relationships and experimental measurements in practically used pipes, concluded in 1944 the configuration of the homonym diagram (Figure 5.8).

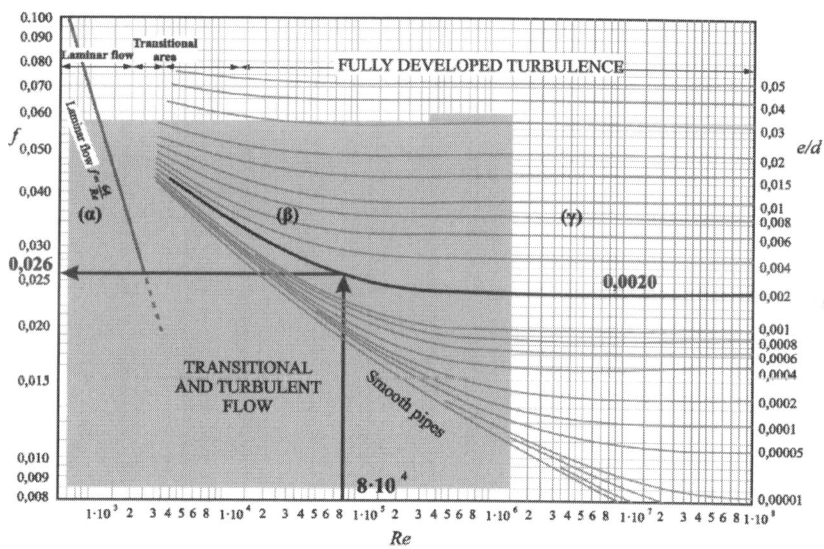

Figure 5.8. *The Moody diagram*

The Moody diagram, as we mentioned in the previous paragraph, has been accepted by all researchers for the calculation of friction coefficient f related to the Reynolds number and relative roughness $\frac{\varepsilon}{d}$, as characteristically shown in Figure 5.8. Mathematically, it is expressed by the relationship:

$$f = F\left(\mathrm{Re} = \frac{Vd}{\nu} \cdot \frac{\varepsilon}{d}\right) \qquad [5.69]$$

Figure 5.8 also shows the way that the diagram is used. Thus, we first calculate the relative roughness $\frac{\varepsilon}{d}$ and find the curve in the diagram that corresponds to it, say $\frac{\varepsilon}{d} = 0{,}0020$. Then, we calculate the Reynolds number as $\mathrm{Re} = 8 \cdot 10^4$. We then find the intersection point of the curve $\frac{\varepsilon}{d}$ and Re and finally read in the diagram the value of the friction coefficient, which in this case is $f = 0{,}026$.

In the Moody diagram in Figure 5.8, three areas are observed:

a) *Area of laminar flow (α),* in which the coefficient f depends exclusively on the Re $(f = 64/\mathrm{Re})$.

b) *Transitional area (b)*, in which the flow is turbulent, but it has not completely developed. Here, the coefficient f depends on both the number Re and the relative roughness $\frac{\varepsilon}{d}$. Now, between areas (a) and (b), there is the *critical area* in which the laminar flow turns gradually turbulent, where the values of the Reynolds number are:

$$2100 < \mathrm{Re} < 4000\,. \qquad [5.70]$$

c) *The turbulent area (c),* in which the flow has turned completely turbulent and the coefficient f practically depends exclusively on the relative roughness $\frac{\varepsilon}{d}$.

Between the two last areas there are no clear limits, and the value given by the relationship is usually considered to be a separator limit:

$$\mathrm{Re} = \frac{3.500}{\varepsilon/d} \qquad [5.71]$$

5.6.4. *Calculation of relative roughness*

In order to have more reliable information and more convenience in the solution of the problems, there is also the following diagram for the calculation of relative roughness depending on the size of the pipe (inches) and the material that is constructed (Figure 5.9).

5.6.5. *Empirical expressions for the friction coefficient*

For the calculation of the coefficient of turbulent flow in pipes with roughness, some remarkable empirical equations have been suggested. As they are complicated, they are not used in a wide range. In the last decades, the use of computers has made them practically usable. Some of the most important equations are presented as follows.

The first is the one that *Prandtl* suggested for *smooth pipes*, which is also used in pipes with low roughness:

$$\frac{1}{f^{1/2}} = 2 \cdot \log\left(\frac{\mathrm{Re} \cdot f^{1/2}}{2,51}\right) = 2 \cdot \log\left(\mathrm{Re} \cdot f^{1/2}\right) - 0,8. \qquad [5.72]$$

We note that in the smooth pipes, the friction coefficient depends exclusively on the Reynolds number.

The second one is that suggested by *von Karman* (1930) for *rough pipes* and high Reynolds numbers for the area of absolute turbulent flow $(y_0 < 0,3 \cdot \varepsilon)$.

$$\frac{1}{f^{1/2}} = 2 \cdot \log\left(\frac{3,7}{\varepsilon/d}\right) = 1,14 - 2 \cdot \log\left(\varepsilon/d\right). \qquad [5.73]$$

This means that in the fully turbulent flow, the friction coefficient depends exclusively on the relative roughness.

The third one is *the equation of Colebrook* (1939), which combines the calculation in smooth pipes with the ones in pipes of high roughness and gives accurate answers in the transitional area as well $(5 \cdot \varepsilon > y_0 > 0,3 \cdot \varepsilon)$:

$$\frac{1}{f^{1/2}} = -2 \cdot \log\left(\frac{2,51}{\mathrm{Re} \cdot f^{1/2}} + \frac{\varepsilon/d}{3,7}\right). \qquad [5.74]$$

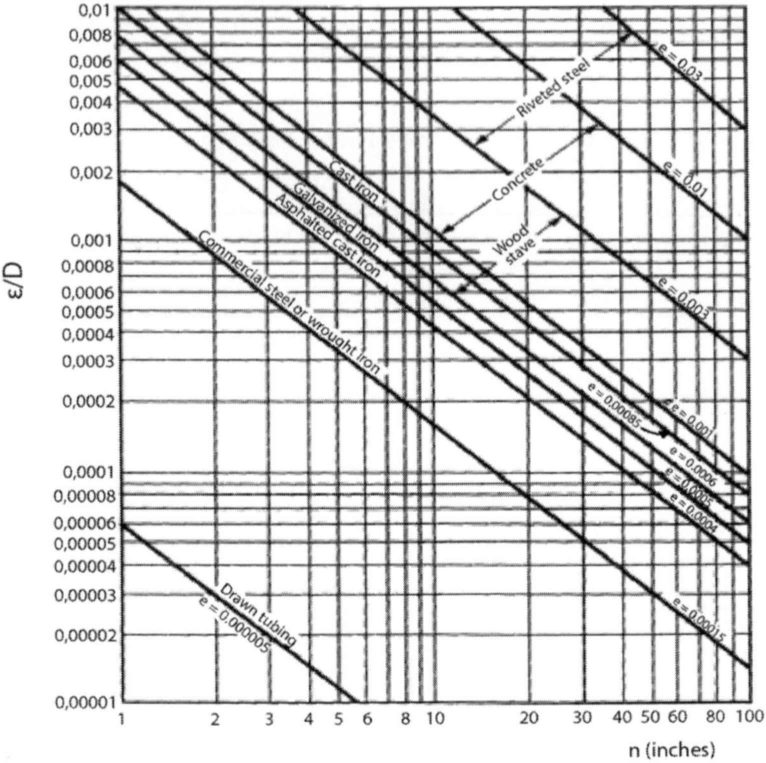

Figure 5.9. *Relative roughness according to the pipe sizes*

Although equation [5.74] gives accurate results, it is difficult to use it without a computer. Therefore, other empirical relationships have been stated, the most important of which is the *Haaland equation* with an error possibility of only 2%:

$$\frac{1}{\sqrt{f}} = -1{,}8\log\left[\frac{6{,}9}{\text{Re}} + \left(\frac{\varepsilon/d}{3{,}7}\right)^{1{,}11}\right] \qquad [5.75]$$

Also the *Swamee–Jain equation* is used (1976), which is:

$$\frac{1}{\sqrt{f}} = -2\log\left(\frac{5{,}74}{\text{Re}^{0{,}9}} + \frac{\varepsilon/d}{3{,}7}\right), \qquad [5.76]$$

as well as the equation:

$$f = \left[A^{12} + \left(B^{-8} + C^{-8} \right)^{-1,5} \right]^{1/12} \qquad [5.77]$$

where:

$$A = \frac{64}{\text{Re}}$$

$$B = \left(\frac{\text{Re}}{12900} \right)^2$$

$$C = \left\{ -0,8687 \cdot \ln \left[\left(\frac{7}{\text{Re}} \right)^{0,9} + 0,27 \left(\frac{\varepsilon}{d} \right) \right] \right\}^{-2}$$

As we said, the above relationships are particularly useful for the solution of flow problems with the assistance of a computer and the relevant software.

5.6.6. *Minor local losses*

The relationships that we mentioned in the previous sections refer to rectilinear pipes without any obstacles and without any variations in the flow. However, it is known that when there is a flow of a real fluid inside a pipe, this flow will either meet prominences from various appliances that connect the pipes during its motion or change direction and go through valves and narrowings, or will even be forced to be compressed in a pipe of a smaller cross-sectional area or released in a pipe with a higher cross-sectional area and many other cases. Every time that even one of the above takes place, there are energy losses, which we must take into account for the correct study in the pipelines. All these losses are expressed individually with a relationship that gives us the local head losses, as follows:

$$h_1 = k \frac{V^2}{2g} \qquad [5.78]$$

with h_1 being the energy losses (e.g. in m) because of the geometrical variations of the flow, $V =$ average flow velocity, $k =$ non-dimensional coefficient, which is different for every kind of obstacle in the flow and is called the *coefficient of local losses*.

The above energy losses expression, when it is added to the energy losses during the fluid flow in a rectilinear part of the pipe, which is given by the relationship:

$$h_1 = f\frac{\ell}{d}\frac{V^2}{2g}, \qquad [5.79]$$

gives us the total of the losses Σh_L of the flow in a pipe. Consequently, the total losses of energy are:

$$\Sigma h = h_1 + \Sigma h_1' = \text{ total losses due to the friction}, \qquad [5.80]$$

where the energy equation is given by the relationship:

$$\frac{p_1}{\rho g} + h_1 + \frac{V_1^2}{2g} + H_A = \frac{p_2}{\rho g} + h_2 + \frac{V_2^2}{2g} + H_T + \Sigma h, \qquad [5.81]$$

where H_A is the head of energy due to the pump, H_T is the head of energy due to the turbine and Σh is the total energy losses given by [5.80].

Instead of the coefficient of the local losses *K*, the concept of the *equivalent length* can be used. This is defined as the ratio of the pipe's length (*l*) to the pipe's diameter (d):

$$\text{Equivalent length } = \frac{\ell}{d}. \qquad [5.82]$$

The local losses have been defined by relationship [5.79] as:

$$h_i = f \cdot \frac{\ell}{d} \cdot \frac{V^2}{2g}. \qquad [5.83]$$

For example, if in a quality table of flow applications, we meet an equivalent length equal to 30, then $\ell/d = 30$. If the (internal) diameter of the pipe is 5 cm, we have $\ell_q = 150 \text{ cm} = 1.5 \text{ m}$. This means that losses will be equal to the linear losses of the pipe of length 1.5 m.

The relationship between the coefficient of the local losses and the equivalent length arises from the definitions of the two values:

$$h_1 = f \cdot \frac{\ell}{d} \cdot \frac{V^2}{2g} = K \cdot \frac{V^2}{2g} \Rightarrow \frac{\ell}{d} = \frac{K}{f}. \qquad [5.84]$$

This means that the equivalent length is equal to the coefficient of the local losses divided by the friction coefficient. The equivalent length is a non-dimensional value (as well as the coefficient of the local losses).

5.6.7. *K values*

The coefficient of the local losses *K* in various flows depends on the shape, the relative roughness of the area and the Reynolds number of the flow. In Table 5.3, the values of the coefficient of the local losses *K* are presented for various types of flows.

5.6.8. *Valves and other devices*

When flow in pipes occurs, besides the inlet in the tank, or the variation of the direction of the flow or the exit from the tank and all the other cases that we mentioned in the previous paragraph, a pipeline contains many other devices, which also influence the flow and cause additional energy losses and must be calculated as well. In general, besides the nature of the obstacles, the *K* value of a flow also depends on the diameter of the pipe and the Reynolds number.

Some of the most important devices that intervene in fluid flows inside pipes are the *valves*, the *tees* and the *spigots*.

Valves, fully open	**Flanged**				**Screwless**				
	0.5″	1″	2″	4″	1″	2″	4″	8″	20″
Spherical	14	8.2	6.9	5.7	13	8.5	6	5.8	5.5
Sliding	0.3	0.24	0.16	0.11	0.80	0.35	0.16	0.07	0.03
Back-flow	5.1	2.9	2.1	2	2	2	2	2	2
Angular	9	4.7	2	1	1	2.4	2	2	2

Table 5.3. *Coefficient of local losses in valves*

Valves are placed in almost all pipelines and are distinguished as spherical, sliding, angular, one-way or back-flow, butterfly valves and others (Figure 5.10). These valves, even if they are completely open, create remarkable local losses, and their coefficient of local losses depends on the type, diameter, roughness and Reynolds number. In fact, the coefficient *K* is higher for valves of lower diameter. Tables 5.3–5.5 present the dependence of *K* on the diameter and pipe geometry changes.

Description of the cause of local loss	**K_i**	
1. Entrance in a pipe that protrudes	0.8	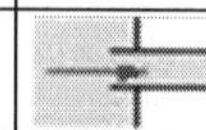
Entrance in a pipe with sharp lips	0.5	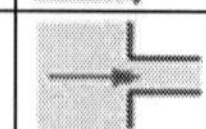
Entrance in a pipe with rounded lips	0.02–0.5	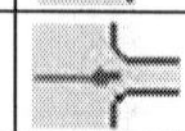
2. Exit from a pipe in a tank	1.0	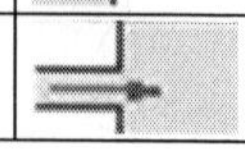
3. Change of direction		
90° angles	1.2	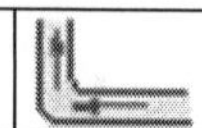
60° angles	0.6	
45° angles	0.4	
Curved 90° angles	0.2–0.7	
Curved 90° angles	2.0	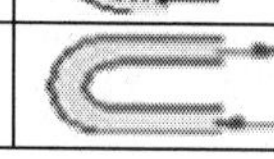

Table 5.4. *K values for pipe scheme changes*

Diameter variation		
Sudden narrowing with $d_2 / d_1 = 0.2$	0.45	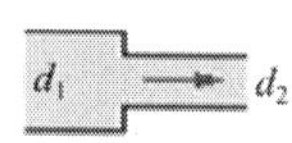
Sudden narrowing with $d_2 / d_1 = 0.4$	0.4	
Sudden narrowing with d_2/d_1=0.6	0.3	
Sudden narrowing with $d_2 / d_1 = 0.8$	0.13	
Gradual narrowing	0.05–0.25	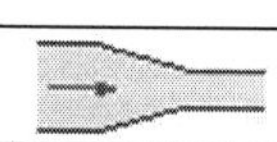
Sudden expansion with $d_1 / d_2 = 0.2$	0.95	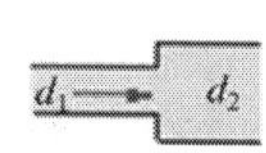
Sudden expansion with $d_1 / d_2 = 0.4$	0.75	
Sudden expansion with $d_1 / d_2 = 0.6$	0.4	
Sudden expansion with $d_1 / d_2 = 0.8$	0.15	
Gradual expansion ($\Theta < 20°$)	0.10–0.60	

Table 5.5. *K values according to pipe diameter*

When a valve is slightly open, the losses are much higher. The coefficient *K* depends on the ratio of the valve's opening *h* to its diameter *d*. In Figure 5.11, we can see this dependence for sliding (1) and spherical (2) valves.

Also, in butterfly valves, the coefficient of losses, whose value is low when the valves are completely open, shows a significant increase when the angle of the flow velocity varies.

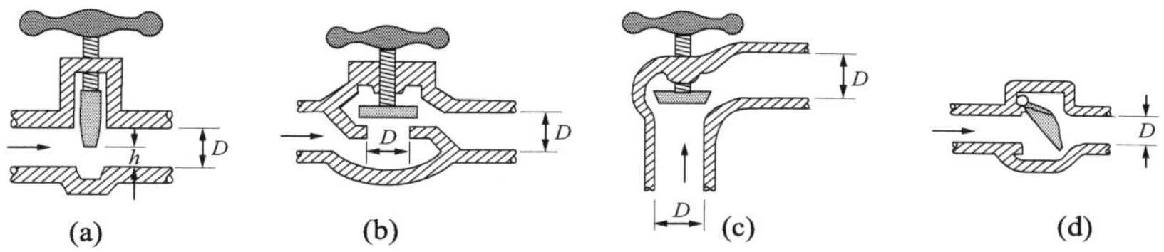

Figure 5.10. *Various types of valves*

Therefore, when $\theta = 90°$, $K = 0,2 - 0,3$ for fully open valves; when $\theta = 60°$, $K = 3 - 4$ and when $\theta = 40°$, $K = 20 - 30$.

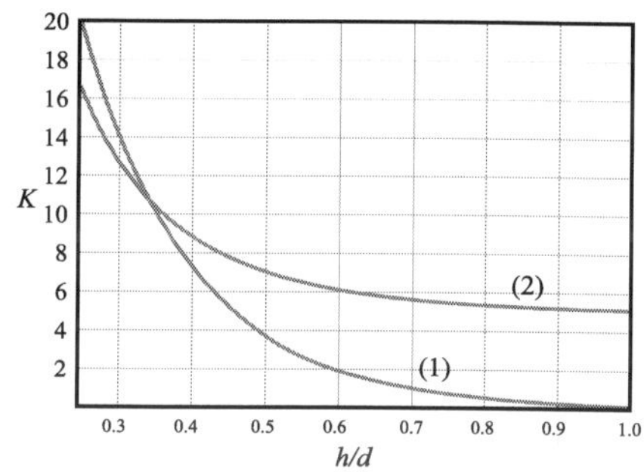

Figure 5.11. *K values as a function of valve ratio (h/d)*

Tee connections, which get their name from their shape (T), are devices that are used for the branches of the pipelines. Their coefficient *K* depends on the diameter of the pipe, the Reynolds number and the direction of the flow.

	Screwless				Flanged				
Tee connections	0.5″	1″	2″	4″	1″	2″	4″	8″	20″
Horizontal flow	0.9	0.9	0.9	0.9	0.24	0.1 9	0.14	0.10	0.07
Vertical flow	2.4	1.8	1.4	1.1	1.0	0.8	0.64	0.58	0.41

Table 5.6. *K values for tee connections*

In Table 5.6, the coefficients of the local losses are presented in *tee* connections.

Concluding in this reference about the values of the coefficients of the local losses of these devices, we must say that they differ much between constructors and the variation reaches 50% often. Hence, for the calculation of various problems, we can use the values given in Table 5.7.

Description of the cause of the local loss	K_i
Spherical valves, open	0.17
Spherical valves, open (1–20 in)	4.0–14
Spherical valves, open a little	4–60
Sliding valves, open (1–20 in)	0.03–0.8
Sliding valves, open a little	0.6–60
Angular valves, open (1–20 cm)	1.0–9.0
One-way valves (back-flow) (1–20 cm)	2.0–2.9
Diaphragm valves, open	2.3–4.3
Butterfly valves, open	0.2–0.3
Butterfly valves, open a little	0.3–60
Valves of type Y	1–6.5
Valves with thread	15
Tee connections of straight flow	0.07–0.09
Tee connections of vertical flow	0.4–2.4
Flange, connection	0.04

Table 5.7. *K values for valves and fittings*

5.6.9. *Total losses*

The total losses because of the frictions Σh, as we mentioned in relationship [5.80], are equal to the sum of the linear (major) and local (minor) losses:

$$\Sigma h = h_f + \Sigma h_i \quad [5.85]$$

According to the equation of Darcy–Weisbach [5.62]:

$$h_f = f \cdot \frac{\ell}{d} \cdot \frac{V^2}{2g}. \quad [5.86]$$

Every local disorder leads to a loss, which is given by relationship [5.78]:

$$h_i = K_i \cdot \frac{V^2}{2g} \qquad [5.87]$$

When the diameter of the pipe does not vary, the velocity remains constant; therefore, using [5.86] and [5.87], relationship [5.85] becomes:

$$\Sigma h = \left[f \frac{\ell}{d} + \Sigma K_i \right] \cdot \frac{V^2}{2g}. \qquad [5.88]$$

ΣK_i is called the total coefficient of local losses. If the diameter varies, then we suppose that different pipes are connected in a row and the calculation for each pipe is carried out separately.

The head losses Σh are significantly influenced by the diameter of the pipe, *d*. More specifically, the linear losses are reversely proportional to the fifth power of the diameter:

$$h_f = \frac{8}{\pi^2 \cdot g} \cdot f \cdot \frac{\ell}{d^5} \cdot Q^2, \qquad [5.89]$$

while the local losses are reversely proportional to the fourth power of the diameter:

$$h_i = \frac{8}{\pi^2 g \cdot d^4} \cdot K_i \cdot Q^2 \qquad [5.90]$$

where the flow rate is $Q = \frac{\pi d^2 V}{4}$.

This means that for a certain flow rate, if we double the diameter, then the local losses reduce $2^4 = 16$ times and linear losses reduce $2^5 = 32$ times approximately.

5.6.10. *Solution of flow problems in pipes*

In order to study the constant flow of an incompressible fluid in a pipe, we may use the following four equations:

a) *Continuity equation:*

$$Q = \frac{\pi \cdot d_i^2}{4} \cdot V_i = \text{const} \qquad [5.91]$$

b) *Energy equation (Bernoulli):*

$$\frac{p_1 - p_2}{\gamma} + \frac{V_1^2 - V_2^2}{2 \cdot g} + (y_1 - y_2) = \Sigma h \qquad [5.92]$$

c) *Equation of losses (Darcy–Weisbach):*

$$\Sigma h = \left(f \cdot \frac{\ell}{d} + \Sigma K_i \right) \cdot \frac{V^2}{2g} \qquad [5.93]$$

or

$$\Sigma h = \frac{8}{\pi^2 \cdot g \cdot d^4} \cdot \left(f \cdot \frac{\ell}{d} \cdot \Sigma K_i \right) \cdot Q^2 \qquad [5.94]$$

d) *Moody diagram:*

$$f = F\left(\text{Re} = \frac{V \cdot d}{\nu}, \frac{\varepsilon}{d} \right) \qquad [5.95]$$

The first two equations resulted from the application of the laws of physics, including the law of conservation of mass and the law of conservation of energy. The third equation of losses, the Darcy–Weisbach equation, resulted empirically from the development of the dimensional analysis and experimental measurements.

The fourth equation is the most peculiar, as it is not an equation, but a diagram relationship. If the flow is laminar, then we can use, instead of the diagram, the equation $(f = 64/\text{Re})$ and the solving procedure is thus simplified. However, in most problems, the flow is turbulent. In fact, there is a possibility of using, instead of the Moody diagram, other empirical relationships that we have mentioned, but those are difficult to use and require computer support.

Therefore, the simplest procedure for the calculation of the friction coefficient f remains *the use of the Moody diagram.*

Especially for the case of laminar flow, as we have already mentioned, the calculations are simplified: instead of the Moody diagram, equation [5.96] is used, as the equation of the losses takes the form [5.97]:

$$f_{lam} = \frac{64}{\text{Re}} \tag{5.96}$$

$$h_{f,lam} = \frac{64}{\text{Re}} \cdot \frac{\ell}{d} \cdot \frac{V^2}{2g} = \frac{32 \cdot \nu \cdot \ell}{g \cdot d^2} \cdot V . \tag{5.97}$$

In this case, the roughness of the pipe does not influence the flow.

5.7. Categories of pipes' flow problems

The following are the three categories of problems in turbulent flow of a real fluid in pipes:

A′ category: given the fluid flow rate in the pipe; the length, diameter and roughness of the pipe and the kinematic viscosity of the flow, it is required to calculate the total head losses (Σh).

B′ category: given the length, diameter, roughness and head of the losses of the pipe and the kinematic viscosity of the flow, it is required to calculate the flow rate of the pipe.

C′ category: given the length, roughness and head of the losses of the pipe and the kinematic viscosity of the flow, it is required to calculate the diameter of the pipe.

In each category, the Darcy–Weisbach relationship of the continuity and the Moody diagram must be used for the definition of the unknown.

5.7.1. *A' category flow problems*

Given: $Q =$ flow rate, $\ell =$ length, $d =$ diameter, $\nu =$ kinematic viscosity, $\varepsilon =$ roughness. Asked: $h_l =$ head of losses

In this case, we proceed as follows:

1) We calculate the ratio $\frac{\varepsilon}{d}$ and the velocity $V = \frac{Q}{S}$.

2) We calculate $\text{Re} = \rho\frac{Vd}{\mu} = \frac{Vd}{\nu} = \frac{Qd}{\nu S} = \frac{Qd}{\nu\frac{\pi d^2}{4}} = \frac{4Q}{\pi\nu d}$

3) From the Re and $\frac{\varepsilon}{d}$, we calculate from the Moody diagram the friction coefficient *f*.

4) With the value of *f* known, we calculate the losses h_l :

$$h_l = f\frac{\ell}{d}\frac{V^2}{2g}$$

5.7.2. *B' category flow problems*

5.7.2.1. *Analytical way*

Given: $h_l =$ head losses, $\ell =$ length, $d =$ diameter, ε = roughness, $\nu =$ kinematic viscosity. Asked: Q = flow rate.

As h_l is given, from the Darcy–Weisbach relationship, we calculate the velocity *V*:

1) We calculate the ratio $\frac{\varepsilon}{d}$ from the given data of the problem.

2) We assume a value for the coefficient *f*, which corresponds approximately to (from the Moody diagram) $\frac{\varepsilon}{d}$.

3) We calculate velocity *V* from the Darcy–Weisbach relationship:

$$h_l = f\frac{\ell}{d}\frac{V^2}{2g}$$

so $V = \sqrt{\frac{2gh_l d}{f\ell}}$

4) We calculate Re with the velocity V.

5) With the value $\frac{\varepsilon}{d}$ known and Re calculated from the Moody diagram, we calculate the friction coefficient f.

6) With the new value of f (if we find it different from the one we assumed), we calculate again from the Darcy–Weisbach relationship velocity V and we repeat steps 4 and 5 until the friction coefficient f agrees with the previous f.

With this final value, the velocity is calculated in step 3.

7) The flow rate is found as follows:

$$Q = VS = \pi \frac{D^2}{4} V$$

5.7.2.2. *Assumption of intense turbulence (AIT)*

1) We assume that we have an absolutely turbulent flow (assumption of intense turbulence – AIT).

2) We calculate the hypothetical f' from the Moody diagram.

3) We calculate the hypothetical flow rate Q' from the losses relationship.

4) We calculate velocity V from the continuity equation, Reynolds number $\left(\mathrm{Re} = \frac{Vd}{v}\right)$ and f.

5) We check the following assumptions:

i) If $f = f' \Rightarrow Q = Q'$, then the problem ends.

ii) If $f \neq f'$, then we go back to step 3.

5.7.3. *C' category flow problems*

Given: $h_l =$ head losses, $Q =$ flow rate, $\ell =$ length, $v =$ kinematic viscosity, $\varepsilon =$ roughness. Asked: $d =$ diameter.

For the solution of these problems where diameter d as well as the factors calculated from it, V, f and Re are unknown, we will use two or three approaching solution circles.

1) First, from the Darcy–Weisbach relationships and the continuity, we express the diameter as a function of friction coefficient f:

$$h_l = f\frac{\ell}{d}\frac{V^2}{2g}, \quad Q = VS = V\pi\frac{d^2}{4} \rightarrow V = \frac{4Q}{\pi d^2}$$

so

$$h_l = \frac{f}{2g}\frac{\ell}{d}\left(\frac{4Q}{\pi d^2}\right)^2 = \frac{16 f\ell Q^2}{2g\pi^2 d^5} = 8\frac{f\ell Q^2}{g\pi^2 d^5}.$$

Solving for d, we get $d = \left(\frac{8\ell Q^2}{\pi^2 g h_l}\right)^{0,2} \cdot f^{0,2}$

2) From the given data of the problem and using the Moody diagram, we assume a value for the f and then we calculate the diameter.

3) Then, knowing the diameter, we calculate the Reynolds number:

$$\mathrm{Re} = \frac{Vd}{\nu} = \frac{4Q}{\pi d\nu}$$

4) As the roughness is one of the given data of the problem and as we have calculated the diameter, we calculate the relative roughness $\frac{\varepsilon}{d}$.

5) Knowing now Re and the ratio $\frac{\varepsilon}{d}$ from the Moody diagram, we find the friction coefficient f.

6) If the value of f that we assumed in step 2 is different from that found in step 5, then with the new value from step 4, we repeat steps $2 \rightarrow 3 \rightarrow 4 \rightarrow 5 \rightarrow 6$ of a new calculating circle until the last value of f in step 5 is the same (approximately) as that in step 2 of the same circle.

7) Finally, the diameter is calculated with the value of f from step 6 of the last circle:

$$d = \left(\frac{8\ell Q^2}{\pi^2 g h_l}\right)^{0,2} f^{0,2}$$

5.8. Pipes' flow problems: numerical work examples

5.8.1. *A' category*

A pipe is made of cast iron, with diameter 10 cm, length 50 m and water flow rate 90 m^3/h. There are four curve bends of 90°.

Calculate the head losses.

Solution

Given: $\ell = 50m$, $d = 10cm$, $\varepsilon = 0,00026m$ (Diagram [5.10]), specific weight $\gamma = 9,81KN/m^3$, $v = 10^{-6} m^2/s$, $Q = 90m^3/h = 0,025m^3/s$

Asked: Σh

Available relationships:

$$\Sigma h = \left(f \cdot \frac{\ell}{d} \Sigma K_i \right) \cdot \frac{V^2}{2g} \qquad [5.98]$$

$$Q = \frac{\pi \cdot d^2}{4} \cdot v \Rightarrow v = \frac{4 \cdot Q}{\pi \cdot d^2}$$

$$f = F\left(\mathrm{Re} = \frac{V \cdot d}{v}, \frac{\varepsilon}{d} \right) \qquad [5.99]$$

The head losses are calculated from the Darcy–Weisbach equation (1). In this relationship, there are four unknowns: $\Sigma h, f, \Sigma K, v$. Therefore, first, we have to calculate the velocity *V*, friction coefficient *f* and the total coefficient of the local losses ΣK_i.

a) Calculation of the velocity V (from the continuity equation).

Relationship (2) gives: $V = \dfrac{4 \cdot 0,025 m^3/s}{\pi \cdot (0,1m)^2} \Rightarrow V = 3,18m/s$

b) Calculation of *f*:

$$\text{Re} = \frac{V \cdot d}{\nu} \Rightarrow \text{Re} = 0,318 \cdot 10^6 = 3,18 \cdot 10^5, \varepsilon / d = 0,0026$$

and from the Moody diagram (Figure 7.10): $f = 0,026$

c) Calculation of ΣK : there are four curve angles of 90°, for which the curving radius is not given. From Table 5.4, the coefficient K varies between 0.2 and 0.7. We take $K_{90} = 0,5$; as we have four curve angles of 90°, we will have:

$$\Sigma K = 4 \cdot K_{90} = 2 \qquad [5.100]$$

Substituting, we have:

$$\Sigma h = \left(0,026 \cdot \frac{50m}{0,1m} + 2\right) \cdot \frac{\left(3,18\, m/s\right)^2}{2 \cdot 9,81\, m/s^2} \Rightarrow$$

$$\Sigma h = 7,73m$$

5.8.2. *B' category*

Ethanol (temperature $20°C$) is transferred from tank A to tank B as shown in Figure 5.12. The pipe has a circular cross-sectional area with diameter 7.5 cm and is made of galvanized cast iron. The two bends of the pipeline are 90° with ratio $\frac{\varepsilon}{d} = 5$. We calculate the flow rate of ethanol.

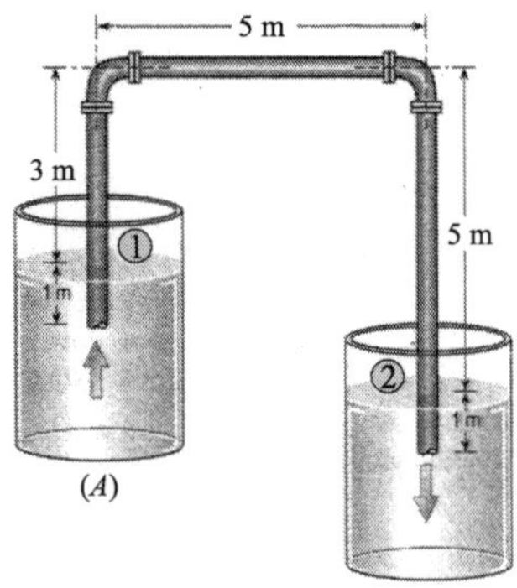

Figure 5.12. *Ethanol flow*

Solution

Given: $d = 7.5$ cm, $\dfrac{\varepsilon}{d} = 5$,

To be calculated: Q

The flow rate Q of ethanol is calculated by the product of the average velocity and the surface cross-sectional area of the pipe:

$$Q = \bar{V}S = \bar{V}\frac{\pi d^2}{4}$$

For the calculation of velocity $\bar{V}$, we use the equation of the mechanical energy, in head form, for an incompressible, steady flow between positions (1) and (2)

$$z_1 + \frac{p_1}{\gamma} + \frac{V_1^2}{2g} = z_2 + \frac{p_2}{\gamma} + \frac{V_2^2}{2g} + \Sigma h \qquad [5.101]$$

where h_i is the head of the total of the energy losses that occur in the pipeline. Because the tanks are big, we can assume that the level of ethanol in them remains almost constant (i.e. $V_1 \approx V_2 \approx 0$). Besides, as both tanks are open to the atmosphere, the pressures p_1 and p_2 are equal, and are also equal to the atmospheric pressure $(p_1 = p_2 = p_0)$. Therefore, equation [5.101] is written as:

$$\Sigma h_i = Z_1 - Z_2 = 2 \text{ m}$$

which means that the head of the losses of the mechanical energy of the flow is equal to the height difference of the level of ethanol between the two tanks. According to equation [5.85], the head Σh is equal to the sum:

$$\Sigma h = h_f + h_{m,ent} + h_{m,90^\circ} + h_{m,ex} \qquad [5.102]$$

where $h_f, h_{m,ent}, h_{m,90^\circ}$ and $h_{m,ex}$ are the heads of the energy losses that occur in the three rectilinear sections, the entrance, the curve of 90° and the exit of the pipeline, respectively. The head h_f is calculated from the Darcy–Weisbach equation (equation [5.62]) as:

$$h_f = f\frac{\ell}{d}\frac{\overline{V}^2}{2g} \qquad [5.103]$$

and the heads $h_{m,ent}, h_{m,90°}$ and $h_{m,ex}$, from equation [5.78]:

$$h_{m,ent} = k_{m,ent}\frac{V^2}{2g} \quad h_{m,90°} = k_{m,90°}\frac{V^2}{2g} \quad h_{m,ex} = k_{m,ex}\frac{V^2}{2g}$$

and therefore equation [5.102] is written as:

$$\Sigma h = \left(f\frac{\ell}{d} + k_{m,ent} + 2k_{m,90°} + k_{m,ex} \right)\frac{V^2}{2g}\,. \qquad [5.104]$$

Solving the last equation for velocity V, we have:

$$V = \sqrt{\frac{2gh_\ell}{f(\ell/d) + k_{m,ent} + 2k_{m,90°} + k_{m,ex}}}$$

The absolute roughness of the pipes made of galvanized cast iron is $e = 0.15$ mm (from Table 5.2), so the roughness $\frac{e}{d}$ is:

$$\frac{e}{d} = \frac{(0.15 \text{ mm})}{75 \text{ mm}} = 0.002\,.$$

The values of the losses coefficient $k_{m,ent}, k_{m,90°}$ and $k_{m,ex}$ are taken from Table 5.3 (for ratio ($\varepsilon/d = 5$ for roughness $e/d = 0,002$) and Table 5.4:

$$k_{m,ent} = 0,78 \quad k_{m,90°} = 0,2 \quad k_{m,ex} = 1,0$$

We already know the values of the quantities that appear in the second term of equation [5.104], apart from the value of the friction coefficient. However, in order to calculate the value of f, the Reynolds number must be known, which in this case is unknown. Therefore, the problem of calculating the velocity (and therefore the flow rate) is now transferred in a problem of calculating the friction coefficient. For the solution of this problem, we apply the iterating method of *trial* and *error*, following the steps that we described for the B′ category problems.

For the initiation of the circular procedure that leads to the calculation of Q, we assume that the friction coefficient of the pipe is:

$$f_1 = 0{,}0235$$

We take this value from the Moody diagram, for the roughness $e/d = 0.002$ as it is a fully developed turbulence, so the Reynolds number is not required. Setting the value f_1 in equation [5.117], we calculate velocity V_1 for ethanol provided the flow is fully developed:

$$V_1 = \sqrt{\frac{2\left(9{,}81m/s^2\right)(2m)}{(0{,}0235)\left[(15m)/(0{,}075m)\right]+(0{,}78)+2(0{,}2)+(1{,}0)}} = 2{,}39m/s\,.$$

Based on the velocity V_1, the Reynolds number, Re_1, is calculated as:

$$\mathrm{Re}_1 = \frac{\rho V_1 d}{\mu} = \frac{\left(788kg/m^3\right)(2{,}39m/s)(0{,}075m)}{\left(1{,}2\times10^{-3}\,Pa\cdot s\right)} = 1{,}18\times10^5\,.$$

From the Moody diagram, for $\mathrm{Re}_1 = 1{,}18\times10^5$ and $e/d = 0{,}002$, we take the value:

$$f_2 = 0{,}0245\,.$$

From the comparison of the above equations, we conclude that $f_2 \neq f_1$. Therefore, the former procedure must be repeated, using now, as initial value of the friction coefficient, the value f_2. From the velocity equation, for $f = f_2$, we have:

$$V_2 = \sqrt{\frac{2\left(9{,}81m/s^2\right)(2m)}{(0{,}0245)\left[(15m)/(0{,}075m)\right]+(0{,}78)+2(0{,}2)+(1{,}0)}} = 2{,}35m/s\,.$$

Based on velocity V_2, the Reynolds number, Re_2, is calculated as:

$$\mathrm{Re}_2 = \frac{\rho V_2 d}{\mu} = \frac{\left(788kg/m^3\right)(2{,}35m/s)(0{,}075m)}{\left(1{,}2\times10^{-3}\,Pa\cdot s\right)} = 1{,}16\times10^5$$

From the Moody diagram, for $\mathrm{Re}_2 = 1.16\times10^5$ and $e/d = 0.002$, we take the value:

$$f_3 = 0,0245 .$$

From the comparison of the above equations, $f_3 = f_2$. Therefore, the value V_2 is the correct value of the velocity of ethanol, and the flow rate is:

$$Q = (\pi/4)(2,35m/s)(0,075m)^2 = 0,010m^3/s .$$

Therefore, ethanol is transferred from tank A to tank B at a flow rate of $10\ell/s$.

5.8.3. *C' category*

A pressure of $2\,bar$ is applied on the free surface of a closed tank, as shown in Figure 5.13. The pipe is made of asphalted cast iron, the length of which is 25 m. Calculate the diameter (to the nearest 0.5 cm) so that the water flow rate is at least 40 m^3/h.

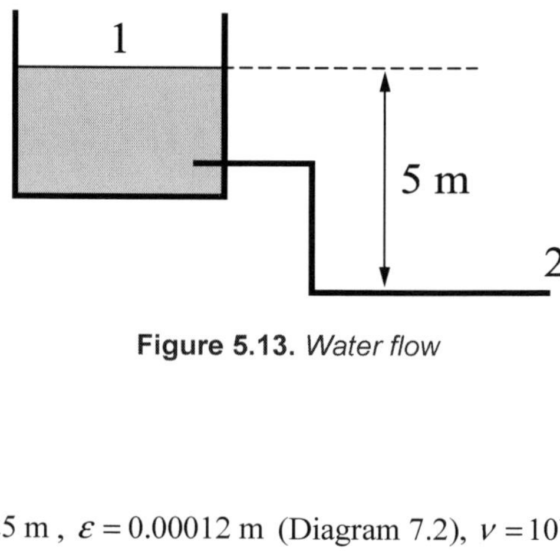

Figure 5.13. *Water flow*

Solution

Given: $\ell = 25$ m, $\varepsilon = 0.00012$ m (Diagram 7.2), $\nu = 10^{-6}$ m^2/s,

$Q \geq 40$ m$^3/h = 40/3,600$ m^3/s, $p_1 = 2bar = 200$ Kpa, $p_2 = 101.3$ Kpa, $y_1 - y_2 = 5$ m. Asked: d .

Available relationships:

$$\Sigma h = \left(f \cdot \frac{\ell}{d} + \Sigma K_i \right) \cdot \frac{V^2}{2g} \qquad [5.105]$$

$$Q = \frac{\pi \cdot d^2}{4} \cdot V \Rightarrow V = \frac{4 \cdot Q}{\pi \cdot d^2} \qquad [5.106]$$

$$f = F\left(\mathrm{Re} = \frac{Vd}{v} \cdot \frac{\varepsilon}{d} \right) \qquad [5.107]$$

$$\frac{p_1 - p_2}{\gamma} + \frac{V_1^2 - V_2^2}{2 \cdot g} + (y_1 - y_2) = \Sigma h \qquad [5.108]$$

In order to use the losses relationship, which is relationship [5.105], we will have to calculate ΣK and losses Σh . Therefore:

a) *For the calculation of* $\mathbf{\Sigma K}$: from Table 5.4, the entrance of the pipe has $K_1 = 0{,}8$ and two simple angles 90°, $K_2 = 2 \cdot 1{,}2$ (from the same table) and after adding $\Sigma K = 3{,}2$.

b) *For the calculation of* $\mathbf{\Sigma h}$: we apply energy equation [5.108] between the points 1 and 2 $(V_1 = 0, V_2 = V)$, and we have:

$$\frac{p_1 - p_2}{\gamma} - \frac{V^2}{2 \cdot g} + (y_1 - y_2) = \Sigma h \Rightarrow$$

$$\Rightarrow \Sigma h = \frac{(200 - 101{,}3)\, KPa}{9{,}81\, KN/m^3} - \frac{V^2}{2 \cdot g} + 5m \Rightarrow$$

$$\Rightarrow \Sigma h = 15{,}06 - \frac{V^2}{2 \cdot g} (SI) \qquad [5.109]$$

Although we do not know the precise value of Σh , relationship [5.109] is enough. Substituting this into [5.105], we have:

$$15{,}06 - \frac{V^2}{2 \cdot g} = \left(f \cdot \frac{L}{d} + \Sigma K_i \right) \cdot \frac{V^2}{2g} \Rightarrow$$

$$\Rightarrow 15,06m = \left(f \cdot \frac{L}{d} + \Sigma K_i + 1 \right) \cdot \frac{V^2}{2g}. \qquad [5.110]$$

By setting randomly a logical value d', the problem of calculating the flow rate occurs. We will solve it by the second way (assumption of intense turbulence):

Assuming that $d' = 10 \text{ cm} = 0.10 \text{ m}$:

$(\varepsilon / d = 0,0012, AIT) \rightarrow f = 0,0205$ (from the Moody diagram)

By substituting it into [5.110], we have:

$V = 5.626$ m/s.

and finally:

$Q' = 159,1 m^3 / h > 40 m^3 / h \Rightarrow d \leq 10 cm$

At this point, we will have to check the assumption of the fully developed turbulence and may correct the value of Q'. However, because the diameter and flow rate are not the final values, we ignore this step. We will apply it only when the values of the flow rate approach the limit of $40 m^3 / h$.

Assuming that $d' = 5cm = 0,05m$:

$(\varepsilon / d = 0,0024, \text{AIT}) \rightarrow f = 0,0246$ (from the Moody diagram).

Therefore:

$V = 4,232 m / s$.

Therefore, we have:

$Q' = 29,9 m^3 / h < 40 m^3 / h \Rightarrow d > 5cm$.

Assuming that $d' = 7cm = 0,07m$

and $(\varepsilon / d = 0,0117, \text{AIT}) \rightarrow f = 0,0225$ (from the Moody diagram).

Therefore, from [5.110] $\Rightarrow V' = 4,915 m/s$

and from [5.106] $\Rightarrow Q' = 68,1 m^3/h > 40 m^3/h \Rightarrow d \leq 7cm$.

Assuming that $d' = 6cm = 0,06m$

$(\varepsilon/d = 0,002, \text{AIT}) \rightarrow f = 0,0234$ (from the Moody diagram).

Therefore, from [5.110] $\Rightarrow V' = 4,60 m/s$

and from [5.106] $\Rightarrow Q' = 46,8 m^3/h$.

Checking the AIT: $\text{Re} = Vd/\nu = 2,8 \cdot 10^5 \rightarrow f = 0,024$ (Moody).

From [5.110] $\Rightarrow V' = 4,56 m/s$ and

from [5.106] $\Rightarrow Q' = 46,4 m^3/h > 40 m^3/h \Rightarrow d \leq 6cm$.

Assuming that $d' = 5,5cm = 0,055m$

$(\varepsilon/d = 0,00218, \text{AIT}) \rightarrow f = 0,024$ (Moody).

From [5.110] $\Rightarrow V' = 4,424 m/s$ and

from [5.106] $\Rightarrow Q' = 37,8 m^3/h$

Checking the AIT: $\text{Re} = Vd/\nu = 2,4 \cdot 10^5 \rightarrow f = 0,0246$ (Moody).

From [5.110] $\Rightarrow V' = 4,38 m/s$ and

from [5.106] $\Rightarrow Q' = 37,47 m^3/h < 40 m^3/h \Rightarrow d > 5,5cm$

Therefore: $5,5cm < d \leq 6cm \Rightarrow d = 6cm$.

For a better control of the results of the testing, we present the following table.

d′ (cm)	ε/d′	f′ (AIT)	V′ (m/s)	Q′	Q′-Q	d
10	0,00120	0,0205	5,626	159,08	119,08	d <= 10
5	0,00240	0,0246	4,232	29,91	-10.09	d > 5
7	0,00171	0,0225	4,915	68,09	28,09	d <=7
6	0,00200	0,0234	4,601	46,83	6,83	d <= 6
5,5	0,00218	0,0240	4,424	37,84	-2,16	d > 5,5

Table 5.8. *Iteration technique results*

5.9. Energy and hydraulic grade lines

As we know, when we have a flow without any friction in a pipe, the energy line of the flow is always parallel to the horizontal surface that we refer to. However, this does not happen in the flow with friction because of the energy losses due to the viscous nature of the fluid. Therefore, during the viscous flow in a pipe, the *total manometer head*, *h*, which is defined by the equation:

$$h^* = z + \frac{p^*}{\gamma} + \frac{V^2}{2g}, \qquad [5.111]$$

always decreases in the direction of the flow. Therefore, the energy line of the viscosity line in a pipe can never be a horizontal or present inclination reversed to the direction of the flow.

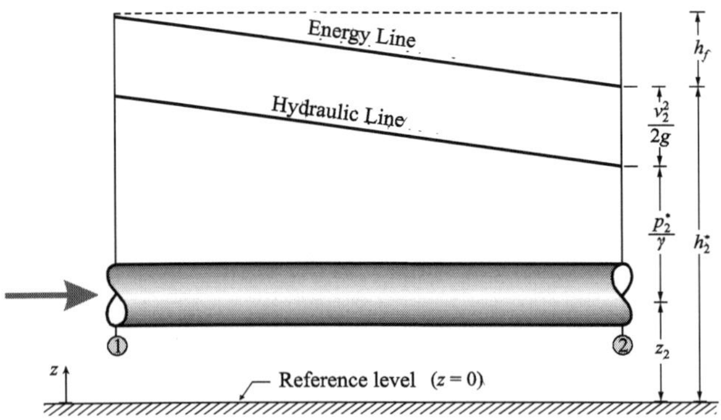

Figure 5.14. *Energy and hydraulic grade lines*

Figure 5.14 shows both the energy line and the hydraulic line of a steady, incompressible viscous flow in a horizontal pipe of constant cross-sectional area.

We note that the two lines are parallel, because the velocity head, $V^2/2g$, of the flow is constant. The vertical drop of the energy line between the cross-sectional areas 1 and 2 shows the *friction head*, h_f, of the flow.

When we want to draw the energy line (EL) and the hydraulic line (HL) of the flows with friction, we must take into account the following observations:

a) The hydraulic line is always under the energy line and at equal distance to the head velocity of the flow. Therefore, if the velocity is zero, as happens for example in a tank, the EL and HL will coincide with the free surface of the flow in the tank (Figure 5.15(a) and (b)).

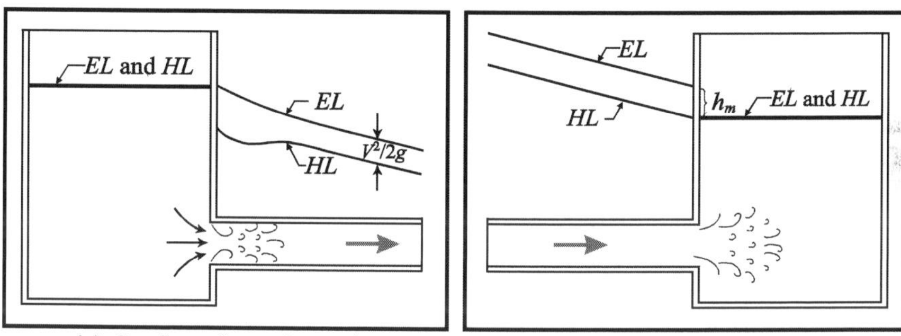

(a) Outflow from the tank (b) Inflow in a tank

Figure 5.15. *Tank – pipe flow*

b) The inclination of the energy line, J_E, which is defined from the ratio:

$$J_E = -\frac{dh^*}{ds} = -\frac{d\left(z + p/\gamma + V^2/2g\right)}{ds}, \qquad [5.112]$$

where dh^* is the variation of the manometer head of the flow that is caused by the elementary length ds of the pipe, which is always positive along the direction of the fluid's motion. The inclination J_E of the EL shows the loss of manometer's head of the flow to the length unit of the pipe. For a steady flow in a pipe with uniformly natural characteristics, namely diameter, roughness, shape, etc., to its whole length, the energy loss of the fluid in the length unit is constant. Therefore, in such a flow, the inclination of the EL along the pipe will be constant. Besides, as in this case the average velocity of the flow is the same everywhere, the HL will be parallel to the EL (Figure 5.14). On the contrary, for a steady flow in a declining or converging parts of pipes, the inclination of the EL and its distance from the HL are not constant along the flow (Figure 5.16).

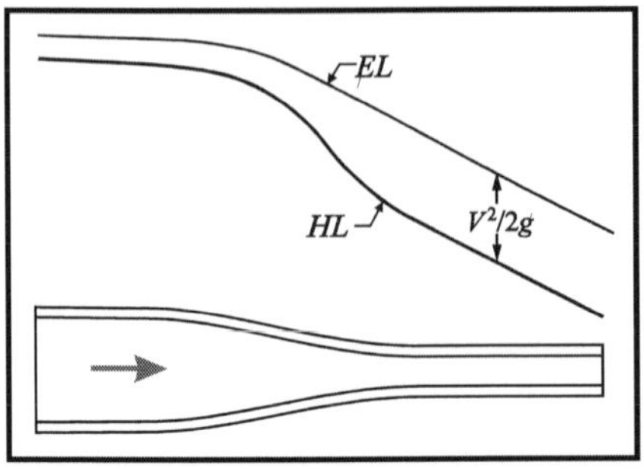

Figure 5.16. *Energy and hydraulic grade lines in a stenosed pipe*

c) When a device is inserted in the fluid's path, it causes a disorder in the flow at some distance before and after the device. This disorder appears with pressure loss and velocity variation of the fluid in the area of the device. This results in a variation of the inclination and the relevant position of the EL and HL.

In the charts of this phenomenon, we simulate this device with a vertical intersection at its position, from which, when the fluid flows through it, energy and pressure losses are caused, and we assume negligible local inclinations between the energy line and the hydraulic line, which are caused by the local variations of the head and the velocity. Therefore, as shown in Figure 5.17(a) and (b), the head of the energy losses, h_m, that the device causes to the flow is equal to the vertical displacement of the energy line at the point where the device lies.

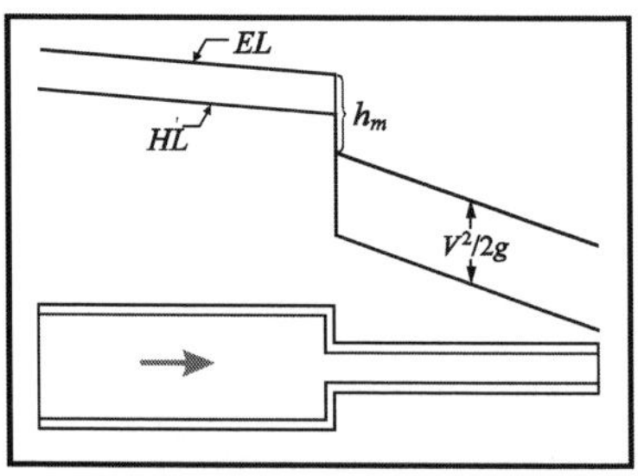

(a) Sudden pipe stenosis

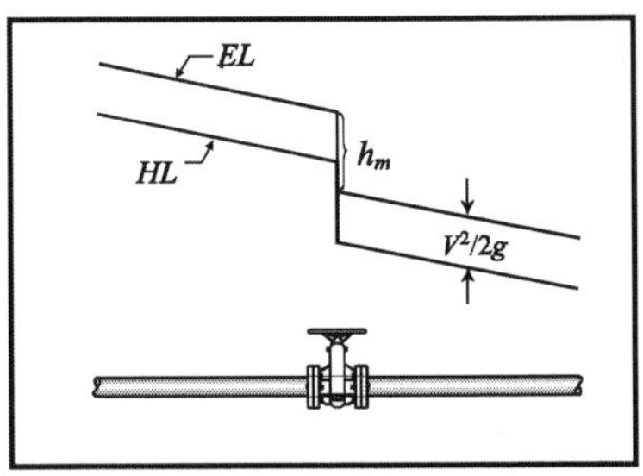

(b) Flow inside a valve

Figure 5.17. *Energy and hydraulic grade lines*

d) As we mentioned in the beginning of the paragraph, when we have a fluid flow inside pipes, the manometer head h is always reduced to the direction of the flow. However, when a pump is inserted, which offers an amount of energy towards the fluid, we have an exception from this rule because we have a sudden rise of the energy line and hydraulic line in the position of the pump because of the amount of energy that the pump offers, as shown in Figure 5.18.

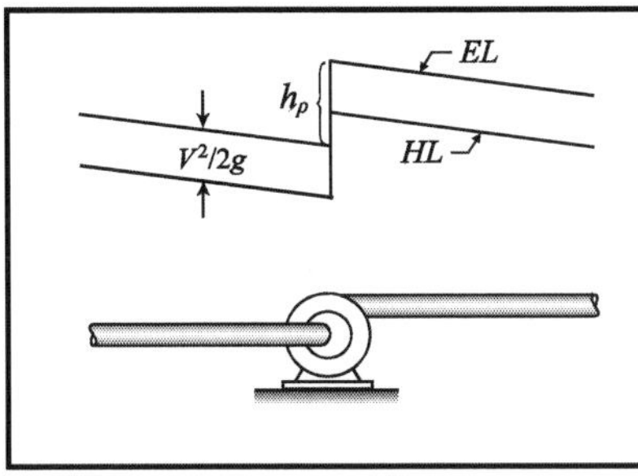

Figure 5.18. *Energy and hydraulic grade lines in pipes with a pump*

This vertical displacement of the energy line that is created by the presence of the pump is called *manometer head of the pump, $h_{\alpha\nu\tau}$*.

e) In the case of turbulent flows, the energy line EL and the hydraulic line HL show a sudden decrease at the position where the vortex occurs because of the energy that the fluid attributes to the vortex as it goes through it. This sudden decrease, h_t, in the total manometer's head of the flow, which causes turbulence, is calculated from the equation:

$$h_t = \frac{P_t}{Qgn_t} \qquad [5.113]$$

where Q is the flow rate, g is the acceleration of gravity, n_t is the degree of distribution and P_t is the axial power of the vortex. The energy losses of the flow that occur during turbulence are contained in the distribution degree n_t.

f) Because the vertical distance of the hydraulic line from the axis of the pipe is equal to the manometer's head of pressure, p^*/γ, of the flow, this line will pass through the points of the pipe where the pressure is equal to the pressure of the environment, so the manometer pressure is $p^* = 0$ (Figure 5.19).

At these points along the hydraulic line under the axis of the pipe, the manometer head of the pressure is negative $\left(p^*/\gamma < 0\right)$. This means that, at these points, the pressure of the fluid is lower than the atmospheric pressure (Figure 5.19).

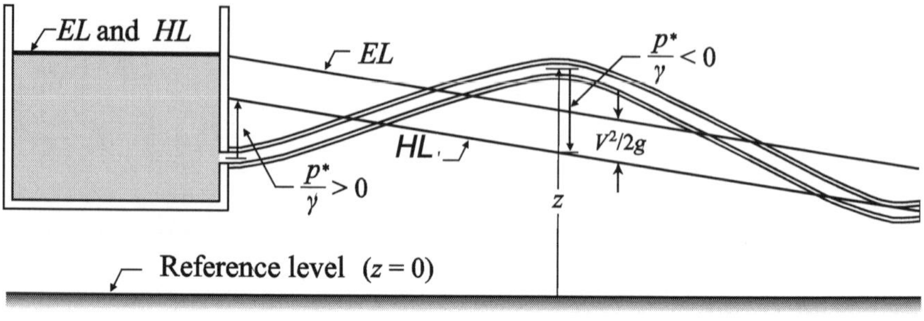

Figure 5.19. *EL and HL under the axis of the pipe*

Finally, we define as *hydraulic inclination* the value:

$$L_{Y.K.} = \frac{d\left(z + p/\gamma\right)}{ds} \qquad [5.114a]$$

Concluding, we have to mention that in many cases, where there are many pipes along, the secondary losses can be considered to be negligible and therefore we do not take them into account. This happens when they are smaller than 5% of the friction losses of the pipe, or they can be included in the equivalent lengths of the pipes and be added to the real length. For these cases, the value of the head of the kinetic energy $V^2/2g$ is low enough, compared to $f(\ell/d)V^2/2g$, to be ignored.

In this special yet very common, case, where the secondary losses are ignored, the *lines of hydraulic inclination and energy* are shown with one line. The only line shown in Figure 5.20 is called the *line of hydraulic inclination.*

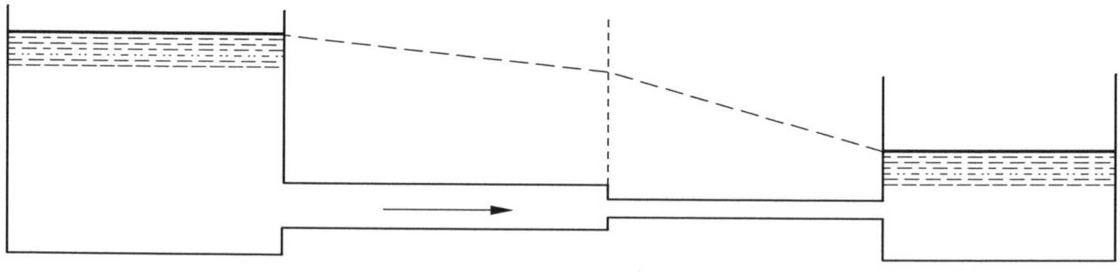

Figure 5.20. *Line of hydraulic inclination*

In the case of long pipes, the *hydraulic inclination* is h_f/ℓ, where h_f is given by the Darcy–Weisbach equation:

$$h_f = f\frac{\ell}{d}\cdot\frac{V^2}{2g}. \qquad [5.114b]$$

Finally, the flow has always the *direction* to which the *energy line* goes through, unless there is a pump.

5.10. Incompressible, viscid flow in connected pipes

The study of flow of an incompressible viscid fluid in pipes has full application in the design of a hydraulic installation, which consists of pipes connected among them as well as with various devices, which eventually form a pipeline. The way of the final pipeline composition arises from the cause that it serves and it is depicted in its design. The basic parameter of this design is the energy study, which has to accompany it in order to estimate its results.

Therefore, in the technical applications, we have the study of pipelines systems that every engineer deals with and which are defined from the power needed for the operation of the pump and, of course, the cost limitations in order to achieve cost-efficient solutions. Because in every single design of a pipeline, the flow rate and the quantities of the fluid are most often known, except for the above values that will have to be calculated. An *energy analysis* of the flow in the pipelines has to be done as well. This is based on the equation of the mechanical energy of the flow.

For the application of this equation, we define a cross-sectional area a at the entrance and another one b at the exit along the pipeline. If the flow in the pipeline is steady and incompressible, with uniform distribution of the velocity, then according to what we have mentioned until now, the total heads of the flow right before the cross sectional area a (head h_a) and right after the cross-sectional area b (head h_β) are:

$$h_a = z_a + \frac{p_a}{\gamma} + \frac{V_a^2}{2g} \qquad [5.115]$$

$$h_b = z_b + \frac{p_b}{\gamma} + \frac{V_b^2}{2g} \qquad [5.116]$$

Under these circumstances, the equation of the mechanical energy is written with the form of heads:

$$h_a + h_w = h_b + h_{tot} \quad [5.117]$$

where h_{tot} is the total head of the energy losses that occur between the cross-sectional areas a and b of the pipeline and h_w is *the head of the mechanical energy* of the flow. The head h_w is given by the relationship:

$$h_w = \frac{P}{gQ} \quad [5.118]$$

where P is the axial power and Q is the flow rate. For the pumps which offer energy to the fluid, the term h_w is positive $(h_w = h_p > 0)$. The total head h_{tot} of the flow is equal to the sum of the head h_f of the energy losses in the parts of the pipeline and the head h_m of the energy losses at various devices that are included in the pipe system, which is:

$$h_{tot} = h_f + h_m = \sum_i f_i \frac{\ell_i}{d_i} \frac{V_i^2}{2g} + \sum_j k_{m,j} \frac{V_j^2}{2g}, \quad [5.119]$$

where f_i, ℓ_i, d_i and V_i are the friction coefficient, length, diameter and average velocity at the $i-$ part of the pipeline and $k_{m,j}$ and V_j are the losses coefficient and the average velocity of the fluid at its $j-$ device. For the application of equation [5.102], experimentally found friction coefficients f_i of the pipes and the losses coefficients $k_{m,j}$ of the devices of the pipeline are of utmost importance.

More specifically, the pipeline networks primarily used in real life applications are listed below, which will be studied analytically:

1) simple pipelines;

2) pipelines in a row;

3) parallel connection of pipelines;

4) mixed connection of pipelines;

5) pipelines branch;

6) pipeline networks.

5.10.1. *Simple pipelines*

We call simple pipelines the completed pipelines, which consist of linear parts of pipes of the same cross-sectional area and are put together in a row with the assistance of suitable fittings. A simple pipeline can contain measurement instruments as well, but no machines support the flow of the fluid, like pumps or systems that produce energy. This means that the variation of the head of the mechanical energy of the flow h_w is equal to zero $(h_w = 0)$. Figure 5.20 presents a simple pipeline with conventional entrance a and conventional exit b. If the flow that occurs inside this pipe is steady and incompressible, then the equation of the mechanical energy of the flow in heads form is written as:

$$h_a = h_b + h_e \qquad [5.120]$$

where h_a is the total head of the flow right after the entrance a, h_β is the total head of the flow right after the exit b and h_e is the total head of the energy losses that occur between the cross-sectional areas a and b.

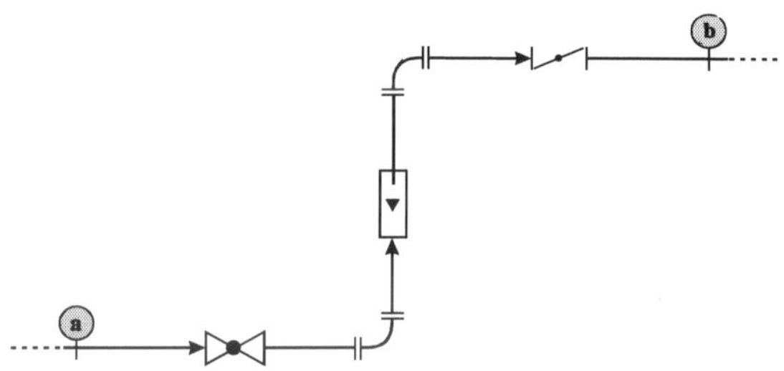

Figure 5.21. *Flow in a pipeline*

As the flow is incompressible and the diameter of the pipe is constant to the whole length of the pipeline, from the continuity equation, it occurs that the average velocity V of the flow will be constant throughout the whole pipeline. Therefore, the total head h_{tot} of the energy losses of the flow in the pipeline will be given by the relationship:

$$h_{tot} = h_i + h_m = \left(f \frac{\ell}{d} + \sum_i k_{m,\ell} \right) \frac{V^2}{2g} \qquad [5.121]$$

where f is the friction coefficient, d is the internal diameter and ℓ is the sum of the lengths of all the straight pipes that consist the pipe system.

In all these simple pipelines, we use the methodology that we developed before for the technical solution of a flow (three categories of problems), which depends on the known parameters in each case. These are the cases that we developed in section 5.7.

5.10.2. *Pipes connected in a row*

This problem is different when we connect two or more different pipes in a row. We consider different pipes with different roughnesses, which means they are made of different materials or mostly have different internal diameters. In this case, we have *pipes connected in a row.*

Therefore, let us assume that pipe 1 is connected to pipe 2 in a row (Figure 5.22). As the fluid is incompressible and the flow is steady, the continuity equation is applicable:

$$Q = Q_1 = Q_2 \qquad [5.122]$$

$$\Rightarrow \frac{\pi \cdot d_1^2}{4} \cdot V_1 = \frac{\pi \cdot d_2^2}{4} \cdot V_2 \Rightarrow$$

$$\Rightarrow d_1^2 \cdot V_1 = d_2^2 \cdot V_2 \Rightarrow \frac{V_1}{V_2} = \left(\frac{d_2}{d_1}\right)^2 \qquad [5.123]$$

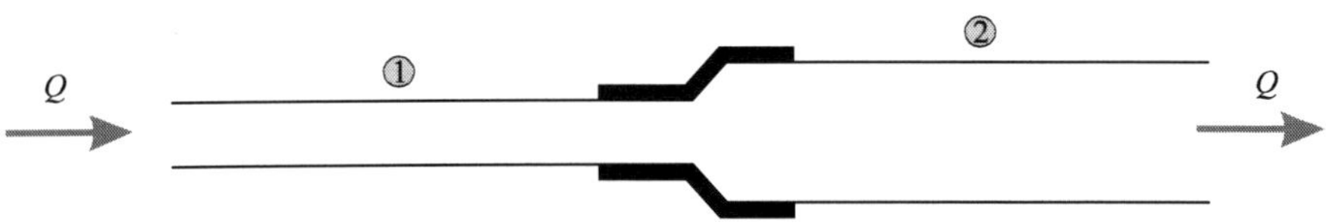

Figure 5.22. *Pipes connected in a row*

The energy losses per weight unit of the fluid that is moving will be added, because every elementary amount of fluid must go through successively. This is easily proved by the Bernoulli equation between points A–B (pipe 1), B–C (pipe 2) and A–C (system of two pipes):

$$\mathrm{A-B}: y_{\mathrm{A}} - y_{\mathrm{B}} + \frac{V_{\mathrm{A}}^2 - V_{\mathrm{B}}^2}{2 \cdot g} + \frac{p_{\mathrm{A}} - p_{\mathrm{B}}}{\gamma} = \Sigma h_1 \qquad [5.124]$$

$$B - C: y_B - y_C + \frac{V_B^2 - V_C^2}{2 \cdot g} + \frac{p_B - p_C}{\gamma} = \Sigma h_2 \quad [5.125]$$

$$A - C: y_A - y_C + \frac{V_A^2 - V_C^2}{2 \cdot g} + \frac{p_A - p_C}{\gamma} = \Sigma h \quad [5.126]$$

but equation [5.126] constitutes the sum of the previous two. Therefore:

$$\Sigma h = \Sigma h_1 + \Sigma h_2 . \quad [5.127]$$

Using the Bernoulli equation between points A and C, the local losses that vary according to the cross-sectional area, by either increasing or decreasing it, were not calculated. In order to have accurate results, they must be calculated and added. The calculation is carried out using the coefficient of the local losses, which refers to the velocity of the pipe with the smaller cross-sectional area and which, according to Figure 5.22, is:

$$h_k = K \cdot \frac{V_1^2}{2g} . \quad [5.128]$$

Therefore, according to the Darcy–Weisbach equation of losses, we have:

$$\Sigma h_1 = \left(f_1 \cdot \frac{\ell_1}{d_1} + \Sigma K_1 \right) \cdot \frac{V_1^2}{2g} = $$
$$= \frac{8}{\pi^2 g} \left(f_1 \cdot \frac{\ell_1}{d_1} + \Sigma K_1 \right) \cdot \frac{Q_1^2}{d_1^4}$$

and

$$\Sigma h_2 = \left(f_2 \cdot \frac{\ell_2}{d_2} + \Sigma K_2 \right) \cdot \frac{V_2^2}{2g} = \frac{8}{\pi^2 g} \left(f_2 \cdot \frac{\ell_2}{d_2} + \Sigma K_2 \right) \cdot \frac{Q_2^2}{d_2^4}$$

and by substituting it into equation [5.127], we get:

$$\Sigma h = \frac{8}{\pi^2 \cdot g} \left(\left(f_1 \cdot \frac{\ell_1}{d_1} + \Sigma K_1 \right) \cdot \frac{Q^2}{d_1^4} + \left(f_2 \cdot \frac{\ell_2}{d_2} + \Sigma K_2 \right) \cdot \frac{Q^2}{d_2^4} \right) . \quad [5.129]$$

If instead of two, we connect three or more pipes in a row, equations [5.128] and [5.129] are generalized. Therefore:

During the connection of pipes in a row:

a) the flow rate is constant to the whole route of the flow,

b) the energy losses are added (possible local losses at the connections are also added):

$$Q = Q_1 = Q_2 = \ldots \qquad [5.130]$$

$$\Sigma h = h_1 + h_2 + \ldots \qquad [5.131]$$

In the calculating procedure of pipes that are connected in a row, except for these two relationships, we use the Bernoulli equation as well, if the energy values are given or asked, as well as, for each pipe, the continuity equation, the Moody diagram and the losses equation. The relevant problems are classified into three categories:

a) calculating problems of the losses head;

b) calculating problems of the flow rate;

c) calculating problems of a diameter.

For the methodological procedure of each problem, we will give numerical examples at the end of this chapter.

5.10.3. *Parallel connection of pipes*

At some points in a pipe, the flow branches into two other sections, and later, the sections converge as one again (bypass). In this case, we talk about a parallel connection of two similar or different pipes.

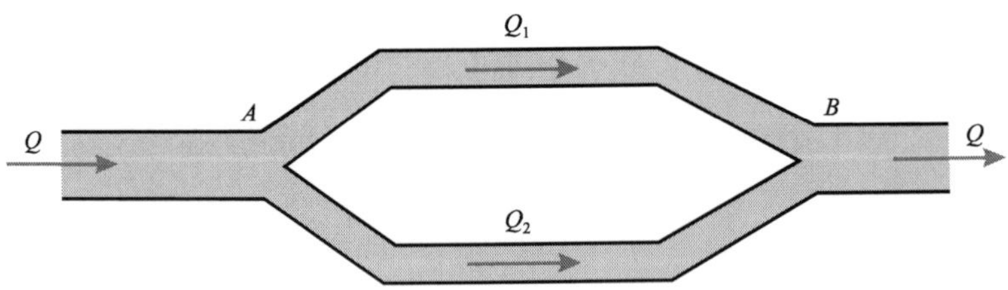

Figure 5.23. *Pipes in parallel*

Pipes 1 and 2 in Figure 5.23 are connected in a parallel way. The general flow, with flow rate Q, branches at point A into two other flows with flow rates Q_1 and Q_2. Point A is called *branch* or *node*.

Applying the continuity equation at the node A, we get:

$$\Sigma Q_{ent} = \Sigma Q_{ex} \Rightarrow$$

$$\Rightarrow Q = Q_1 + Q_2 \qquad [5.132]$$

At point B, the two pipe branches are connected again. Point B is also a node, and the continuity equation will give:

$$\Sigma Q_{ent} = \Sigma Q_{ex} \Rightarrow$$

$$Q_1 + Q_2 = Q,$$

which is a relationship identical to [5.119].

Substitute the flow rates with the velocities and diameters, [5.119] becomes:

$$\frac{\pi d^2}{4} \cdot V = \frac{\pi d_1^2}{4} \cdot V_1 + \frac{\pi d_2^2}{4} \cdot V_2 \Rightarrow$$

$$d^2 V = d_1^2 V_1 + d_2^2 V_2 \qquad [5.133]$$

Applying now the Bernoulli equation in pipe 1 between points A and B, we get:

$$y_A - y_B + \frac{p_A - p_B}{\gamma} + \frac{V_A^2 - V_B^2}{2g} = \Sigma h_1 \qquad [5.134]$$

Similarly, for pipe 2:

$$y_A - y_B + \frac{p_A - p_B}{\gamma} + \frac{V_A^2 - V_B^2}{2g} = \Sigma h_2 \qquad [5.135]$$

These relationships have their first members equal; thus, it occurs that:

$$\Sigma h = \Sigma h_1 = \Sigma h_2 \qquad [5.136]$$

This means that the energy losses per weight unit of the fluid are the same in the two connected in parallel.

Substituting from the losses equation, it occurs that:

$$\left(f_1 \cdot \frac{\ell_1}{d_1} + \Sigma K_1\right) \cdot \frac{Q_1^2}{d_1^4} = \left(f_2 \cdot \frac{\ell_2}{d_2} + \Sigma K_2\right) \cdot \frac{Q_2^2}{d_2^4} \quad [5.137]$$

If the local losses are negligible, which means that $\Sigma K_1 = \Sigma K_2 \simeq 0$, then [5.137] becomes:

$$f_1 \cdot \frac{\ell_1}{d_1} \cdot Q_1^2 = f_2 \cdot \frac{\ell_2}{d_2} \cdot Q_2^2 \quad [5.138]$$

and by using coefficients α_i, we have:

$$a_1 \cdot Q_1^2 = a_2 \cdot Q_2^2 \quad [5.139]$$

Equations [5.132] and [5.136] are generalized in the case where between the two points (nodes) A and B, we have more than two pipes connected in parallel:

$$Q = Q_1 + Q_2 + \ldots \quad [5.140]$$

$$\Sigma h = h_1 = h_2 = \ldots \quad [5.141]$$

Consequently, when two or more pipes are connected in parallel:

a) the total flow rate is equal to the sum of the individual flow rates;

b) the losses per weight unit are the same for all the pipes.

As we will see through the examples that follow, a direct solution is possible only when the head loss is given, which is common for the pipes connected in parallel, and the flow rate has to be calculated.

5.10.4. *Mixed pipe connection*

In some cases, we deal with a mixed pipe connection, as the one shown in Figure 5.24. The relationships arise if we take into account everything that is applicable for the connection in a row and in parallel. Therefore, as far as the flow rate of the connection in Figure 5.24 is concerned and taking into account that:

$$Q_1 = Q_2 + Q_3 \text{ and } Q_3 = Q_4 + Q_5, \quad [5.142]$$

we have:

$$Q = Q_1 = Q_2 + Q_3 = Q_2 + (Q_4 + Q_5) = Q_6 = Q_7 \quad [5.143]$$

Similarly, for the losses, taking into account that:

$$\Sigma h_4 = \Sigma h_5 \text{ and } \Sigma h_2 = \Sigma h_3 + \Sigma h_4,$$

we have:

$$\Sigma h = \Sigma h_1 + \Sigma h_2 + \Sigma h_6 + \Sigma h_7 \quad [5.144]$$

The relationships of the losses and the flow rates are very useful. In the mixed connection, it is better to use the losses equation (Darcy–Weisbach) in equivalent forms:

$$h_i = a_i \cdot Q_i^2 \Leftrightarrow Q_i = a_i^{-0,5} \cdot h_i^{0,5} \quad [5.145]$$

Therefore, for example, for pipes 3, 4 and 5 in Figure 5.24, it is applicable that:

$$Q_3 = Q_4 + Q_5 \Rightarrow Q_3 = \left(a_4^{0,5} + a_5^{0,5}\right) h_4^{0,5} \quad [5.146]$$

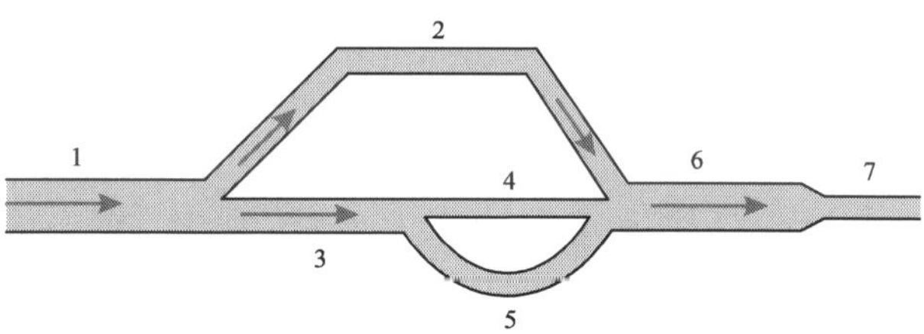

Figure 5.24. *Pipes in mixed connection*

If we take into account that a_i can be calculated approximately by using the assumption of the developed turbulence, we note that the three unknowns of relation [5.142] (Q_3, Q_4, Q_5) have been reduced to (Q_3, h_4) in relation [5.146].

The solution of the problems of the mixed pipeline is based on the methodology of the connections in a row and in parallel. Therefore, if for the pipelines in Figure 5.24 we have enough data for the parallel connection of pipes (4) and (5), we are able to calculate it. Afterward, we calculate the connection of those two pipes to pipe 3 in a row, then the parallel connection of pipe 2 to the system (3,4,5) and finally, the connection of pipes 2, 6 and 7 in a row to the system (2,3,4,5). However, if the given data enforce it, the route is inversed. If, for example, in the assembly of Figure 5.24 we know the total head losses and the flow rate, we will calculate the losses in pipes 1, 6 and 7. From equation [5.144], we will calculate the losses in pipe 2 and then its flow rate. From the relation of the first node, we will calculate the flow rate of pipe 3 and its losses as well. This way, we reach a parallel connection of pipes 4 and 5, for which we know the losses and therefore we can calculate the flow rates or even a diameter.

5.10.5. *Pipe branches*

Studying the parallel connection, we deal with the branch of the flow into two or more individual flows. The branch (or node) is very common to the pipe networks.

In this paragraph, we will refer to the branch of three pipes (Figure 5.25).

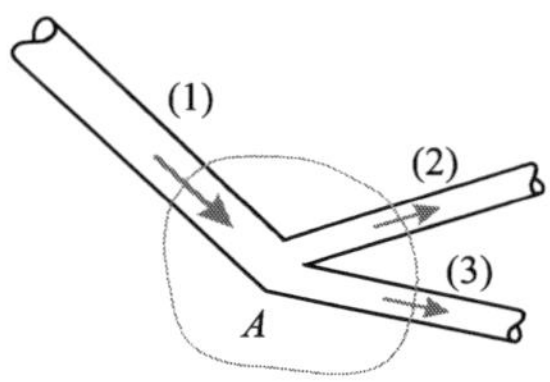

Figure 5.25. *Pipe branches*

Applying the continuity equation in the control volume that contains node A, we have:

$$\Sigma Q_{ent} = \Sigma Q_{ex} \quad [5.147]$$

If we know the direction of the flow in the three pipes, equation [5.147] gets a more specific form. Therefore, as shown in Figure 5.25, flow rate Q_1 is getting in the control volume that is node A, while flow rates Q_2 and Q_3 are coming out. Then, equation [5.147] becomes:

$$Q_1 = Q_2 + Q_3 \quad [5.148]$$

However, if the direction of flow in pipe (2) is opposite to that in the figure, which means that the flow rate Q is getting in the control volume that contains node A; then, equation [5.148] becomes:

$$Q_1 + Q_2 = Q_3 \quad [5.149]$$

In many cases, we do not know the direction of the flow at some pipes of the branch. This problem can be dealt with the following *rule of the pre-marking flow rates*:

a) When the fluid *comes in* the node, the flow rate is considered to be *positive*.

b) When the fluid *comes out* of the node, the flow rate is considered to be *negative*.

Using this rule, we can generalize the relation of the flow rates with the form:

$$\Sigma Q = 0 \Rightarrow Q_1 + Q_2 + Q_3 = 0 \qquad [5.150]$$

Relationship [5.150], in which the flow rates are pre-marked, is always applicable in the node of the three pipes. The rule of pre-marking flow rates could be exactly the opposite as long as opposite signs for inflow and outflow are used.

Although the relation of the flow rates of the branches is reached easily, the relation of the head losses of the three pipes presents significant difficulties and it has to be studied for the whole system and not only for the node. This case will be dealt in the chapter involving pipe networks. We will now examine the case where three pipes end up in three tanks, with their free surfaces at different altitudes. This problem is known as the *problem of the three tanks*.

5.10.5.1. *Problem of the three tanks*

Pipes (1), (2) and (3) of the branch are connected to tanks A, B and C, respectively, with atmospheric pressure on their free surfaces and with altitudes y_1, y_2 and y_3 (Figure 5.26).

If tank A lies at the highest level and tank C at the lowest level, then we have:

$$y_1 > y_2 > y_3$$

It is obvious that there will be outflow of fluid from tank A, due to gravity, whereas at tank C there will be inflow of the fluid. However, it is not certain what we will have at the tank of the intermediate altitude B. First, we can conclude that it depends on its position: if the altitude of the free surface of y_2 is near the altitude of A (y_1), then it will behave like A, which means that we will have outflow from B. However, if it is near the altitude of C (y_3), then we will have inflow at B. In both cases, the equations of the node flow rates will be applicable:

$$Q_1 + Q_2 + Q_3 = 0 \qquad [5.151]$$

In this relation, if we put (+) sign for the outflow flow rate from the tank, which means inflow at the node, then inflow will be set (−) sign for the inflow flow rate of the fluid in the tank, when we have outflow from the node.

If we use a non-pre-marked flow rate, and if $y_1 > y_2 > y_3$, then we have:

$$Q_1 \pm Q_2 - Q_3 = 0 \qquad [5.152]$$

In order to now calculate the relation of the head losses in this case of the three tanks, we use an auxiliary pressure gauge tube as shown in Figure 5.26. With its installation in this system, the fluid will rise in it to a head Λ, which will be smaller than y_1 and bigger than y_3, which is:

$$y_1 > \Lambda > y_3 \qquad [5.153]$$

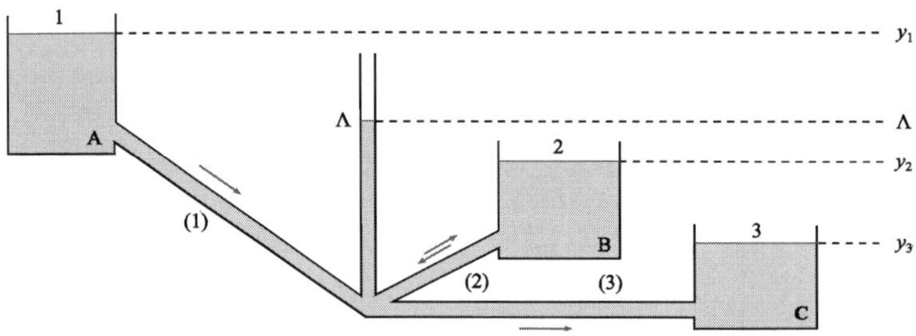

Figure 5.26. *Three-tank arrangement*

Applying the Bernoulli equation between points (1) and (Λ) and taking into account that the velocity is zero in the pressure gauge, as the pressure is atmospheric at the surface of the tank, we have:

$$\text{if } \Sigma h_1 = h_1 \Rightarrow y_1 - \Lambda + \frac{V_1^2 - V_P^2}{2g} + \frac{p_1 - p_P}{\gamma} = h_1 \Rightarrow y_1 - \Lambda = h_1$$

$$\Rightarrow y_1 - h_1 = \Lambda \qquad [5.154]$$

Similarly, the Bernoulli equation between (Λ) and (3) $(\Lambda > y_3)$ gives:

$$\Lambda - y_3 = h_3 \Rightarrow h_3 + y_3 = \Lambda . \qquad [5.155]$$

For the tank with the intermediate height, we distinguish two cases:

a) If the free surface of B is higher than Λ $(y_2 > \Lambda)$, then we have outflow from B and the Bernoulli equation is applied between (2) and (Λ), which gives:

$$y_2 - \Lambda = h_2 \Rightarrow y_2 - h_2 = \Lambda . \qquad [5.156]$$

b) If the free surface of B is lower than Λ $(y_2 < \Lambda)$, then we have inflow to B and the Bernoulli equation is applied between (Λ) and (2), which gives:

$$\Lambda - y_2 = h_2 \Rightarrow h_2 + y_2 = \Lambda . \qquad [5.157]$$

In the special case in which the free surface of the intermediate tank is at the level Λ $(y_2 = \Lambda)$, the Bernoulli equation gives:

$$y_2 - \Lambda = h_2 \Rightarrow h_2 = 0 \Rightarrow Q_2 = 0 .$$

Equations [5.154–5.156] have the second members equal, and they can be written with the general relation:

$$y_1 - h_1 = y_2 \mp h_2 = y_3 + h_3 = \Lambda . \qquad [5.158]$$

Completing the analysis, we have to mention that equations [5.148] and [5.158] are only applicable for the case where the heights of the tank are:

$$y_1 > y_2 > y_3$$

Finally, in the problems of the three tanks, we have the following three groups of values:

a) The values of the three pipes $d_i, \ell_i, \varepsilon_i$.

b) The heights of the three tanks y_1, y_2, y_3.

c) The flow rates Q_1, Q_2, Q_3.

The problems of the three tanks are ranked into three categories, in which the values of the pipes are known and two altitudes and one flow rate or three flow rates are unknown. For the solution of these problems, we use relationships [5.138] and [5.145], combining them with the pipe calculation ones: the *continuity equation*, the *Moody diagram* and the *losses equation*.

5.10.5.2. *General problem of the three tanks*

We consider now the case where the three pipes of the branch do not end up in three open tanks, but the whole system that we examine is defined from the cross-sectional areas of the pipes at points (1), (2) and (3), as shown in Figure 5.27.

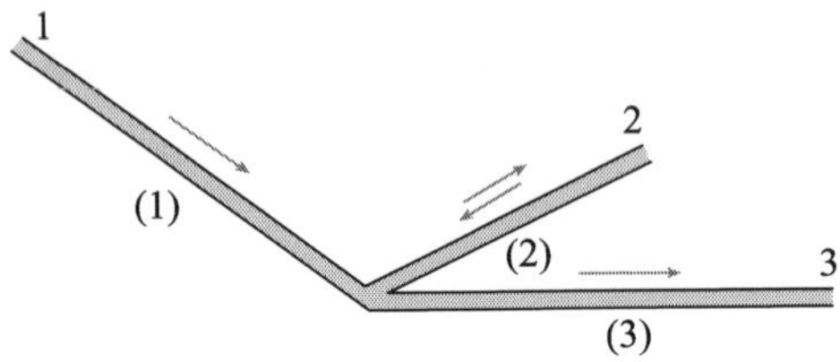

Figure 5.27. *Pipe branches*

In the previous case of the three tanks, the surfaces, where the three pipes ended up, were the free surfaces of the tanks, which are characterized energetically only from their altitudes y_1, y_2, y_3.

But now that the system of Figure 5.27 is defined from the cross-sectional areas of the pipes at points (1), (2) and (3), the energy head at their points is:

$$H_1 = y_1 + \frac{V_1^2}{2g} + \frac{p_1}{\gamma} \quad [5.159]$$

$$H_2 = y_2 + \frac{V_2^2}{2g} + \frac{p_2}{\gamma} \quad [5.160]$$

$$H_3 = y_3 + \frac{V_3^2}{2g} + \frac{p_3}{\gamma} \quad [5.161]$$

The equation of the flow rates [5.150] is applicable in this case, too. However, in the losses equation [5.158] in the position of the head of dynamic energy y_1, we put the energy heads H_i. If we know that pipe (1) inserts in the node and pipe (3) outflows from it, equation [5.158] becomes:

$$H_1 + h_1 = H_2 \mp h_2 = H_3 + h_3 = \Lambda ,$$

where H_1, H_2 and H_3 are calculated from equations [5.159], [5.160] and [5.161], respectively.

It is obvious that the problems become more complicated because in the calculations other values are inserted, such as velocities and pressures. However, methodologically, we work with the same ways that we saw in the problems of the three tanks.

Even more general is the case where in the node more than three pipes participate. Equation [5.150] takes the form:

$$\sum Q_i = 0 \qquad [5.162]$$

In this equation, in the flow rates we set conventional signs (+) for the inflow and (−) for the outflow. Similarly, we also set at the losses equation (+) for the inflow, (−) for the outflow conventional signs, so it becomes:

$$H_i - h_i = \Lambda . \qquad [5.163]$$

For the solution, more data are needed, so that we can place the problem in a proportional category to the previous ones that we examined.

5.11. Simple applications of pipeline networks

5.11.1. *Simple pipeline*

Problem: in a hydraulic installation, water comes out in the atmosphere at a flow rate of 0.06 m^3/s. The point of outflow lies 5 m lower from the free surface of the water of the tank, as shown in Figure (1) below. The transition pipe of the water ends up to a nozzle whose diameter is 10 cm and whose internal diameter is 20 cm. If the energy losses are considered to be negligible:

a) What kind of machine is *M*, pump or turbine?

b) Calculate the power of *M*, if the efficiency rate is 85%.

c) Calculate the pressure at point A, if the barometric pressure is 100 KPa.

Solution

We will use the equation of the mechanical energy in the head form, which is:

$$h_a + h_w = h_\beta + h_\ell \,. \tag{1}$$

In order to find if *M* is a pump or a turbine, we will have to check if the head h_w, which is between the two positions *a* and *b*, is positive or negative, and if we find that $h_w > 0$, then the machine is a pump, whereas if $h_w < 0$, then it will be a turbine. Therefore:

We apply equation (1) between positions 1 and 2, replacing the total heads of the flow in these positions with the respective sums of the pressure heads, level and velocities. Therefore, we have:

$$z_1 + \frac{p_1}{\gamma} + \frac{\upsilon_1^2}{2g} + h_w = z_2 + \frac{p_2}{\gamma} + \frac{\upsilon_2^2}{2g} + h_\ell \tag{2}$$

This equation is simplified significantly, if we take into account that: (1) pressures p_1 and p_2 are equal $\left(p_1 = p_2 = 100kPa\right)$, (2) the velocity $V_1 \approx 0$ and (3) the height $h_\ell = 0$, because the energy losses of the flow are considered to be negligible. Therefore, from equation (2), for the head h_w, the following relationship holds:

$$h_w = \frac{V_2^2}{2g} - \left(z_1 - z_2\right) \tag{3}$$

Velocity V_2 at the exit of the nozzle is calculated from the relation:

$$V_2 = \frac{Q_2}{S_2} = \frac{4Q}{\pi d_2^2} = \frac{4\left(0,06m^3/s\right)}{\pi\left(0,10m\right)^2} = 7,64m/s \,. \tag{4}$$

Therefore, we can calculate, from equation (3) the head h_w:

$$h_w = \frac{\left(7,64m/s\right)^2}{2\left(9,81m/s^2\right)} - \left(5m\right) = -2,02m \tag{5}$$

Thus, as the head $h_w < 0$, machine *M* is a *turbine*.

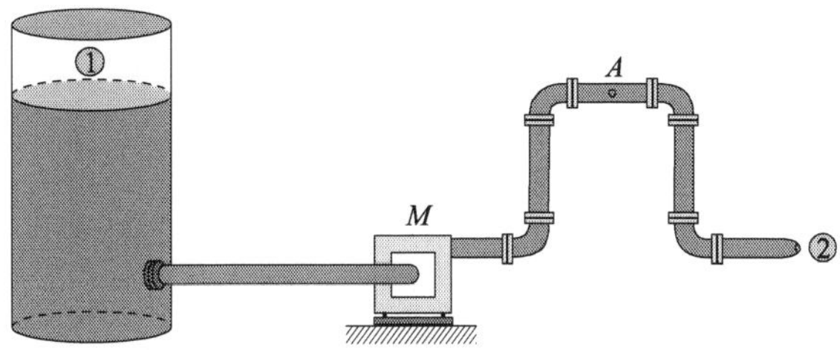

Figure 5.28. *Hydraulic system*

b) Power P_t of the turbine is calculated by the equation:

$$P_t = n_t g \rho Q h_t = (0,85)(9,81 m/s^2)(998 kg/m^3)(0,06 m^3/s)(2,02 m) = 1008\, W. \quad (6)$$

Thus, the axial power of the turbine is approximately 1 KW.

c) For the calculation of pressure p_A, we apply the equation of the mechanical energy, in head form, between positions 1 and A:

$$z_1 + \frac{p_1}{\gamma} + \frac{V_1^2}{2g} = z_2 + \frac{p_A}{\gamma} + \frac{V_A^2}{2g} + h_t + h_\ell \quad (7)$$

or, because the head $h_\ell = 0$, altitude $z_1 = z_A$ and velocity $V_1 \approx 0$:

$$p_A = p_1 - \rho g h_t - \rho \frac{V_A^2}{2}. \quad (8)$$

The velocity V_A of water is calculated from the equation:

$$V_A = \frac{Q}{S} = \frac{4Q}{\pi d_A^2} = \frac{4(0,06 m^3/s)}{\pi (0,20 m)^2} = 1,91 m/s. \quad (9)$$

We can already calculate, from equation (8), pressure p_A:

$$p_A = (100 \times 10^3 Pa) - (998 kg/m^3)\left[(9,81 m/s^2)(2,02 m) - 0,5(1,91 m/s)^2\right] =$$

$$= 82000 Pa = 82 kPa \quad (10)$$

which means that, at point A, the pressure of water is smaller than the atmospheric pressure.

5.11.2. *Pipes in a row*

5.11.2.1. *Calculation of head losses*

Pipes AB and BC in Figure 5.29, which are made of cast iron, are of length 100 and 120 m and diameter 12 and 10 cm, respectively. The total coefficient of the local losses for each pipe is equal to 3. Calculate the total losses Σh if the water flow rate is 60 m^3/h Also, the altitude range H of the two tanks needs to be calculated.

Solution

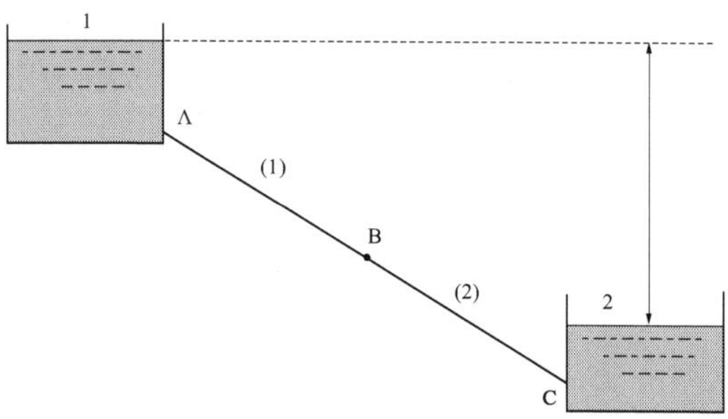

Figure 5.29. *Piping system*

The connection is in a row, so relationships [5.128] and [5.129] are applicable:

$$Q_1 = Q_2 = Q \tag{1}$$

$$\Sigma h = \Sigma h_1 + \Sigma h_2 . \tag{2}$$

We examine whether the data allow us to calculate one of the two pipes. We note that for both pipes the flow rate and $\ell_i, d_i, \varepsilon_i$ are known. We can therefore calculate the losses of each pipe. Thus, for pipe (1), we have:

$$V_1 = \frac{4 \cdot Q_1}{\pi \cdot d_1^2} \Rightarrow V_1 = 1,47 m/s$$

$$\left(\mathrm{Re}_1 = \frac{V_1 d_1}{\nu} \Rightarrow \mathrm{Re}_1 = 1{,}8 \cdot 10^5 \, , \, \varepsilon / d_1 = 0{,}00217\right) \rightarrow f_1 = 0{,}025 \text{ (Moody)}$$

$$\Sigma h_1 = \left(f_1 \cdot \frac{\ell_1}{d_1} + \Sigma K_1 \right) \cdot \frac{V_1^2}{2g} \Rightarrow \Sigma h_1 = 2{,}62 m \, .$$

Similarly, for pipe (2), we have:

$$V_2 = \frac{4 \cdot Q_2}{\pi \cdot d_2^2} \Rightarrow V_2 = 2{,}12 m / s$$

$$\mathrm{Re}_2 = \frac{V_2 d_2}{\nu} \Rightarrow \mathrm{Re}_2 = 2{,}1 \cdot 10^5 , \varepsilon / d_2 = 0{,}00216 \rightarrow f_2 = 0{,}0258 \text{ (Moody)}$$

$$\Sigma h_2 = \left(f_2 \cdot \frac{\ell_2}{d_2} + \Sigma K_2 \right) \cdot \frac{V_2^2}{2g} \Rightarrow \Sigma h_2 = 7{,}80 m \, .$$

The total losses are equal to the sum of the losses, so from (2), we have: $\Sigma h = 10{,}5 m$.

For the calculation of the altitude range of the free surfaces of the two tanks, we apply the Bernoulli equation:

$$y_1 - y_2 + \frac{V_1^2 - V_2^2}{2 \cdot g} + \frac{p_1 - p_2}{\gamma} = \Sigma h \Rightarrow H = y_1 - y_2 = \Sigma h \Rightarrow H = 10{,}5 m$$

5.11.2.2. *Calculation of the flow rate*

If in the previous problem, all the data remain the same, calculate the flow rate if the height range of the two tanks is 15 m.

Solution

We have two pipes connected in a row. So, the following relationships are applicable:

$$Q_1 = Q_2 = Q \tag{1}$$

$$\Sigma h = \Sigma h_1 + \Sigma h_2 \tag{2}$$

As the altitude range $H = y_1 - y_2$ is given, we apply the Bernoulli equation and we calculate the total losses:

$$y_1 - y_2 + \frac{V_1^2 - V_2^2}{2 \cdot g} + \frac{p_1 - p_2}{\gamma} = \Sigma h \Rightarrow H = y_1 - y_2 = \Sigma h \Rightarrow \Sigma h = 15m .$$

We note that we know the sum of the losses of the two pipes, but we do not know the losses of each one of them individually. As we do not know the flow rate as well, it is impossible to calculate each pipe directly. There are two ways to deal with the problem:

First method (analytical)

We assume a random, but logical, flow rate Q', calculate the losses of each pipe, $\Sigma h_1'$ and $\Sigma h_2'$ (like in example 1), and continue with the methodology of testing for the value of the flow rate. Therefore:

First test: assume that $Q' = 100m^3 / h = 100/3600m^3 / s$

Calculation for pipe (1):

$$V_1 = \frac{4Q}{\pi d_1^2} \Rightarrow V_1 = 2,46m/s ,$$

$$\left(\mathrm{Re} = \frac{V_1 d_1}{\nu} \Rightarrow \mathrm{Re}_1 = 3 \cdot 10^5, \varepsilon / d_2 = 0,00217 \right) \rightarrow f_1 = 0,0245 \text{ (Moody)},$$

$$\Sigma h_1 = \left(f_1 \frac{\ell_1}{d_1} + \Sigma K_1 \right) \cdot \frac{V_1^2}{2g} \Rightarrow \Sigma h_1 = 7,2m .$$

Calculation for pipe (2):

$$V_2 = \frac{4Q_2}{\pi d_1^2} \Rightarrow V_2 = 3,54m/s$$

$$\left(\text{Re}_2 = \frac{V_2 d_2}{\nu} \Rightarrow \text{Re}_2 = 3{,}5 \cdot 10^5, \varepsilon / d_2 = 0{,}0026\right) \rightarrow f_2 = 0{,}0256 \text{ (Moody)}$$

$$\Sigma h_2 = \left(f_2 \frac{\ell_2}{d_2} + \Sigma K_2 \right) \cdot \frac{V_2^2}{2g} \Rightarrow \Sigma h_2 = 21{,}48m .$$

Thus, if $Q' = 100m^3 / h$, then the losses will be:

$$\Sigma h' = 7{,}2 + 21{,}48 = 28{,}67m$$

$$\Sigma h = 15m \text{ (fact)} \Rightarrow \Sigma h' > \Sigma h \Rightarrow Q' > Q \Rightarrow Q < 100m^3 / h .$$

Second test: assume that $Q' = 50m^3 / h = 50 / 3600m^3 s$

We repeat the calculation and we find:

$$\Sigma h' = 1{,}83 + 5{,}44 = 7{,}27m \Rightarrow \Sigma h' < \Sigma h \Rightarrow Q' < Q \Rightarrow Q > 50m^3 / h .$$

Third test: assume that $Q' = 75m^3 / h = 75 / 3600m^3 / s$

We calculate:

$$\Sigma h' = 4{,}07 + 12{,}13 = 16{,}21m \Rightarrow \Sigma h' > \Sigma h \Rightarrow Q' > Q \Rightarrow Q < 75m^3 / h .$$

Fourth test: assume that $Q' = 70m^3 / h = 70 / 3600m^3 / s$

$$\Sigma h' = 3{,}55 + 10{,}58 = 14{,}14m \Rightarrow \Sigma h' < \Sigma h \Rightarrow Q' < Q \Rightarrow Q > 70m^3 / h .$$

Fifth test: assume that $Q' = 72m^3 / h = 72 / 3600m^3 / s$

$$\Sigma h' = 3{,}76 + 11{,}19 = 14{,}95m \Rightarrow \Sigma h' \cong \Sigma h \Rightarrow Q' \approx Q \Rightarrow Q = 72m^3 / h .$$

So: $Q = 72m^3 / h$

If we are not satisfied with the approach, we continue.

The relevant table is reported and listed with the results of the successive tests:

Q' v_1	f_1	Σh_1	v_2	f_2	Σh_2	$\Sigma h'$	Q	
m^3/h	m/s		m	m/s		m	m	m^3/h
100	2,46	0,0245	7,20	3,54	0,0256	21,48	28,67>15	Q<100
50	1,23	0,0250	1,83	1,77	0,0259	5,44	7,27<15	Q>50
75	1,84	0,0247	4,07	2,65	0,0257	12,13	16,21>15	Q<75
70	1,72	0,0247	3,55	2,48	0,0257	10,58	14,14<15	Q>70
72	**1,77**	**0,0247**	**3,76**	**2,55**	**0,0257**	**11,19**	**14,95≅ 15**	**Q≅ 72**

Table 5.9. *Iteration techniques results*

Second method: assumption of intense turbulence (AIT)

The direct solution of the problem is prevented from the fact that we cannot calculate f_1 and f_2, while for each pipe, we know the relative roughness, but we do not know the Reynolds number, so here we can also use the assumption of the intense turbulence (AIT) as in the case of the simple pipe and take from the Moody diagram values for f_1 and f_2.

From equation [5.129], we have:

$$\Sigma h = \frac{8}{\pi^2 y}\left[\left(f_1 \frac{\ell_1}{d_1} + \Sigma K_1\right)\cdot\frac{1}{d_1^4} + \left(f_2 \frac{\ell_2}{d_2} + \Sigma K_2\right)\cdot\frac{1}{d_2^4}\right]\cdot Q^2 \qquad (1)$$

Here, the only unknown is Q, which we calculate.

Afterward, we make use of the AIT by calculating successively the V, the Re and the values of f_1 and f_2 from the Moody diagram. If at least one of f_1 and f_2 is very different from the one we assumed, then we repeat its solution [5.118] with the new value. To make the sums easier, we simplify the losses equation of the pipe with replacement of the values in the system SI. Therefore:

$$\Sigma h_i = a_1 \cdot Q_i^2 \text{ (SI)} \qquad (2)$$

$$\text{where } a_i = \frac{8}{\pi^2 g}\left(f_i \frac{\ell_i}{d_i} + \Sigma K_i\right)\frac{1}{d_i^4} \text{ (SI)}, \qquad (3)$$

with units of a_i, $m^{-5} \cdot s^2$.

Thus, [5.131] becomes:

$$\Sigma h = (a_1 + a_2) \cdot Q^2, \tag{4}$$

from which occurs:

$$Q = \left(\frac{\Sigma h}{a_1 + a_2} \right)^{1/2}. \tag{5}$$

Therefore, by assuming the AIT, we successively produced the following results:

Pipe (1):

$$(\varepsilon/d_1 = 0{,}0021, \text{AIT}) \rightarrow f_1 = 0{,}024 \text{ (Moody)}.$$

$$\text{From (3)} \Rightarrow a_1 = 9141\ (SI) \Rightarrow \Sigma h_1 = 9141 \cdot Q^2\ (SI)$$

Pipe (2):

$$(\varepsilon/d_2 = 0{,}0026, \text{AIT}) \rightarrow f_2 = 0{,}025 \text{ (Moody)}.$$

$$\text{From (3)} \Rightarrow a_2 = 27410\ (SI) \Rightarrow \Sigma h_2 = 27410 \cdot Q^2\ (SI).$$

$$\text{From (4)} \Rightarrow 15 = 36551 \cdot Q^2\ (SI) \Rightarrow Q = 0{,}02026 m^3/s = 72{,}9 m^3/h.$$

Assumption control:

$$V_1 = \frac{4 \cdot Q}{\pi \cdot d_1^2} \Rightarrow V_1 = 1{,}8 m/s \Rightarrow \text{Re}_1 = 2{,}15 \cdot 10^5 \rightarrow f_1 = 0{,}0247 \text{ (Moody)}$$

$$V_2 = \frac{4 \cdot Q}{\pi \cdot d_2^2} \Rightarrow V_2 = 1{,}8 m/s \Rightarrow \text{Re}_2 = 2{,}15 \cdot 10^5 \rightarrow f_2 = 0{,}0257 \text{ (Moody)}$$

As we see the differences of the f values are approximately of class 3%, the approximation can be considered to be satisfactory. If we want higher accuracy, we give new values for f so we have:

$$a_1 = 9387\ (SI) \qquad a_2 = 2.7970\ (SI)$$

and from (4), we have:

$$Q = 0,02004m^3 / s = 72,1m^3 / h$$

Third problem: diameter calculation

We want to connect two water tanks that are separated by a distance of 40 m and have a height difference of 5 m. The pipe which we have in order to achieve this connection is made of asphalted cast iron of length 25 m and diameter 12 cm. If we want to complete this connection so that the flow rate is at least 50 m^3/h, calculate in (mm) the minimum diameter that the pipe of the same material must have, if the secondary losses are considered to be negligible.

Solution

First, as we have energy data, the altitude range $y_1 - y_2$, we apply the Bernoulli equation:

$$y_1 - y_2 + \frac{V_1^2 - V_2^2}{2 \cdot g} + \frac{p_1 - p_2}{\gamma} = \Sigma h \Rightarrow H = y_1 - y_2 = \Sigma h \Rightarrow \Sigma h = 5m .$$

The losses are connected with the flow rate, with equation [5.131]:

$$\Sigma h = \frac{8}{\pi^2 \cdot g} \cdot \left[\left(f_1 \cdot \frac{\ell_1}{d_1} + \Sigma K_1 \right) \cdot \frac{1}{d_1^4} + \left(f_2 \cdot \frac{\ell_2}{d_2} + \Sigma K_2 \right) \cdot \frac{1}{d_2^4} \right] \cdot Q^2 .$$

In this equation, unknowns are diameter d_2, which needs to be calculated, and the friction coefficients f_1 and f_2. Regarding f_2, no certain information is provided by the assumption of the totally developed turbulence, as we do not know the relative roughness ε_2 / d_2. Therefore, the only way is to assume a value d_2 for the diameter, to calculate Q' that occurs and to compare it with the given minimum $Q_{\min}$. Therefore:

a) If $Q' \leq Q_{\min} \Rightarrow d_2 > d_2'$, d_2 is rejected.

b) If $Q' > Q_{\min} \Rightarrow d_2 \leq d_2'$, d_2 is not rejected, but there might be a smaller diameter that satisfies the system.

The minimum diameter is the one for which we will have:

$Q' \cong Q_{\min}$.

For the calculation, we will apply the assumption of the intense turbulence (AIT). Therefore, we have:

First test: assume that $d_2' = 100mm = 0,100m$

By supposing an intensely turbulent flow, we calculate the friction coefficients and flow rate:

Pipe (1):

$(\varepsilon/d_1 = 0,001, \text{AIT}) \rightarrow f_1 = 0,0196$ (Moody) and from:

$(3) \Rightarrow a_1 = 1630\,(SI) \Rightarrow \Sigma h_1 = 1630 \cdot Q^2\,(SI)$.

We note that the successive tests that will follow do not influence Pipe (1), so a revision of the above calculation is not needed.

Pipe (2):

$(\varepsilon/d_2 = 0,0012, \text{AIT}) \rightarrow f_2 = 0,0205$ (Moody) and from:

$(3) \Rightarrow a_2 = 2545\,(SI) \Rightarrow \Sigma h_2 = 2545 \cdot Q^2\,(SI)$.

Flow rate calculation: $Q = \left(\dfrac{\Sigma h}{a_1 + a_2} \right)^{1/2}$ (5)

(5) $Q' = 0,0346 m^3/s = 124,5 m^3/h$

$Q' > Q_{\min} \Rightarrow d_2 \le d_2' = 100mm$.

Second test: assume that $d_2' = 50mm = 0,050m$

Pipe (2):

$(\varepsilon/d_2 = 0,0024, \text{AIT}) \rightarrow f_2 = 0,0245$ (Moody) and from:

$(3) \Rightarrow a_2 = 97560\,(SI) \Rightarrow \Sigma h_2 = 97560 \cdot Q^2\,(SI)$.

Flow rate calculation (5) $Q' = 0,0071 m^3/s = 25,5 m^3/h$, so

$$Q' < Q_{\min} \Rightarrow d_2 > d_2' = 50mm$$

We present the following testing table:

d_2' mm	ε/d_2	f_2	a_2 m^3/s	$a_1 + a_2$ m^3/h	Q'	Q'	Comparison	d_2
100	0.0012	0.0205	2545	4175	0.0346	124.6	$Q' > Q_{min}$	$d_2 \leq 100$
50	0.0024	0.0246	97559	99189	0.0071	25.6	$Q' > Q_{min}$	$d_2 > 50$
70	0.0017	0.0225	16584	18214	0.0166	59.6	$Q' > Q_{min}$	$d_2 \leq 70$
60	0.0020	0.0234	37330	38960	0.0113	40.8	$Q' > Q_{min}$	$d_2 > 60$
65	0.0018	0.0229	24493	26123	0.0138	49.8	$Q' > Q_{min}$	$d_2 > 65$
66	**0.0018**	**0.0228**	**22602**	**24232**	**0.0144**	**51.7**	**$Q' > Q_{min}$**	**$d_2 \leq 66$**

Table 5.10. *Results*

From this testing, we conclude that the minimum diameter lies between 65 mm and 66 mm. This means that:

$$65mm < d_2 \leq 66mm.$$

However, before we conclude this, we must check the assumption of the developed turbulence only for the minimum accepted diameter of the table and if needed correct it. Therefore:

AIT control

$$V_1 = \frac{4Q}{\pi \cdot d_1^2} \Rightarrow V_1 = 1,25m/\sec \Rightarrow$$

$$\Rightarrow \text{Re}_1 = 1,5 \cdot 10^5 \rightarrow f_1 = 0,0213 \text{ (Moody)}$$

and

$$V_2 = \frac{4Q}{\pi \cdot d_2^2} \Rightarrow V_2 = 4,12m/\sec \Rightarrow \text{Re}_2 = 2,7 \cdot 10^5 \rightarrow f_2 = 0,0235 \text{ (Moody)}$$

Correction: $a_1 = 1767\,(SI)$

$$a_2 = 23263(SI)$$

$$(5)\ Q' = 0{,}0141 m^3/s = 50{,}9 m^3/h \quad \left(> Q_{\min} = 50 m^3/h\right)$$

So: $d_2 = 66mm$ and:

$$Q = 50{,}9 m^3/h.$$

5.11.3. *Pipes in parallel*

5.11.3.1. *Calculation of head losses*

We have three pipes made of cast iron which are connected in a parallel way as shown in Figure 5.30 and which are of diameter 3 cm, 4 cm and 5 cm. The lengths of these pipes are 60 m, 40 m and 60 m respectively. Calculate the head losses from point A to point B and the flow rate in each pipe, if the total flow rate of water is $Q = 0{,}015 m^3/\sec$. Finally, it is given that $\Sigma K_1 = 3, \Sigma K_2 = 1, \Sigma K_3 = 3$.

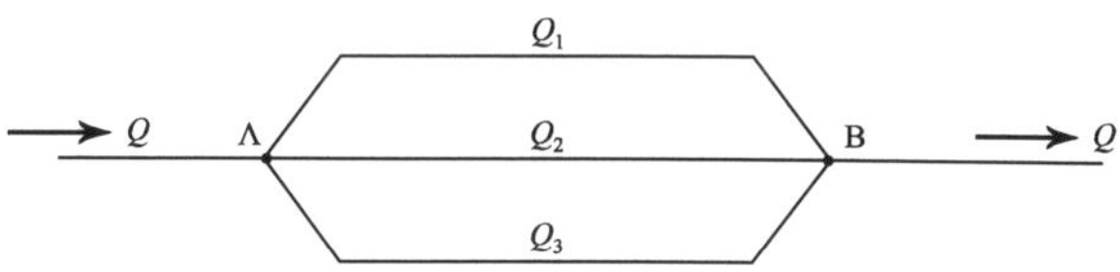

Figure 5.30. *Pipes in parallel connection*

Solution

We have a parallel connection, so equations [5.140] are [5.141] applicable:

$$[5.140] \Rightarrow Q = Q_1 + Q_2 + Q_3 \tag{1}$$

$$[5.141] \Rightarrow \Sigma h = \Sigma h_1 = \Sigma h_2 = \Sigma h_3 \tag{2}$$

We note that we have not obtained sufficient data to calculate some parameters of the three pipes, as for each one of them the flow rate and the head losses are unknown. Thus, there is not any direct solution. In this case, there are two ways of solution with assumptions:

First method (analytical):

We assume head losses $\Sigma h'$. This, according to equation (2), is common for all three pipes. For each pipe, we calculate the supposed Q_i'. We use relation (1):

If $Q' > Q \Rightarrow \Sigma h' > \Sigma h$, we repeat with a smaller $\Sigma h'$.

If $Q' < Q \Rightarrow \Sigma h' < \Sigma h$, we repeat with a bigger $\Sigma h'$

If $Q' \approx Q \Rightarrow \Sigma h' \approx \Sigma h$, the problem ends.

The examples given above show that this method is accurate but time-consuming and therefore we will proceed to the solution with the second method, assumption of developed turbulence, which we recommend.

Second method (AIT):

We know the relative roughness of the pipes connected in parallel; therefore, assuming a totally developed turbulence, we first calculate the friction coefficients from the Moody diagram, and then calculate a_i.

Pipe (1):

$$\left(\varepsilon/d_1 = 0{,}0087, \text{AIT}\right) \rightarrow f_1 = 0{,}036 \text{ (Moody)}$$

$$a_1 = 7677922 (SI) \Rightarrow \Sigma h_1 = \Sigma h = 7677922 \cdot Q_1^2 \, (SI). \tag{3}$$

Pipe (2):

$$\left(\varepsilon/d_2 = 0{,}0065, \text{AIT}\right) \rightarrow f_2 = 0{,}033 \text{ (Moody)}$$

$$a_2 = 1097388 (SI) \Rightarrow \Sigma h_2 = \Sigma h = 1097388 \cdot Q_2^2 \, (SI). \tag{4}$$

Pipe (3):

$$\left(\varepsilon/d_3 = 0{,}0052, \text{AIT}\right) \rightarrow f_3 = 0{,}0307 \text{ (Moody)}$$

$$a_3 = 527193 (SI) \Rightarrow \Sigma h_3 = \Sigma h = 527193 \cdot Q_3^2 \, (SI). \tag{5}$$

From relationships (3), (4) and (5), we calculate Q_1, Q_2, Q_3 to Σh as required. Therefore, we have:

$$Q_1 = \sqrt{\Sigma h} \cdot \sqrt{1/a_1} \Rightarrow Q_1 = 0{,}000361 \cdot \Sigma h \; (SI) \tag{6}$$

$$Q_2 = \sqrt{\Sigma h} \cdot \sqrt{1/a_2} \Rightarrow Q_2 = 0{,}000955 \cdot \Sigma h \; (SI) \tag{7}$$

$$Q_3 = \sqrt{\Sigma h} \cdot \sqrt{1/a_3} \Rightarrow Q_3 = 0{,}001377 \cdot \Sigma h \; (SI) \tag{8}$$

Replacing Q_i in the relation of the flow rates (1), an equation occurs with only one unknown, the head losses Σh :

$$\begin{aligned}
&Q = Q_1 + Q_2 + Q_3 \Rightarrow \\
&\Rightarrow Q = 0{,}000361 \cdot \sqrt{\Sigma h} + 0{,}000955 \cdot \sqrt{\Sigma h} + 0{,}001377 \cdot \sqrt{\Sigma h} \Rightarrow (SI) \\
&\Rightarrow Q = 0{,}002693 \cdot \sqrt{\Sigma h} \Rightarrow (SI) \\
&\Rightarrow \Sigma h = 31m \ .
\end{aligned}$$

From relationships (6), (7) and (8), we calculate the flow rates in each pipe:

$$(6) \Rightarrow Q_1 = 0{,}00201 m^3 / s$$

$$(7) \Rightarrow Q_2 = 0{,}00532 m^3 / s$$

$$(8) \Rightarrow Q_3 = 0{,}00767 m^3 / s \, .$$

AIT control: before we accept the results, we must check the assumptions of the developed turbulence. If at least one of the f 's has a value different from the one we used, we repeat the calculation with new values:

Pipe (1): $V_1 = 2{,}84 m/s \Rightarrow \mathrm{Re}_1 = 8{,}5 \cdot 10^4 \rightarrow f_1 = 0{,}0369$ (Moody)

Pipe (2): $V_2 = 4{,}23 m/s \Rightarrow \mathrm{Re}_2 = 1{,}7 \cdot 10^5 \rightarrow f_2 = 0{,}0335$ (Moody)

Pipe (3): $V_3 = 2{,}84 m/s \Rightarrow \mathrm{Re}_3 = 8{,}5 \cdot 10^4 \rightarrow f_3 = 0{,}0312$ (Moody).

We note that the differences are smaller than 2%, so they are considered to be satisfactory and correction is not needed. Therefore:

$$\Sigma h = 31m$$

$$Q_1 = 0,00201 m^3 / s$$

$$Q_2 = 0,00532 m^3 / s$$

$$Q_3 = 0,00767 m^3 / s$$

5.11.3.2. *Calculation of the flow rate*

The pipes in Figure 5.31 are connected in a parallel way; pipe (1) has a length of 100 m and diameter of 8 cm; pipe (2) has a length of 150 m and diameter of 6 cm and pipe (3) has a length of 80 m and diameter of $4cm$. The roughness values of the three pipes are 0.24, 0.12 and 0.20 mm, respectively. At point A, the pressure is $150KPa$ higher than the pressure at point B, and the A lies 5 m higher than B. If minor losses are considered to be negligible, calculate the flow rates of the water at the above pipes, in m^3/h.

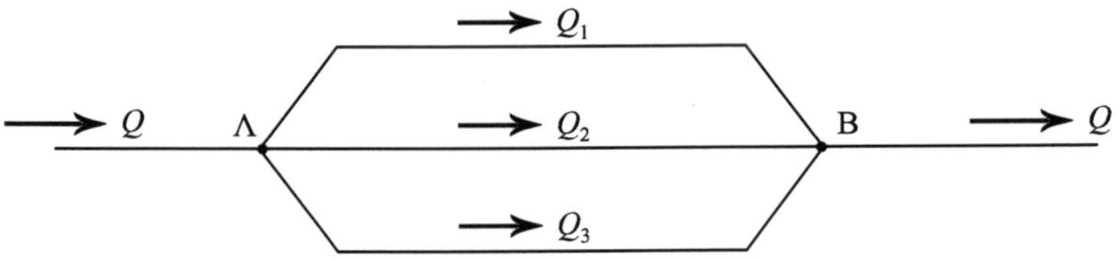

Figure 5.31. *Flow rate calculation in the pipe network*

Solution

The connection is parallel, so equations [5.140] and [5.141] will be applicable:

$$[5.140] \Rightarrow Q = Q_1 + Q_2 + Q_3 \qquad (1)$$

$$[5.141] \Rightarrow \Sigma h = \Sigma h_1 = \Sigma h_2 = \Sigma h_3 \qquad (2)$$

Because we have an altitude range between points A and B, we apply the Bernoulli equation and we calculate the head losses from A to B:

$$\text{A} - \text{B} : y_A - y_\text{B} + \frac{V_A^2 - V_B^2}{2g} + \frac{p_\text{A} - p_\text{B}}{\gamma} = \Sigma h \Rightarrow$$

$$\Rightarrow y_\text{A} - y_\text{B} + \frac{p_\text{A} - p_\text{B}}{\gamma} = \Sigma h \Rightarrow \Sigma h = 20,3m\ .$$

As the head losses are common (relation (2)), we have enough data to calculate the flow rate of each pipe.

Pipe (1):

$$\left(\varepsilon/d_1 = 0,003, \text{AIT}\right) \rightarrow f_1 = 0,0265 \text{ (Moody)}$$

$$a_1 = 65977\,(SI) \Rightarrow \Sigma h_1 = \Sigma h = 65977 \cdot Q_1^2\,(SI) \Rightarrow$$

$$\Rightarrow Q_1 = 0,0174 m^3/s = 62,8 m^3/h.$$

AIT control:

$$V_1 = 3,5 m/s \Rightarrow \text{Re}_1 = 2,8 \cdot 10^5 \rightarrow f_1 = 0,0267 \approx 0,0265 \text{ (Moody).}$$

Pipe (2):

$$\left(\varepsilon/d_2 = 0,002, \text{AIT}\right) \rightarrow f_2 = 0,0237 \text{ (Moody)}$$

$$a_2 = 378077\,(SI) \Rightarrow \Sigma h_2 = \Sigma h = 378077 \cdot Q_2^2\,(SI) \Rightarrow$$

$$\Rightarrow Q_2 = 0,00733 m^3/s = 26,4 m^3/h.$$

AIT control:

$$V_2 = 2,59 m/s \Rightarrow \text{Re}_2 = 1,6 \cdot 10^5 \rightarrow f_2 = 0,0245 \text{ (Moody).}$$

Correction: $Q_2 = 0,00721 m^3/s = 26,0 m^3/h$

Pipe (3):

$$\left(\varepsilon/d_3 = 0,005, \text{AIT}\right) \rightarrow f_3 = 0,0307 \text{ (Moody)}$$

$$a_3 = 1966744\,(SI) \Rightarrow \Sigma h_3 = \Sigma h = 1966744 \cdot Q_3^2\,(SI) \Rightarrow$$

$$\Rightarrow Q_3 = 0{,}00321 m^3/s = 11{,}6 m^3/h.$$

AIT control:

$$V_3 = 2{,}56 m/s \Rightarrow \text{Re}_3 = 1 \cdot 10^5 \rightarrow f_3 = 0{,}0313 \text{ (Moody)}.$$

Correction: $Q_3 = 0{,}00317 m^3/s = 11{,}4 m^3/h$

So: $Q_1 = 62{,}8 m^3/h$, $Q_2 = 26{,}0 m^3/h$ and $Q_3 = 11{,}4 m^3/h$.

5.11.3.3. *Diameter calculation*

Between points A and B, which have a distance of 8 m between them, of a straight pipe made of commercial cast iron with diameter 10 cm through which water flows, we want to connect in a parallel way another pipe of the same material with length 10 m, through which water flows at a rate of 15 m^3/h, as shown in Figure (5.32). For the flow in the main pipe not to be higher than 60 m^3/h, calculate the diameter of the pipe that we want to connect, if $\Sigma K_1 = 0, \Sigma K_2 = 5$.

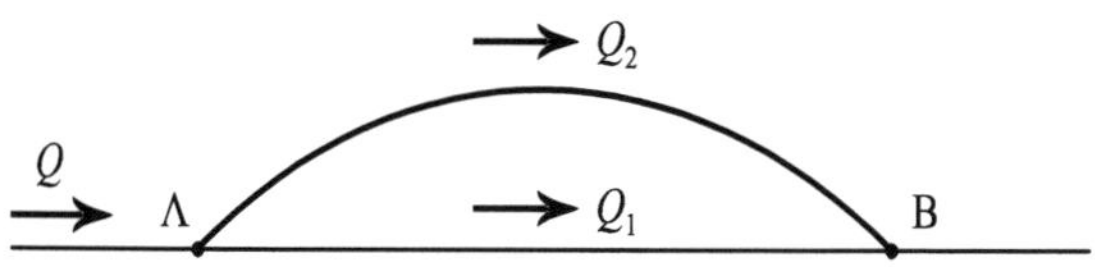

Figure 5.32. *Pipe network for diameter calculation*

Solution

When in a pipe, the maximum flow rate is given, indirectly the maximum head of the allowed losses is given, as the losses of the pipe are proportional to the square of the flow rate:

$$\Sigma h_{1,\max} = a_1 \cdot Q_{1,\max}^2 \tag{1}$$

Thus, we can calculate the highest losses that are allowed in pipe (1) from A to B:

Calculations of losses in pipe (1) for $Q_1 = Q_{1,\max} = 60/3600 m^3/s$:

$$V_1 = \frac{4 \cdot Q}{\pi \cdot d_1^2} \Rightarrow V_1 = 2{,}12 m/s \ \ \mathrm{Re}_1 = \frac{V_1 \cdot d_1}{\nu} \Rightarrow \mathrm{Re}_1 = 2{,}1 \cdot 10^5$$

$$\left(\mathrm{Re}_1 = 2{,}1 \cdot 10^5, \varepsilon/d_1 = 0{,}00046\right) \rightarrow f_1 = 0{,}018 \text{ (Moody)}$$

$$\Sigma h_1 = \left(f_1 \cdot \frac{\ell_1}{d_1} + \Sigma K_1\right) \cdot \frac{V_1^2}{2g} \Rightarrow \Sigma h_1 = \Sigma h_2 = \Sigma h = 0{,}337 m$$

Thus, the highest losses that are allowed are: $\Sigma h_{\max} = 0{,}337 m$

The problem is transformed into a problem of calculation of the pipe's diameter, when we know the maximum head of the allowed losses. Therefore, we will solve it with tests.

First test: assume that $d_2' = 100 mm = 0{,}1 m$

$$V_2 = \frac{4 \cdot Q_2}{\pi \cdot d_2^2} \Rightarrow V_2 = 0{,}53 m/s \ \ \mathrm{Re}_2' = \frac{V_2 \cdot d_2}{\nu} \Rightarrow \mathrm{Re}_2' = 5{,}3 \cdot 10^4 .$$

$$\left(\mathrm{Re}_2' = 5{,}3 \cdot 10^4, \varepsilon/d_2' = 0{,}00046\right) \rightarrow f_2 = 0{,}022 \text{ (Moody)}$$

and because $\Sigma h_2 = \left(f_2 \cdot \frac{\ell_2}{d_2} + \Sigma K_2\right) \cdot \frac{V_2^2}{2g} \Rightarrow \Sigma h_2' = 0{,}103 m$

Thus: $\Sigma h' < \Sigma h_{\max} \Rightarrow d_2' \geq d_2 \Rightarrow d_2 \leq 100 mm$.

Second test: we choose a diameter smaller than $100 mm$ and repeat with successive tests, listing the relative table that follows:

Assume that $d_2' = 50 mm = 0{,}05 m \Rightarrow \Sigma h_2' = 2{,}138 m > \Sigma h_{\max} \Rightarrow d_2 > 50 mm$.

d_2 mm	***e/d_2***	***V_2 m/s***	***Re_2'***	***f_2'***	***$\Sigma h_2'$ m***	***Comparison with Σh_{max}***	***d_2 mm***
100	0.00046	0.53	$5.3 \cdot 10^4$	0.0219	0.103	< 0.337	$d_2 \leq 100$
50	0.00092	2.12	$1.1 \cdot 10^5$	0.0216	2.138	> 0.337	$d_2 > 50$
70	0.000657	1.08	$7.6 \cdot 10^5$	0.0214	0.481	> 0.337	$d_2 > 70$

80	0.000575	0.83	$6.6 \cdot 10^5$	0.0215	0.269	<0.337	$d_2 \leq 80$
75	0.000613	0.94	$7.1 \cdot 10^5$	0.0215	0.356	>0.337	$d_2 > 75$
77	0.000597	0.89	$6.9 \cdot 10^5$	0.0215	0.318	<0.337	$d_2 \leq 77$
76	**0.000605**	**0.92**	$\mathbf{7.0 \cdot 10^5}$	**0.0215**	**0.336**	**<0.337**	$\mathbf{d_2 \leq 76}$

Table 5.11. *Results*

Thus, the minimum diameter that satisfies the requirements of the problem is $d_2 = 76mm$.

5.11.4. *Mixed pipe network*

Problem: the mixed assembly shown in Figure 5.33 consists of six pipes made of asphalted cast iron, and their dimensions are as follows: pipe (1): length $\ell_1 = 30m$ and diameter $d_1 = 12cm$; pipe (2): length $\ell_2 = 15m$ and diameter $d_2 = 6cm$; pipe (3): length $\ell_3 = 8m$ and diameter $d_3 = 6cm$; pipe (4): length $\ell_4 = 4m$ and diameter $d_4 = 3cm$; pipe (5): length $\ell_5 = 7m$ and diameter d_5 = 3 cm; and pipe (6): length $\ell_6 = 20m$ and diameter $d_6 = 10cm$. If all these pipes lie on the same level and the flow rate of pipe (1) is $Q_1 = 60m^3/h$, calculate the flow rates in all the other pipes and the pressure drop from point A to point B with the condition that minor losses are negligible.

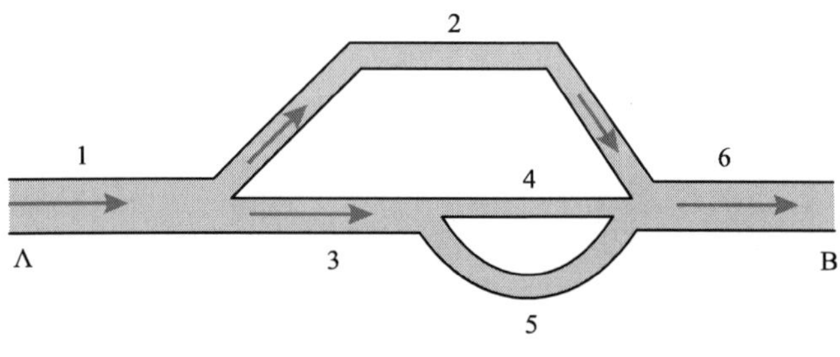

Figure 5.33. *Mixed pipe network*

Solution

As shown in Figure (6), the flow rates are expressed by the relationships:

$$Q_1 = Q_2 + Q_3 = Q_6 \tag{1}$$

and $Q_3 = Q_4 + Q_5$. (2)

While the losses:

$$h_{AB} = h_1 + h_2 + h_6 \tag{3}$$

$$h_2 = h_3 + h_4 \tag{4}$$

and $h_4 = h_5$. (5)

Because for pipes (1) and (6) we have that $Q_1 = Q_6$, this is enough to calculate the losses h_1 and h_6. Therefore, we have:

Calculation for pipe (1):

$$V_1 = \frac{4 \cdot Q_1}{\pi \cdot d_1^2} \Rightarrow V_1 = 1,47 m/s$$

$$\left(\mathrm{Re}_1 = \frac{V_1 \cdot d_1}{v} \Rightarrow \mathrm{Re}_1 = 1,8 \cdot 10^5 \ , \ \varepsilon/d_1 = 0,001 \right) \rightarrow f_1 = 0,021 \text{ (Moody)}$$

$$h_1 = f_1 \cdot \frac{\ell_1}{d_1} \cdot \frac{V_1^2}{2g} \Rightarrow h_1 = 0,58 m \ .$$

Calculation for pipe (2):

$$V_6 = \frac{4 \cdot Q_6}{\pi \cdot d_6^2} \Rightarrow V_6 = 2,12 m/s$$

$$\left(\mathrm{Re}_6 = \frac{V_6 \cdot d_6}{v} \Rightarrow \mathrm{Re}_6 = 2,1 \cdot 10^5 \ , \varepsilon/d_6 = 0,0012 \right) \rightarrow f_6 = 0,0216 \text{ (Moody)}$$

$$h_6 = f_6 \cdot \frac{\ell_6}{d_6} \cdot \frac{V_6^2}{2g} \Rightarrow h_6 = 0,99 m \ .$$

For the pipes with unknown flow rates, e.g. (2), (3), (4) and (5), we assume totally developed turbulent flow, and from the equation of losses, we form the relationships of losses and flow rates (all the relationships in the SI):

Pipe (2):

$$\left(\varepsilon_2/d_2 = 0,002\,\text{AIT}\right) \rightarrow f_2 = 0,0237 \text{ (Moody)}$$

$$[5.131] \Rightarrow a_2 = 37808 \Rightarrow h_2 = 37808 \cdot Q_2^2 \qquad (6)$$

Pipe (3):

$$\left(\varepsilon_3/d_3 = 0,002\,\text{AIT}\right) \rightarrow f_3 = 0,0237 \text{ (Moody)}$$

$$[5.131] \Rightarrow a_3 = 20164 \Rightarrow h_3 = 20164 \cdot Q_3^2 \qquad (7)$$

Pipe (4):

$$\left(\varepsilon_4/d_4 = 0,004\,\text{AIT}\right) \rightarrow f_4 = 0,0287 \text{ (Moody)}$$

$$[5.131] \Rightarrow a_4 = 390564 \Rightarrow h_4 = 390564 \cdot Q_4^2 \qquad (8)$$

Pipe (5):

$$\left(\varepsilon_5/d_5 = 0,004\,\text{AIT}\right) \rightarrow f_5 = 0,0287 \text{ (Moody)}$$

$$[5.131] \Rightarrow a_5 = 683487 \Rightarrow h_5 = 683487 \cdot Q_5^2 \,. \qquad (9)$$

We try to replace as many unknown flow rates as possible. We start from the parallel connection of pipes (4) and (5) and proceed to the total pipeline:

$$h_4 = h_5 \Rightarrow 390564 \cdot Q_4^2 = 390564 Q_5^2 \Rightarrow$$

$$\Rightarrow Q_4 = 1323 \cdot Q_5 \qquad (10)$$

$$Q_3 = Q_4 + Q_5 \Rightarrow Q_3 = 1.323 \cdot Q_5 + Q_5 \Rightarrow Q_3 = 2.323 \cdot Q_5$$

$$h_2 = h_3 + h_5 \Rightarrow 37808 \cdot Q_2^2 = 20164 \cdot Q_3^2 + 683487 Q_3^2 \Rightarrow \tag{11}$$

$$37808 \cdot Q_2^2 = 20164 \cdot (2{,}323 \cdot Q_5)^2 + 683487 Q_5^2 \Rightarrow$$

$$37808 Q_2^2 = 792287 \cdot Q_5^2 \Rightarrow Q_2 = 4{,}578 Q_5 \tag{12}$$

$$Q_1 = Q_2 + Q_3 \Rightarrow Q_1 = 4{,}578 \cdot Q_5 + 2{,}323 \cdot Q_5 \Rightarrow$$

$$Q_1 = 6{,}9 Q_5 . \tag{13}$$

However, Q_1 is known, thus:

$$(10) \Rightarrow Q_5 = 0{,}00242 m^3 / s = 8{,}70 m^3 / h$$

$$(7) \Rightarrow Q_4 = 0{,}00319 m^3 / s = 11{,}50 m^3 / h$$

$$(8) \Rightarrow Q_3 = 0{,}00561 m^3 / s = 20{,}20 m^3 / h$$

$$(9) \Rightarrow Q_2 = 0{,}01106 m^3 / s = 39{,}80 m^3 / h$$

At this point, we check the assumptions of the intense turbulence for the four pipes. There is a remarkable inclination only at pipe (3) $(f_2 = 0{,}0248)$. The repetition of the procedure brings small changes to the calculation of the flow rates:

$$Q_5 - 0{,}00241 m^3 / s = 8{,}66 m^3 / h$$

$$Q_4 = 0{,}00319 m^3 / s = 11{,}50 m^3 / h$$

$$Q_3 = 0{,}00560 m^3 / s = 20{,}16 m^3 / h$$

$$Q_2 = 0{,}01107 m^3 / s = 39{,}84 m^3 / h .$$

– Calculation of total losses:

$$h_{\text{AB}} = h_1 + h_2 + h_6$$

$$h_2 = \frac{8}{\pi^2 g} f_2 \cdot \frac{\ell_2}{d_2^5} \cdot Q_2^2 \Rightarrow h_2 = 4,71m$$

$$h_{AB} = 6,29m \,.$$

We apply the Bernoulli equation between points A and B and calculate the pressure drop:

$$y_A - y_B + \frac{V_A^2 - V_B^2}{2g} + \frac{p_A - p_B}{\gamma} = h_{AB} \Rightarrow \frac{p_A - p_B}{\gamma} = h_{AB} - \frac{V_A^2 - V_B^2}{2g} \Rightarrow$$

$$\Rightarrow p_A - p_B = 61,7KPa \,.$$

5.11.5. *Problem of the three tanks*

A′ problem: at the A′ problem of the three tanks, the values of the three pipes, the altitudes of the two tanks or their altitude range and the flow rate of one tank with known head are given and the height of the third tank and the other two flow rates are asked.

The pipes in Figure 5.34 are made of cast iron with the following parameters:

Diameter	**Length**
$d_1 = 30cm$	$\ell_1 = 500m$
$d_2 = 15cm$	$\ell_2 = 250m$
$d_3 = 25cm$	$\ell_3 = 200m$

The heights of the free surfaces of tanks A and B are 60 m and 45 m, respectively. In tank B, water flows in at $100m^3/s$. Calculate the height range between A and C and the unknown flow rates. The secondary losses are negligible.

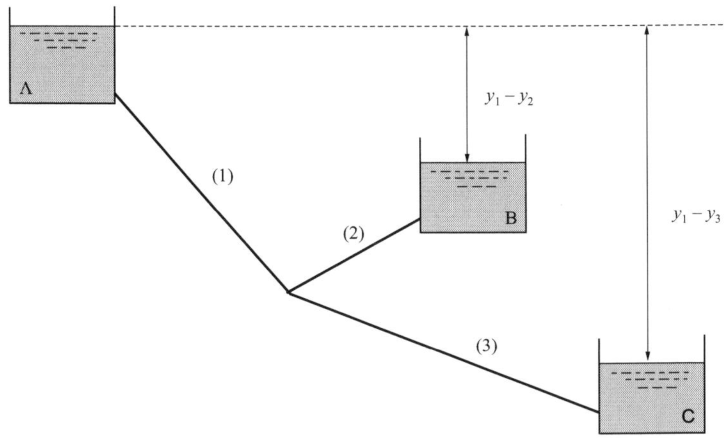

Figure 5.34. *Three-tank pipe network*

Solution

In a pipe, the flow rate is connected with the losses, with the equation of the losses: $h_i = a_1 \cdot Q_i^2$. Therefore, when we know the values of a pipe and the flow rate, we can calculate the head losses. Therefore, we have enough data to calculate the losses at pipe (2).

Step 1: we calculate the losses of pipe (2)

$$V_2 = \frac{4 \cdot Q_2}{\pi \cdot d_2^2} \Rightarrow V_2 = 1,57m/s$$

$$\left(\mathrm{Re}_2 = \frac{V_2 \cdot d_2}{\nu} \Rightarrow \mathrm{Re}_2 = 2,36 \cdot 10^5, \varepsilon_2/d_2 = 0,0017 \right) \rightarrow f_2 = 0,0233 \text{ (Moody)}$$

$$h_2 = f_2 \cdot \frac{\ell_2}{d_2} \cdot \frac{V_2^2}{2g} \Rightarrow h_2 = 4,9m$$

Step 2: we calculate the losses and the flow rate of pipe (1)

From equation [5.158], we can calculate the losses from pipe (1) because we know its altitude. Therefore, we have:

$$y_1 - h_1 = y_2 + h_2 \Rightarrow h_1 = y_1 - y_2 - h_2 \Rightarrow$$

$$\Rightarrow h_1 = 10,1m. \quad (1)$$

If the result of relation (1) was negative, then the assumption that tank A lies higher than B, as we do not know if C is higher or lower, would be wrong and we would have to make a correction iterating with another assumption.

Returning to the problem, as we know the losses, we can calculate the flow rate.

$$\left(\varepsilon_1/d_1 = 0,00087\, AIT\right) \rightarrow f_1 = 0,0193 \text{ (Moody)}$$

$$h_1 = \frac{8}{\pi^2 \cdot g} \cdot f_1 \cdot \frac{\ell_2}{d_2} \cdot Q_1^2 \Rightarrow Q_1 = 0,175 m^3/s = 630 m^3/h.$$

AIT control: $V_2 = 2,48 m/s \Rightarrow \text{Re}_1 = 7,5 \cdot 10^5 \rightarrow f_1 = 0,0193$ (Moody).

So: $Q_1 = 0,175 m^3/s = 630 m^3/h$.

Step 3: calculation of the flow rate, losses and height of pipe (3). Because we know the two flow rates and the equation of the node [5.151], we calculate the flow rate of the third pipe.

$$[5.151] \Rightarrow Q_1 + Q_2 + Q_3 = 0 \Rightarrow \left(+630 m^3/h\right) + \left(-100 m^3 h\right) + Q_3 = 0 \Rightarrow$$

$\Rightarrow Q_3 = -530 m^3/h$ or simpler

$Q_3 = 530 m^3/h$ (inflow in Γ)

– Calculation of losses:

$$V_3 = \frac{4 \cdot Q_3}{\pi \cdot d_3^2} \Rightarrow V_3 = 3,0 m/s$$

$$\left(\text{Re}_3 = \frac{V_3 \cdot d_3}{\nu} \Rightarrow \text{Re}_3 = 7,5 \cdot 10^5, \varepsilon_3/d_3 = 0,001\right) \rightarrow f_3 = 0,0202 \text{ (Moody)}$$

$$h_3 = f_3 \cdot \frac{\ell_3}{d_3} \cdot \frac{V_3^2}{2g} \Rightarrow h_3 = 7,41m .$$

– Calculation of height:

$$[5.158] \Rightarrow y_1 - h_1 = y_3 + h_3 \Rightarrow y_1 - y_3 = h_1 + h_3 \Rightarrow y_1 - y_3 = 17,51m .$$

$$(y_3 = 42,49m)$$

B′ problem: in the B′ problem of the three tanks, the *values* of the three pipes, the *heights of the two tanks*, their height range and the *flow rate* of the tank of the unknown heights are *given* and the height of the tank and the other *two flow rates* are *asked.*

The three pipes, as shown in Figure 5.35, are made of asphaltic cast iron with diameters $d_1 = 6in, d_2 = 4in$ and $d_3 = 5in$ and lengths $\ell_1 = 60m, \ell_2 = 20m$ and $\ell_3 = 30m$, respectively.

If the free surface of the tank ς B lies 5 m higher than Γ and the water outflow from the tank A is $150m^3 / h$, calculate the altitude range between the free surfaces of the tanks A and Γ and the flow rates Q_2 and Q_3 (in m^3 / h).

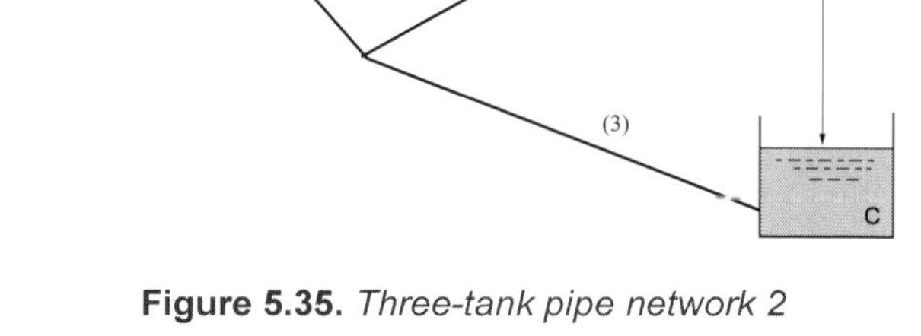

Figure 5.35. *Three-tank pipe network 2*

Solution

Compared to the data of the previous example, the first difference lies in the fact that the heights of the two tanks as well as the height range are not given. However, this has no influence on the method that we use, as even in the previous example, the required data was the height range. Moreover, if we consider the free surface of C as the reference level, we get $y_3 = 0m, y_2 = 5m$.

The important methodological difference lies in the fact that tank A, whose flow rate is given, has unknown height. There are enough data to calculate the losses in pipe (1). However, equation [5.158] does not allow us go to another pipe.

Step 1: calculation of the losses of pipe (1)

$$V_1 = \frac{4 \cdot Q_1}{\pi \cdot d_1^2} \Rightarrow V_1 = 2,28m/s$$

$$\left(\mathrm{Re}_1 = \frac{V_1 \cdot d_1}{\nu} \Rightarrow \mathrm{Re}_1 = 3,5 \cdot 10^5, \varepsilon_1/d_1 = 0,00079 \right) \rightarrow f_1 = 0,0194 \text{ (Moody)}$$

$$h_1 = f_1 \cdot \frac{\ell_1}{d_1} \cdot \frac{V_1^2}{2g} \Rightarrow h_1 = 2,04m.$$

At this point, the similarity with the previous problem stops. Equation [5.158] does not allow the losses calculation h_2 or h_3, because every equation that arises from it has two unknowns:

$$[5.158] \Rightarrow y_1 = h_1 = y_2 \mp h_2 \text{ (unknowns: } y_1, h_2\text{)}.$$

$$[5.158] \; y_1 - h_1 = y_3 + h_3 \text{ (unknowns: } y_1, h_3\text{)}.$$

So we are obliged to proceed in tests.

Because we know one of the three (Q_1), if we assume a second one (Q_2') from equation [5.151], we calculate the third one (Q_3) as well. By assuming Q_2', Q_3', we calculate losses h_2' and h_3'.

From equation [5.158], we form the *criterion for the control and correction* for the initial assumption for (Q_2'). But because we do not know the signs, we usually start the first test for $Q_2' = 0$, so the loss $h_2' = 0$.

First test: if $Q_2' = 0 \Rightarrow h_2' = 0$.

From equation [5.151], we have:

$$Q_3' = Q_1' = 150/3600 m^3/\sec \text{ (inflow in C).}$$

We calculate the losses in pipe (3):

$$V_3 = \frac{4 \cdot Q_3}{\pi \cdot d_3^2} \Rightarrow V_3 = 3,29 m/s$$

$$\left(\text{Re}_3 = \frac{V_3 \cdot d_3}{\nu} \Rightarrow \text{Re}_3 = 4,2 \cdot 10^5, \varepsilon_1/d_1 = 0,00094 \right) \to f_1 = 0,02 \text{ (Moody)}$$

$$h_2 = f_2 \cdot \frac{\ell_2}{d_2} \cdot \frac{V_2^2}{2g} \Rightarrow h_3' = 2,65 m$$

Control [5.156] $\Rightarrow y_2 \mp h_2 = y_3 + h_3 \Rightarrow y_2 - y_3 = h_3 \pm h_2 = 5m$

In our test:

$$h_3' \pm h_2' = h_3' = 2,65m < 5 \Rightarrow h_3' < h_3 \Rightarrow Q_3' < Q_3 \Rightarrow Q_3 > 150 m^3/h \Rightarrow Q_3 > Q_1.$$

Because the inflow in C is larger than the outflow of A, we have the *outflow from tank* **B** *as well.* Equations [5.149] and [5.156] take the form:

$$Q_1 + Q_2 = Q_3 \qquad (1)$$

$$y_1 - h_1 = y_2 - h_2 = y_3 + h_3 = \Lambda \qquad (2)$$

From relation (2), we form the *control criterion* of the tests:

$$y_2 - y_3 = h_3 + h_2 \Rightarrow h_3 + h_2 = 5m. \qquad (3)$$

Second test: assume that $Q_2' = 50 m^3/h$ (outflow) $\Rightarrow Q_3' = 200 m^3/h$

Applying the first problem of the pipe (as above), we calculate the losses h_2' and h_3':

$$h_2' = 0,64m, h_3' = 4,61m.$$

Control:

$$h_2' + h_3' = 5,25m \Rightarrow h_2' + h_3' > h_3 + h_2 \Rightarrow Q_1' > Q \Rightarrow Q < 50m^3 / h .$$

We proceed with a new assumption of 50 m^3/h, as presented in the following table:

Q'_2 m^3/h	Q'_3 $(Q_1 + Q_2)$	f'_2	h'_2 $(a_2 \times Q_2^2)$	f'_3	h'_3 $(a_3 \times Q_3^2)$	$h'_2 + h'_3$	Criterion $h_3 + h_5 = 5$	Q_2 m^3/h
0	150	0.0000	0	0.0201	2.61	2.61	2.61 < 5	$Q_2 > 0$
50	200	0.0218	0.64	0.0199	4.61	5.25	5.25 > 5	$Q_2 < 50$
40	190	0.0220	0.42	0.0199	4.16	4.58	4.58 < 5	$Q_2 > 40$
45	195	0.0219	0.52	0.0199	4.38	4.90	4.90 < 5	$Q_2 > 45$
47	197	0.0218	0.57	0.0199	4.47	5.04	5.04 > 5	$Q_2 < 47$
46	196	0.0219	0.54	0.0199	4.43	4.97	4.97 < 5	$Q_2 > 46$
46.4	**196.4**	**0.0218**	**0.55**	**0.0199**	**4.44**	**5.00**	**5.00 = 5**	$\mathbf{Q_2 = 46.4}$

Table 5.12. *Results*

Therefore:

$$Q_2 = 46,4m^3 / h \text{ (outflow)}$$

$$Q_3 = 196,4m^3 / h \text{ (inflow)}$$

$$h_2 = 0,55m$$

$$h_3 = 4,44m$$

Finally, the height difference between the free surfaces of tanks A and C is calculated from equation (2), which is a similar equation to [5.156].

Therefore, from (2), we have:

$$y_1 - y_3 = h_1 + h_3 = (2,04 + 4,44)m \Rightarrow y_1 - y_3 = 6,48m .$$

C′ Problem: in the C′ problem of the three tanks, the *values* of the three pipes and the *heights of the three tanks* or their height difference are *given* and the *three flow rates* are *asked*.

In these problems, we cannot calculate some of the parameters of the three pipes, because for all three of them, the flow rate and their losses are unknown. For the solution of these problems, we will use the assumption method. Therefore, we will make use of one of the relationships of the three tanks for the assumptions and the other one for the formation of the control criterion.

In Figure 5.36, pipe (1) is made of cast iron with $\ell_1 = 450m$, $\varepsilon_1 = 0,00026m$ and $d_1 = 30cm$. Pipes (2) and (3) are made of asphalted cast iron with $\varepsilon_2 = 0,00012m$, $d_2 = 0,15m$, $\ell_2 = 250m$ and $\varepsilon_3 = 0,00012m$, $d_3 = 0,20m$ and $\ell_3 = 350m$, respectively. The surface of tank B is 6 m below the surface of tank A and the surface of tank C is 18 m below the surface of tank A. Calculate the flow rates in the three pipes if the kinematic viscosity of the flow is $v = 10^{-6} m^2 / \sec$.

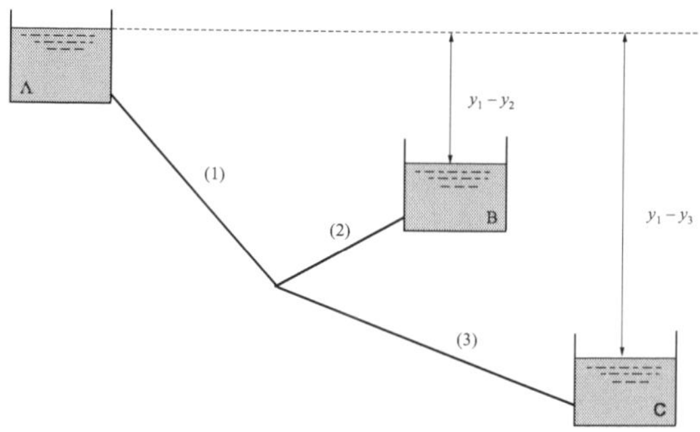

Figure 5.36. *Three tanks problem C*

Taking the level of the free surface of the lower tank usually as reference level, to avoid the negative altitudes, we can assume the altitudes of the free surfaces of the three tanks. Therefore, if we set $y_3 = 0m$, we will have:

$$y_3 = 0 \Rightarrow y_1 = 18m \text{ and } y_2 = 12m$$

Therefore, by using the assumption of intense turbulence without being interested in the accuracy of the values of f, because the results are based on the assumption of some values of Λ, which will change, we are not interested in the accuracy of its value. We will examine the assumptions of the intense turbulence only for the value of Λ, which we will accept. Therefore:

First test: assume that $\Lambda' = y_2 = 12m$

$$[5.156] \Rightarrow h_2' = 0 \Rightarrow Q_2' = 0$$

$$[5.156] \Rightarrow y_1 - h_1' = \Lambda' \Rightarrow h_1' = 6m$$

$$y_3 + h_3' = \Lambda' \Rightarrow h_3' = 12m$$

– Calculation of the flow rate of pipe (1):

$$(\varepsilon_1/d_1 = 0{,}00087\,\text{AIT}) \rightarrow f_1 = 0{,}0193 \text{ (Moody)}$$

$$a_1 = 295\,(SI) \Rightarrow h_1 = 295 \cdot Q_1^2 \;(SI) \Rightarrow Q_1' = 0{,}143m^3/s = 513{,}6m^3/h\,.$$

– Calculation of the flow rate of pipe (3):

$$(\varepsilon_3/d_3 = 0{,}0006\,\text{AIT}) \rightarrow f_3 = 0{,}0177 \text{ (Moody)}$$

$$a_3 = 1600\,(SI) \Rightarrow h_3 = 1600 \cdot Q_3^2 \;(SI) \Rightarrow Q_3' = 0{,}0866m^3/s = 318{,}8m^3/h\,.$$

– Control:

$$Q_1' \pm Q_2' - Q_3' = Q_1' - Q_3' = 201{,}8m^3/h \Rightarrow Q_1' > Q_3' \Rightarrow$$

$\Rightarrow$ Inflow in the B $\Rightarrow \Lambda > y_2$

Relationship formation:

$$[5.149] \Rightarrow Q_1 - Q_2 - Q_3 = 0 \Leftrightarrow Q_1 = Q_2 + Q_3 \text{ (criterion)} \tag{1}$$

$$[5.156] \Rightarrow y_1 - h_1 = y_2 + h_2 = y_3 + h_3 = \Lambda \tag{2}$$

$$(2) \Rightarrow h_1 = y_1 - \Lambda \tag{2a}$$

$$h_2 = \Lambda - y_2 \tag{2b}$$

$$h_3 = \Lambda - y_3\,. \tag{2c}$$

The first test allowed us to conclude that in pipe B there is inflow, and therefore to form equations (1) and (2).

We proceed with a new test with $\Lambda > y_2$:

Second test: assume that $\Lambda' = 15m$.

– Calculation of the flow rate of pipe (C): for pipe (1), the equation that occurred in the first test is applicable (with the assumption of intense turbulence):

$$h_1 = 295 \cdot Q_1^2 \ (SI) \tag{3a}$$

$$(2a) \Rightarrow h_1' = 3m$$

$$(3a) \Rightarrow Q_1' = 0{,}101 m^3 / \sec = 363{,}1 m^3 / h .$$

– Calculation of the flow rate of pipe (2):

$$\left(\varepsilon_2 / d_2 = 0{,}0008\,\text{AIT}\right) \rightarrow f_2 = 0{,}0189 \text{ (Moody).}$$

$$a_2 = 5.144 (SI) \Rightarrow h_2 = 5.144 \cdot Q_2^2 \ (SI) \tag{3b}$$

$$(2b) \Rightarrow h_2' = 3m$$

$$(3\beta) \Rightarrow Q_2' = 0{,}024 m^3 / \sec = 86{,}9 m^3 / h .$$

– Calculation of the flow rate of pipe (3): from the first testing, it is applicable:

$$h_3 = 1600 \cdot Q_3^2 \ (SI) \tag{3c}$$

$$(2c) \Rightarrow h_3' = 15m$$

$$(3\gamma) \Rightarrow Q_3' = 0{,}097 m^3 / \sec = 348{,}6 m^3 / h .$$

Control:

$$363{,}1 m^3 / h < \left(86{,}9 + 348{,}6\right) m^3 / h \Rightarrow Q_1' < Q_1 \Rightarrow h_1' < h_1 \Rightarrow \Lambda < 15m$$

We present the relevant table to proceed with the tests:

P' *m*	*h'₁* *(y₁-Λ)*	*h'₂* *(Λ-y₂)*	*h'₃* *(Λ-y₃)*	*Q'₁* *m³/h*	*Q'₂* *m³/h*	*Q'₃* *m³/h*	*Criterion* *Q₁-Q₂-Q₃=0*	*Λ* *m*
12,00	6,00	0,00	12,00	513,6	0,0	311,8	201,8	Λ>12
15,00	3,00	3,00	15,00	363,1	86,9	348,6	-72,4	Λ<15
13,00	5,00	1,00	13,00	468,8	50,2	324,5	94,1	Λ>13
14,00	4,00	2,00	14,00	419,3	71,0	336,7	11,6	Λ>14
14,5	3,50	2,50	14,50	392,2	79,4	342,7	-29,8	Λ<14,5
14,10	3,90	2,10	14,10	414,0	72,7	338,0	3,3	Λ>14,1
14,15	3,85	2,15	14,15	411,4	73,6	338,6	-0,8	Λ<14,15
14,14	**3,86**	**2,14**	**14,14**	**411,9**	**73,4**	**338,4**	**0,05**	**Λ=14,14**

Table 5.13. *Results*

Before we accept the above results, we must check the assumption of intense turbulence and confirm or alter the values of f and consequently the coefficients $\boldsymbol{a}$ of relation (3) and the values of Q:

Control of the AIT:

– Pipe (1): $V_1 = 1{,}62 m/s \Rightarrow \text{Re}_1 = 4{,}9 \cdot 10^5 \Rightarrow f_1 = 0{,}0195 \approx 0{,}0193$

– Pipe (2): $V_2 = 1{,}15 m/s \Rightarrow \text{Re}_2 = 1{,}7 \cdot 10^5 \Rightarrow f_2 = 0{,}0203 \neq 0{,}0189$

Correction: $a_2 = 5514 (SI) \Rightarrow h_2 = 5514 \cdot Q_2^2 \, (SI) \Rightarrow Q_2 = 71{,}1 m^3/h$.

– Pipe (3): $V_3 = 3 m/s \Rightarrow \text{Re}_3 = 6 \cdot 10^5 \Rightarrow f_3 = 0{,}018 \approx 0{,}0177$.

Smoothing of the results: the following flow rates are obtained:

$$Q_1 = 411{,}9 m^3/h$$

$$Q_2 = 71{,}1 m^3/h$$

$$Q_3 = 338{,}4 m^3/h$$

However, the continuity equation for incompressible fluids defines:

$$Q_1 = Q_2 + Q_3 .$$

Thus, from the above approximate values, we have:

$$411{,}9m/h \neq 71{,}1m^3/h + 338{,}4m^3/h = 409{,}5m^3/h\,.$$

The difference that arises from the application of the continuity equation, which is 2,4 m^3/h, is equally divided between the inflow and the outflow from the node. Therefore, we finally have:

$$Q_1 = 410{,}7m^3/h$$

$$Q_2 = 71{,}4m^3/h$$

$$Q_3 = 339{,}3m^3/h$$

This accuracy is satisfactory because even if we return to the testing table and continue the calculation with rectified relationships (2), the corrections are so small that they do not surmount the 0.5%.

5.12. Formulae

1) Reynolds number:

$$\mathrm{Re} = \frac{\rho V L}{\mu} = \frac{VL}{\nu} = \text{const}$$

where ρ is the fluid density, V is the fluid velocity, L is the characteristic length of the flow, μ is the fluid viscosity and ν is the kinematic viscosity $\left(\nu = \frac{\mu}{\rho}\right)$.

2) Reynolds number in cylindrical pipes:

$$\mathrm{Re} = \frac{\rho V d}{\mu} = \frac{Vd}{\nu}$$

where d is the internal diameter of the cylindrical pipe.

3) Reynolds number in non-cylindrical pipes:

$$\mathrm{Re} = \frac{4r_h \rho V}{\mu} = \frac{4r_h V}{\nu}$$

where r_h is the hydraulic radius, which is defined by the relation:

$$r_h = \frac{S}{P}$$

where S is the surface area of the pipe's cross-sectional area and P is the length of the pipe's perimeter.

4) Local velocity for a laminar flow in a cylindrical pipe:

$$u = u_{\max}\left(1 - \frac{r^2}{R^2}\right)$$

where $u_{\max}$ is the flow velocity at the center of the pipe, r is the distance from the center of the circular cross-sectional area and R is the distance from the internal wall of the pipe.

5) Local velocity for a turbulent flow in a cylindrical pipe:

$$u = u_{\max}\left(1 - \frac{r}{R}\right)^n$$

where the exponent n has values $1/9 \leq n \leq 1/5$.

6) Empirical type for the average velocity of a flow in cylindrical pipes:

$$V = \frac{m}{\rho S}$$

where m is the fluid mass, ρ is the fluid density and S is the surface area of the pipe's cross-sectional area.

7) Relation between the average and the local velocity:

$$V = \frac{1}{S}\int_S u \cdot dS$$

where $V = 0,5u_{\text{max}}$ for a laminar flow and $0,5u_{\text{max}} \leq V \leq 0,8u_{\text{max}}$ for a turbulent flow.

8) Shear stress in a horizontal cylindrical pipe:

$$r = \frac{r}{r\ell}(p_1 - p_2)$$

where r is the radiant distance from the main pipe, S is the surface area, $\ell = \ell_2 - \ell_1$ is the length of the pipe and $p_1 - p_2$ is the pressure difference between two points of the pipe.

9) Pressure drop in a horizontal cylindrical pipe:

$$\Delta p = p_1 - p_2 = 4\frac{\ell}{d} \cdot \tau_0$$

where ℓ is the pipe's length and τ_0 is the shear stress on the pipe's walls.

10) Pressure drop in a horizontal non-cylindrical pipe:

$$\Delta p = \frac{\ell p}{S} \cdot \tau_0$$

where ℓ is the pipe's length, S is the surface area of the pipe's cross-sectional area, P is the length of the pipe's perimeter and τ_0 is the shear stress on the pipe's walls.

11) Pressure drop of a real fluid in a cylindrical pipe:

$$\Delta p = \frac{f}{2} \cdot \frac{\ell}{d} \cdot \rho \cdot V^2$$

where f is the friction coefficient of the flow and ℓ, d, ρ, V are the pipe's length, pipe's diameter, fluid density and flow velocity, respectively.

12) Shear stress and friction coefficient:

$$f = \frac{8\tau_0}{\rho V^2}$$

where f is the friction coefficient, τ_0 is the shear stress on the pipe's walls, ρ is the fluid density and V is the flow velocity.

13) Pressure drop and friction coefficient:

i) Cylindrical pipe:

$$\Delta p = \frac{f}{2} \cdot \frac{\ell \rho V^2}{d}$$

ii) Non-cylindrical pipe:

$$\Delta p = \frac{f}{8} \cdot \frac{P\ell}{S} \cdot \rho V^2$$

where all the values in both relationships express known concepts.

14) Continuity equation in cylindrical pipes:

$$Q = \frac{\pi \cdot d^2}{4} \cdot V = \text{const}$$

where Q is the flow rate, d is the pipe's diameter and V is the flow velocity.

15) Bernoulli equation for a cylindrical pipe:

$$\frac{p_1 - p_2}{\gamma} + \frac{V_1^2 - V_2^2}{2 \cdot g} + (h_1 - h_2) = \Sigma h + h_t - h_p$$

where p_1 is the fluid's pressure in cross-sectional area (1), p_2 is the fluid's pressure in cross-sectional area (2), γ is the specific weight of the fluid, V_1 is the flow velocity in cross-sectional area (1), V_2 is the flow velocity in cross-sectional area (2), h_1,h_2 are head losses in cross-sectional areas (1) and (2), respectively, Σh are total head losses, h_t is the energy head of a turbine or a pump and h_p is the energy head of the pump.

16) Flow velocity max value in a horizontal cylindrical pipe in relation to the pressure drop:

$$V_{\max} = \frac{R^2}{4\mu\ell}(p_1 - p_2)$$

where R is the pipe's radius, μ is the fluid viscosity, ℓ is the pipe's length and p_1, p_2 are pressures of the fluid in cross-sectional areas (1) and (2), respectively.

17) Average flow velocity in a horizontal cylindrical pipe in relation to V_{max}:

$$V_{md} = \frac{R^2}{8\mu\ell}(p_1 - p_2) = \frac{V_{\max}}{2}$$

where all the variables have known values.

18) Friction coefficient of a laminar flow in a horizontal cylindrical pipe:

$$f = \frac{64}{\text{Re}}$$

where Re = Reynolds number $= \frac{\rho \cdot dV_{md}}{\mu}$.

19) Hagen–Poiseuille equation for a laminar flow in a horizontal pipe:

$$V_{md} = (p_1 - p_2) \cdot \frac{d^2}{32\mu\ell}$$

where all the variables have known values.

20) Head losses h_l in a flow of a smooth horizontal cylindrical pipe:

$$h_1 = \frac{f\ell V_{md}}{d2g}$$

where f is the friction coefficient, ℓ is the total length of the pipe, V_{md} is the average velocity of flow, d is the pipe's diameter and g is the acceleration of gravity.

21) Friction coefficient f in a turbulent flow of a smooth horizontal cylindrical pipe:

$$f = \frac{0{,}316}{\mathrm{Re}^{0{,}25}}$$

where Re = Reynolds number $= \dfrac{\rho \cdot dV_{md}}{\mu}$.

22) Friction coefficient f in smooth pipes:

$$\frac{1}{\sqrt{f}} = 0{,}87\ell n\left(\mathrm{Re}\sqrt{f}\right) - 0{,}8$$

where f is the friction coefficient and Re is the Reynolds number.

23) Relative roughness, $\varepsilon_{\sigma x}$, in non-smooth pipes:

$$\varepsilon_{\sigma x} = \frac{\varepsilon}{d}$$

where ε is the average velocity of the pipe and d is the pipe's diameter.

24) Mathematical expression of the Moody diagram for the calculation of the friction coefficient *f*:

$$f = F\left(\mathrm{Re} = \frac{V \cdot d}{\nu} \cdot \frac{\varepsilon}{d}\right)$$

where Re is the Reynolds number.

25) Empirical equations for the friction coefficient *f*:

i) Prandtl equation for smooth pipes:

$$\frac{1}{f^{1/2}} = 2\log\left(\frac{\mathrm{Re}\cdot f^{1/2}}{2,51}\right) = 2\log\left(\mathrm{Re}\cdot f^{1/2} - 0,8\right)$$

where *f* is the friction coefficient and Re is the Reynolds number.

ii) von Karman equation for rough pipes:

$$\frac{1}{f^{1/2}} = 2\log\left(\frac{3,7}{\varepsilon/d}\right) = 1,14 - 2\log\left(\varepsilon/d\right)$$

where *f* is the friction coefficient and ε/d is the relative roughness of the pipe.

iii) Colebrook equation for pipes:

$$\frac{1}{f^{1/2}} = -2\log\left(\frac{2,51}{\mathrm{Re}\cdot f^{1/2}} + \frac{\varepsilon/d}{3,7}\right)$$

where *f* is the friction coefficient, Re is the Reynolds number and ε/d is the relative roughness.

iv) Haaland equation:

$$\frac{1}{f^{1/2}} = -1,8\log\left[\frac{6,9}{\mathrm{Re}} + \left(\frac{\varepsilon/d}{3,7}\right)^{1,11}\right]$$

where the variables are defined as above.

v) Swamee–Jain equation:

$$\frac{1}{f^{1/2}} = -2\log\left(\frac{5,74}{\mathrm{Re}^{0,9}} + \frac{\varepsilon/d}{3,7}\right)$$

where the variables are defined as above.

vi) Finally, the following equation can also be used:

$$f = \left[A^{1,2} + \left(B^{-8} + C^{-8}\right)^{-1,5}\right]^{1/12}$$

where $A = \frac{64}{\mathrm{Re}}$, $B = \left(\frac{\mathrm{Re}}{12.900}\right)^2$, $C = \left\{-0,8687 \cdot \ell n\left[\left(\frac{7}{\mathrm{Re}}\right)^{0,9} + 0,27 \cdot \left(\frac{\varepsilon}{d}\right)\right]\right\}^{-2}$

where the variables are defined as above.

26) Minor losses in cylindrical pipes:

$$h_l' = k\frac{V^2}{2g}$$

where h_1' are energy minor losses, V is the average velocity of the flow, k is the non-dimensional coefficient (coefficient of the local losses).

27) Energy equation with total energy losses:

$$\frac{p_1}{\rho g} + h_1 + \frac{V_1^2}{2g} + H_A = \frac{p_2}{\rho g} + h_2 + \frac{V_2^2}{2g} + H_T + H_L$$

where p_1, p_2 are pressures in cross-sectional areas (1) and (2), respectively; h_1, h_2 are head losses in cross-sectional areas (1) and (2), respectively; V_1, V_2 are average velocities; ρ is the density of the fluid; g is the acceleration of gravity and H_L is the total energy loss, which is defined by the relation:

$$H_L = h_1 + h_1'$$

with h_1' being energy minor losses and h_1 being energy losses through the flow; H_A is the head of energy due to the pump and H_T is the head of the energy due to the turbine.

28) Equivalent length:

$$\text{Equivalent length} = \frac{\ell}{d}$$

where ℓ is the length of a pipe and d is the pipe's diameter.

29) Relation between the coefficient of minor losses and coefficient of the equivalent length:

$$\frac{\ell}{d} = \frac{K}{f}$$

where ℓ is the pipe's length, d is the pipe's diameter, K is the coefficient of minor losses and f is the friction coefficient.

30) Equation of total losses, Σh:

$$\Sigma h = \left\{ f \cdot \frac{\ell}{d} + \Sigma K_i \right\} \cdot \frac{V^2}{2g}$$

or

$$\Sigma h = \frac{8}{\pi^2 \cdot g \cdot d^4} \cdot \left\{ f \cdot \frac{\ell}{d} + \Sigma K_i \right\} \cdot Q^2$$

where f is the friction coefficient (which is given by the Moody diagram, according to the relationship: $f = F\left(\text{Re} = \frac{Vd}{\nu} \cdot \frac{\varepsilon}{d}\right)$), ℓ is the pipe's length, ΣK_i is the total losses coefficient, V is the flow velocity, g is the acceleration of gravity and Q is the flow rate.

The second equation occurs because:

Head losses h_f are given by the relation:

$$h_f = \frac{8}{\pi^2 g} \cdot f \cdot \frac{\ell}{d^5} \cdot Q^2$$

Moreover, *minor losses* from:

$$h_i = \frac{8}{\pi^2 g d^4} \cdot K_i \cdot Q^2$$

with $Q = \frac{\pi d^2 V}{4}$ and $\sum h = h_f + h_i$.

31) System of equations for solving problems of flows in pipes:

i) Continuity equation:

$$Q = \frac{\pi \cdot d_i^2}{4} \cdot V_1 = \text{const}$$

ii) Energy equation (Bernoulli):

$$\frac{p_1 - p_2}{\gamma} + \frac{V_1^2 - V_2^2}{2g} + (y_1 - y_2) = \Sigma h$$

iii) Losses equation (Darcy–Weisbach):

$$\Sigma h = \frac{8}{\pi^2 \cdot g \cdot d^4} \cdot \left(f \cdot \frac{\ell}{d} + \Sigma K_i\right) Q^2$$

iv) Moody diagram:

$$f = F\left(\mathrm{Re} = \frac{V \cdot d}{\nu} \cdot \frac{\varepsilon}{d}\right)$$

where all the variables have known values.

v) Equation of total losses of a laminar flow Σh_{lam}:

Because $$f_{lam} = \frac{64}{\mathrm{Re}}$$

for a laminar flow, we have:

$$h_{f\,lam.} = \frac{64}{\mathrm{Re}} \cdot \frac{\ell}{d} \cdot \frac{V^2}{2g} = \frac{32 \cdot \nu \cdot \ell}{g \cdot d^2} \cdot V$$

32) Total manometer head h^*:

$$h^* = z + \frac{p}{\gamma} + \frac{V^2}{2g}$$

where Z is the distance from the energy line to the reference level, p is the fluid's pressure, γ is the specific weight, V is the flow velocity and y is the acceleration of gravity.

33) Inclination of the energy line JE:

$$J_E = -\frac{dh^*}{ds}$$

where dh^* is the variation of the manometer head of the flow that is caused by the elementary length ds of the pipe.

34) Decrease of the total manometer head h_t due to the presence of a turbine:

$$h_t = \frac{P_t}{Qgn_t}$$

where P_t is the axial power of a turbine, Q is the fluid's flow rate, g is the acceleration of gravity and n_t is the distribution rate.

35) Equation of mechanical energy in the form of head energy losses:

$$h_a + h_w = h_b + h_{tot}$$

where h_{tot} is the total head of energy losses and h_w is the variation of the head of the mechanical energy, which is given by the relationship:

$$h_w = \frac{P}{gQ}$$

where P is the axial power, Q is the flow rate, g is the acceleration of gravity and h_a, h_b are total heads right before cross-sectional area (a) and right after cross-sectional area (b), respectively, which are given by the relationships:

$$h_a = z_a + \frac{p_a}{\gamma} + \frac{C_a^2}{2g}$$

$$h_b = z_b + \frac{p_b}{\gamma} + \frac{V_b^2}{2g}$$

where the variables have known values.

36) Total head of energy losses of a pipeline, h_{tot}:

$$h_{tot} = h_t + h_m = \sum_i f_i \frac{\ell_i}{d_i} \frac{V_i^2}{2g} + \sum_j K_{mj} \frac{V_j^2}{2g}$$

where f_i, ℓ_i, d_i and V_i are the coefficient of friction, length, diameter and average velocity at the *i*-linear part of the pipeline, respectively, and $K_{m,j}$ and V_j are the coefficient of the losses and the average velocity of the adduction or abduction of the fluid at its *j*-appliance, respectively.

37) Simple pipeline:

Total head of energy losses

$$h_{tot} = h_i + h_m = \left(f \frac{\ell}{d} + \sum_i K_{m,\ell} \right) \frac{V^2}{2g}$$

where f is the friction coefficient, d is the internal diameter, ℓ is the length sum of all the straight pipes that constitute the pipeline, V is the flow velocity, g is the acceleration of gravity and $\sum_i K_{m,\ell}$ is the sum of the losses coefficients at the pipeline.

38) Pipes connected in a row:

Equation system for two pipes:

i) Continuity equation:

$$\frac{V_1}{V_2} = \left(\frac{d_2}{d_1} \right)^2$$

ii) Equation of head losses:

$$\Sigma h = \Sigma h_1 + \Sigma h_2 = \frac{8}{\pi^2 g} \left(f_1 \cdot \frac{\ell_1}{d_1} + \Sigma K_1 \right) \cdot \frac{Q^2}{d_1^4} + \frac{8}{\pi^2 g} \left(f_2 \cdot \frac{\ell_2}{d_2} + \Sigma K_2 \right) \cdot \frac{Q^2}{d_2^4}$$

where all the variables have known values because $Q = Q_1 = Q_2$.

In the calculation procedure, except for the above two equations, we also use the following:

iii) Bernoulli equation if energy and values are given or asked

iv) Continuity equation for each pipe

v) Moody diagram for each pipe

vi) Losses equation for each pipe

39) Parallel connection of pipes:

Equation system for two pipes:

i) Continuity equation:

$$Q = Q_1 + Q_2 \Rightarrow d^2 V = d_1^2 V_1 + d_2^2 V_2$$

ii) Equation of head losses:

$$\Sigma h = \Sigma h_1 \pm \Sigma h_2 \Rightarrow f_1 \frac{\ell_1}{d_1^5} \cdot Q_1^2 = f_2 \frac{\ell_2}{d_2^5} \cdot Q_2^2$$

or $$a_1 \cdot Q_1^2 = a_2 \cdot Q_2^2$$

where all the variables have known values.

40) Mixed pipe connection

Equation system for seven pipes:

i) Continuity equation:

$$Q = Q_1 = Q_2 + Q_3 = Q_2 + (Q_4 + Q_5) = Q_6 = Q_7$$

ii) Equation of head losses:

$$\Sigma h = \Sigma h_1 + \Sigma h_2 + \Sigma h_6 + \Sigma h_7$$

because according to Figure (5.24), we have:

$$\Sigma h_4 = \Sigma h_5$$ and $$\Sigma h_2 = \Sigma h_3 + \Sigma h_4$$

and $$Q_1 = Q_2 + Q_3$$ and $$Q_3 = Q_4 + Q_5$$

$$Q_3 = Q_4 + Q_5 \Rightarrow Q_3 = \left(a_4^{0,5} + a_5^{0,5}\right) h_4^{0,5}$$

where all the variables have known values.

41) Pipe branch:

Equation system for three pipes:

Continuity equation:

$$\Sigma Q = 0 \Rightarrow Q_1 + Q_2 + Q_3 = 0$$

with the condition of the **rule of flow rates pre-marking**, according to which:

i) If the fluid **comes into** the node, then the flow rate is considered to be **positive**.

ii) If the fluid **comes out** of the node, then the flow rate is considered to be **negative**.

42) Problem of the three tanks:

Equations system:

if $y_1 > y_2 > y_3$

i) Flow rates equation:

$$Q_1 + Q_2 + Q_3 = 0$$

ii) Losses equation:

$$y_1 - h_1 = y_2 \mp h_2 = y_3 + h_3 = \Lambda$$

where Q_1, Q_2, Q_3 are flow rates of the three pipes, y_1, y_2, y_3 are heights of the three tanks and h_1, h_2, h_3 are energy losses of the three pipes.

iii) Continuity equation.

iv) Moody diagram.

v) Losses equation.

43) General problem of the three tanks:

i) If the three pipes do not end up in tanks but end up in their cross-sectional areas (1), (2) and (3) (Figure 5.27), then we have:

a) Energy head:

$$H_1 = y_1 + \frac{V_1^2}{2g} + \frac{p_1}{\gamma}$$

$$H_2 = y_2 + \frac{V_2^2}{2g} + \frac{p_2}{\gamma}$$

$$H_3 = y_3 + \frac{V_3^2}{2g} + \frac{p_3}{\gamma}$$

b) Losses equation:

$$H_1 + h_1 = H_2 \mp h_2 = H_3 + h_3 = \Lambda$$

where all the variables have known values, and the energy heads H_1, H_2, H_3 are defined by the previous relationships.

c) Flow rates equation:

$$Q_1 + Q_2 + Q_3 = 0$$

ii) If there are more than three pipes, then we have:

a) Flow rates equation:

$$\Sigma Q_i = 0$$

b) Losses equation:

$$\boxed{H_i - h_i = \Lambda}$$

5.13. Questions

1) What are the *physical values* that characterize the flow of an incompressible real fluid in closed pipes?

2) How is the *Reynolds number* defined in cylindrical and non-cylindrical pipes for laminar and turbulent flows?

3) How is the *velocity distribution* defined in a cylindrical pipe for laminar and turbulent flows, and which expression gives the *average velocity* of this flow?

4) Calculate the *shear stress* in a horizontal cylindrical pipe.

5) Calculate the *pressure drop* in a horizontal cylindrical pipe and in a non-cylindrical pipe.

6) State the relation that connects the *shear stress* τ_0 with the *friction coefficient* f in the flow of an incompressible fluid in a cylindrical horizontal pipe.

7) Which relation connects the *pressure drop* with the *friction coefficient* for cylindrical and non-cylindrical horizontal pipes in the flow of a real incompressible fluid inside them?

8) What is the expression of the *continuity equation* and the Bernoulli equation in cylindrical pipes of an incompressible real fluid?

9) What are *energy heads*, how many kinds are there and what are they, and which relationships express them?

10) What are *head losses* and from which heads do the total head losses Σh consist of in a flow in cylindrical pipes?

11) Which are the forms of the *Bernoulli equation* that are used for the study of flow in cylindrical pipes?

12) What is *entrance length* ℓ for the flow of a real incompressible fluid in a horizontal cylindrical pipe and which relation is attributed to Boussinesq and which to Langhaar and for which form of the flow are they applicable inside the pipe? What happens for the other form of the flow?

13) From the relation $\Delta p = \frac{f}{2} \cdot \frac{\ell}{d} \cdot \rho \cdot V^2$, derive the *Darcy–Weisbach* relation, where f is the friction coefficient, ℓ, d are the length and the diameter of the pipe, respectively, and ρ, V are the density and the velocity of the fluid, respectively.

14) What causes the *turbulent flow* inside a closed cylindrical horizontal pipe, and which relation gives the *velocity* in this flow and what is its *average value*?

15) Draw the graph for the velocity variation V with respect to the time t in a turbulent flow.

Give schematically the form of the velocity distribution in the turbulent flow, compared to that of the laminar flow. What do you notice and why?

16) What are *smooth pipes* and which is the relation that gives the friction coefficient for a turbulent flow in smooth pipes according to Blasius and which according to Nikuradse for the distribution of the velocity inside them?

17) How is the roughness of a pipe expressed and what do you know about the Moody diagram; how is it expressed mathematically and which flow areas are clearly expressed in it? Give these areas schematically.

18) What are local energy losses in the flow of a real incompressible fluid in cylindrical pipes, which relation expresses them and which is the expression of the equation of the mechanical energy (Bernoulli equation) in connection with the energy heads and the total energy losses?

19) What is *equivalent length*, what is its relation with the coefficient of the local losses and how is it proved?

20) Calculate the *total losses coefficient* with the assistance of the Darcy–Weisbach equation and what is the total coefficient of local losses?

21) What are *energy line* and *hydraulic line* of a flow with friction? Draw them for the flow of an incompressible real fluid in pipes.

22) What influences the head losses Σh in the flow in cylindrical pipes more and what is its effect on the local losses as well?

23) Which and which form of *equations* are available for the solution of the flow in pipes?

24) How many and which *problem categories* are in the fluid's flow in pipes, and which quantities are given for each category?

25) What is the *assumption of intense Turbulence* (AIT), in which problems in pipes is it used and what is the methodology of this assumption?

26) What is *inclination of the energy line*, which equation expresses it and what do we have to know in order to draw the energy line and the hydraulic line? What happens when a device is inserted in the fluid's route and how is the inclination varied by it and the relevant position of the E.L. and the H.L.?

27) What happens if a pump is inserted in the flow with the E.L. and H.L., what happens when the fluid goes through a turbine and how does the *manometer head* vary?

28) What is *pipe connection* in the flow of a real incompressible fluid, which has a full application, from which factors is it limited and on which equation is it based?

29) Which are the total heads of the flow in two cross-sectional areas of a pipeline and how is the equation of the mechanical energy written with the form of these heads?

30) Which equation gives the total head of the losses in a pipeline on which lies both a pump and a turbine?

31) How many categories of pipeline nets do we have and where do they appear in practice?

32) What is the *form of the mechanical energy* in a simple pipeline? Give a sketch of such a pipeline. When is a pipeline called *simple*?

33) What is *connection of pipes in a row* and which equations define it and in which form?

34) How many and which problems appear during the study of flow in pipes connected in a row?

35) What is *parallel connection* of pipes, which equations define it and how many and which problem categories appear? What is a *node* or *branch*?

36) When is a pipe connection called *mixed*, which equations system define it and which is the methodology for solving the problems of a flow in such a connection?

37) What is pipe *branch*, which equations define the flow in a branch of three pipes and what is the *problem of the three tanks*?

38) Which equations define the flow in the problem of the three tanks, which values groups do we meet in these problems and in how many and which categories are they classified?

5.14. Problems with solutions

1) In a pipe of diameter 8 cm, two fluids flow: water at temperature 10°C and oil SAE at 20°C. Calculate the highest velocities that both these fluids can have, so that their flow remains laminar, if the kinematic viscosity of the water is $v_N = 1 \cdot 10^{-6} m^2 / \sec$ and that of the oil is $v_\Lambda = 120 \cdot 10^{-6} m^2 / \sec$.

Solution

We know that the Reynolds number for pipes is given by the relation:

$$\mathrm{Re} = \frac{V \cdot d}{v} \quad (1)$$

We also know that the highest value of the Reynolds number in order to have laminar flow is:

$$\mathrm{Re}_N = \mathrm{Re}_\Lambda = 2.100$$

If we solve (1) for V and substitute, we have:

$$V_N = \frac{\mathrm{Re} \cdot v_N}{d} = 0,026 m / \sec$$

and $V_\Lambda = \dfrac{\mathrm{Re} \cdot v_\Lambda}{d} = 3,15 m / \sec$

2) A pipe is connected to a tank with water flow at velocity 3.5 m/sec. If the length of the pipe is 20 m and the diameter is 3 cm, calculate the transitional length of the entrance from the tank to the pipe $\left(v = 10^{-6} m^2 / \sec\right)$.

Solution

The Reynolds number of the flow is:

$$\mathrm{Re} = \frac{V \cdot d}{v} \Rightarrow \mathrm{Re} = 105000 = 1,05 \cdot 10^5$$

Therefore, the flow is turbulent. However, we know that the ratio $\ell_\varepsilon / d = F(\mathrm{Re})$, thus:

$$\ell_\varepsilon / d = 4{,}4 \cdot \mathrm{Re}^{1/6} \Rightarrow \ell_\varepsilon / d = 30{,}2 \Rightarrow \ell_\varepsilon = 0{,}91m \,.$$

We note that the length ℓ_ε is often not negligible compared to the total length of the pipe (in this example, it constitutes 4.5%).

The most important consequence of the existence of this inlet area is the increased losses due to the turbulences.

3) The velocity distribution for a turbulent flow in a pipe is given approximately by the relation $V(r) = V_{\max}\left(1-\frac{r}{R}\right)^{1/7}$, where $V_{\max}$ is the maximum velocity of the flow. Calculate its average velocity.

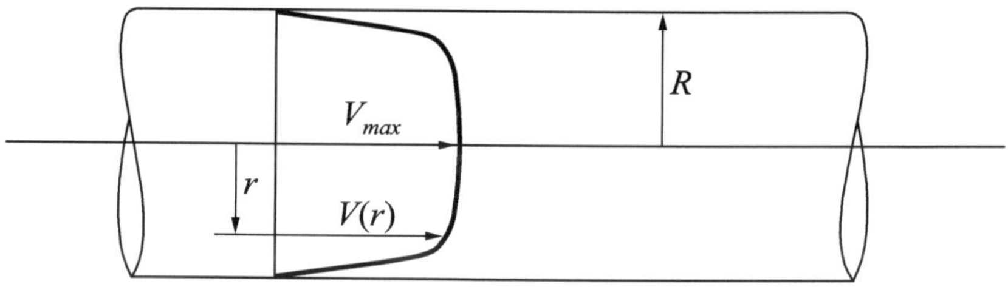

Figure (3)

Solution

The average velocity is given by the relation:

$$V_{md} = \frac{1}{S}\int V dS = \frac{1}{\pi R^2}\int_0^R V_{\max}\left(1-\frac{r}{R}\right)^{1/7} 2\pi r dr = \frac{2\pi V_{\max}}{\pi R^2}\int_0^R \left(1-\frac{r}{R}\right)^{1/7} r dr \,.$$

In order to calculate this integral, we substitute:

$$y = 1 - \frac{r}{R} \text{ or } dy = -\frac{dr}{R},$$

so $r = R(1-y)$ and $dr = -Rdy$.

By substituting, we get:

$$V_{md} = \frac{2V_{\max}}{R^2}\int_1^0 y^{1/7} R(1-y)\,Rdy = \frac{2V_{\max}}{R^2}R^2\int_0^1 (1-y)\,y^{1/7}dy =$$

$$= 2V_{\max}\int_0^1 y^{1/7}dy - 2V_{\max}\int_0^1 y^{8/7}dy = \frac{14V_{\max}}{8} - \frac{14V_{\max}}{15} = 14V_{\max}\left(\frac{1}{8}-\frac{1}{15}\right) =$$

$$14V_{\max}\left(\frac{15-8}{120}\right) = \frac{98}{120}V_{\max} = \frac{49}{60}V_{\max}$$

This means that the average velocity for a turbulent flow, where $V_{\max}$ is the maximum velocity, is given by the relation:

$$V = \frac{49}{60}V_{\max}$$

4) In the pipe shown in Figure (4), oil flows with a constant flow rate $Q = 180m^3/h$ and kinematic viscosity $\nu = 6\cdot 10^5\, m^2/\sec$. If the diameters of pipes (1), (2) and (3) are $d_1 = 4in, d_2 = 7in$ and $d_3 = 2in$, respectively, calculate:

i) The velocities

ii) The kind of the flow in pipes (1), (2) and (3).

Figure (4)

Solution

i) From the relation $Q = \pi d_i^2 \cdot V_i$

we have $V_i = \dfrac{4Q}{\pi \cdot d_i}$

so, by substituting for each pipe, we get:

$$V_1 = 1,54 m/\sec$$

$$V_2 = 0,50 m/\sec$$

and $V_3 = 6,17/\sec$.

ii) The kind of the flow is defined by the Reynolds number. Therefore, by substituting in the relation:

$$\mathrm{Re} = \frac{V \cdot d}{v},$$

we have:

$$\mathrm{Re}_1 = 2607 = \text{transitional flow}$$

$$\mathrm{Re}_2 = 1482 < 2100 = \text{ laminar flow}$$

and $\mathrm{Re}_3 = 5223 > 4000 =$ turbulent flow.

5) A pipe transfers water from quite a big tank to another tank that is 20 m below it. If the water flow rate from the pipe is $0,001 m^3/\sec$, calculate the losses in head values (which means Σ_h in meters).

Solution

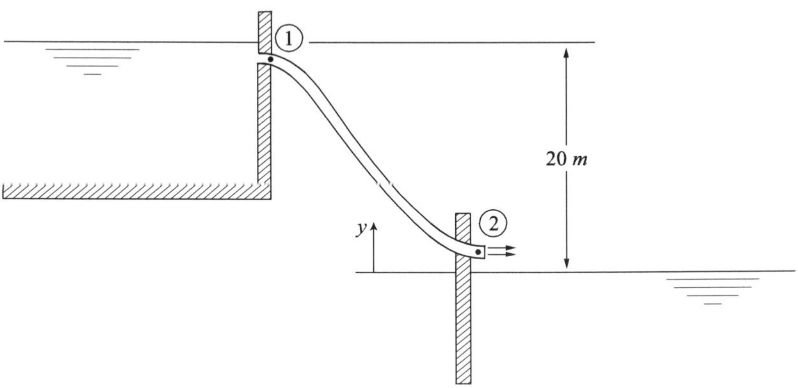

Figure (5)

From the energy equation, we have:

$$\frac{p_1}{\gamma}+h_1+\frac{V_1^2}{2g}=\frac{p_2}{\gamma}+h_2+\frac{V_2^2}{2g}+\Sigma h .$$

Moreover, because $p_1 = p_2 = p_{at}$ and the pipe is considered to have a constant cross-sectional area, $S_1 = S_2$ and Q = flow rate $= V_1S_1 = V_2S_2$, so $V_1 = V_2$, where V_1 and V_2 are the average velocities at points (1) and (2).

The energy equation finally becomes:

$$\boxed{\begin{array}{l} h_1 - h_2 = \Sigma h \\ \Sigma h = 20m \end{array}}$$

6) In a rectilinear cylindrical pipe with diameter 6 cm and length 25 m, water flows at 60 m3/h. Calculate the head losses if the friction coefficient is equal to 0.03.

Solution

From the continuity equation, we calculate the velocity

$$V=\frac{4\cdot Q}{\pi\cdot d^2}\Rightarrow V=\frac{4\cdot\left(60/3600m^3/s\right)}{\pi\cdot(0,06)^2}\Rightarrow V=5,7m/\sec .$$

By substituting in the equation of the linear losses, we calculate the linear losses:

$$h_f=f\cdot\frac{\ell}{d}\cdot\frac{V^2}{2g}\Rightarrow h_f=0,03\cdot\frac{25m}{0,06m}\cdot\frac{(5,7m/s)^2}{2\cdot 9,81m/s^2}\Rightarrow h_f=22,14m .$$

7) In a horizontal pipe of length $\ell = 100\,ft$ and diameter $d = 2in$, oil flows at a relative density $\rho_{rel} = 0,87$ with viscosity $0,1kg/m\cdot\sec$. If the oil flow rate is $60ft^3/\min$, calculate the head losses.

Solution

We calculate the flow velocity:

$$V = \frac{4 \cdot Q}{\pi \cdot d^2} \Rightarrow V = 3,4m/s$$

We calculate the Reynolds number to identify the kind of the flow:

$$\mathrm{Re} = \frac{V \cdot d}{\nu} = \frac{V \cdot d \cdot \rho}{\mu} \Rightarrow \mathrm{Re} = 1543 .$$

Therefore, the flow is laminar $(\mathrm{Re} = 1543 < 2300)$ and for the head of the linear losses, the following relation is applicable:

$$h_f = \frac{32 \cdot \nu \cdot \ell}{g \cdot d^2} \cdot V = \frac{32 \cdot (0,1/870) \cdot 30,48}{9,81 \cdot (0,0508)^2} \cdot 3,5m \Rightarrow h_f = 15,5m ,$$

because: $\nu = \mu/\rho = (0,1Kg/ms)/(870Kg/m^3)$.

8) Calculate the losses during a fluid flow at $0,3m^3/\sec$ with kinematic viscosity $\nu = 9 \times 10^{-6} m/\sec$ in a horizontal pipe made of cast iron of length 500 m and diameter 20 cm.

Solution

Given the length and diameter of the pipe, from the roughness table, we have:

$$\varepsilon = 0,0259cm , \tag{1}$$

$$\text{so } \frac{\varepsilon}{d} = \frac{0,0259cm}{20cm} = 0,001295 . \tag{2}$$

We calculate the Reynolds number as:

$$\mathrm{R_e} = \frac{Vd}{\nu} . \tag{3}$$

From the continuity equation, we have:

$$Q = VS \text{ and } V = \frac{Q}{S} = \frac{4Q}{\pi d^2} \tag{4}$$

Substituting (4) into (3) gives:

$$R_e = \frac{4Qd}{\nu \pi d^2} = \frac{4Q}{\nu \pi d} = \frac{4 \times 0,3}{9 \times 10^{-6} \times \pi \times 0,2} = \frac{1,2}{1,8\pi} \times 10^{-6}$$

and $\mathrm{Re} = 2,12 \times 10^5$. (5)

Finally, from the Moody diagram for $\frac{\varepsilon}{d} = 0,001295$ and $\mathrm{Re} = 2,12 \times 10^5$, we find $f = 0,0215$, so:

$$h_1 = f \frac{\ell}{d} \frac{V^2}{2g} = 0,0215 \times \frac{500}{0,2} \times \left(\frac{4Q^2}{\pi d^2} \right) \frac{1}{2 \times 9,81} =$$

$$= \frac{0,0215 \times 500 \times 16 \times 0,09}{0,2 \times \pi^2 \times (0,2)^4 \times 2 \times 9,81} m = 249,8m. \tag{6}$$

9) Water at temperature 15.5°C flows in a horizontal cylindrical pipe made of concrete with absolute roughness $\varepsilon = 0.1$ cm and diameter 40 cm. If the flow losses are 150 m for every 1,000 m, calculate with relevant accuracy the flow rate in m^3/min.

Solution

We calculate the quotient $\frac{\varepsilon}{d}$

$$\frac{\varepsilon}{d} = \frac{0,1cm}{40cm} = 2,5 \times 10^{-3} = 0,0025. \tag{1}$$

For $\frac{\varepsilon}{d} = 0,0025$, let us take $f = 0,025$ because in the Moody diagram for a fully developed turbulent flow, the line $\frac{\varepsilon}{d} = 0,0025$ extended segments of the axis of f at point $f = 0,025$.

From the Darcy–Weisbach relation, we have:

$$h_1 = f\frac{\ell}{d}\frac{V^2}{2g}. \tag{2}$$

Solving for velocity V, we get:

$$V = \sqrt{\frac{2gh_1 d}{f\ell}} = \sqrt{\frac{2\times 9,81\times 150\times 0,4}{0,025\times 1000}\frac{m}{\sec}} \Rightarrow \tag{3}$$

$$V = 6,86\frac{m}{\sec}. \tag{4}$$

Knowing the velocity, we calculate the Reynolds number (where v is the kinematic viscosity of the water, at 15.5°C given by relating tables) which is $v = 1,12\times 10^{-6}\, m^2/\sec$, so:

$$R_e = \frac{Vd}{v} = \frac{6,86\times 0,4}{1,12\times 10^{-6}} = 2,45\times 10^6. \tag{5}$$

With $\frac{\varepsilon}{d} = 0,0025$ and $\mathrm{Re} = 2,45\times 10^6$ from the Moody diagram, we find the coefficient f.

Indeed, the values of f that correspond to the relative roughness $0,0025$ and $\mathrm{Re} = 2,45\times 10^6$ are the same as the ones we assumed, which is $f = 0,025$.

So, the flow rate is:

$$Q = VS = \frac{\pi d^2}{4}V \Rightarrow$$

$$\Rightarrow Q = \frac{\pi\times (0,4)^2\times 6,86}{4}\frac{m^3}{\sec} = \frac{\pi\times 0,16\times 6,86}{4}\times 60\frac{m^3}{\min}.$$

Therefore: $Q = 51,72\dfrac{m^3}{\min}$. (6)

10) We consider a pipe made of asphalted cast iron with diameter $d = 6cm$ and inside it water flows at 20°C. If the flow rates are 20, 40 and $80m^3 / s$, calculate for these flow rates the friction coefficients (given: $\nu_{H_2O} = 10^{-6} m^2 / \sec$).

Solution

Friction coefficient f is calculated using relative roughness ε/d and Reynolds number as:

$$f = F(\mathrm{Re}, \varepsilon/d)$$

From the relevant table, we get the roughness of the asphalted cast iron: $\varepsilon = 0,012cm$ and calculate the relative roughness:

$$\varepsilon/d = 0,002 . \quad (1)$$

Afterward:

a) We calculate the velocities from the continuity equation:

$$V_i = \frac{4 \cdot Q}{\pi \cdot d_i^2} \Rightarrow V_i = \frac{4 \cdot (20/3600)}{\pi \cdot 0,06^2} m/s \Rightarrow$$

$$V_1 = 1,965 m/s$$

$$V_2 = 3,93 m/s$$

$$V_3 = 7,86 m/s .$$

b) We calculate the Reynolds numbers: $\mathrm{Re}_i = \frac{V_i \cdot d}{\nu} \Rightarrow$

$$\mathrm{Re}_1 = 1,18 \cdot 10^5$$

$$\mathrm{Re}_2 = 2,36 \cdot 10^5$$

$$\mathrm{Re}_3 = 4,72 \cdot 10^5 .$$

With the defined pairs $(\mathrm{Re}, \varepsilon/d)$, we look up the Moody diagram and calculate the respective *f*:

$$\left(\mathrm{Re}_1 = 1{,}18 \cdot 10^5, \varepsilon/d = 0{,}002\right) \xrightarrow{Moody} f = 0{,}025 \text{ (turbulent flow).}$$

$$\left(\mathrm{Re}_2 = 2{,}36 \cdot 10^5, \varepsilon/d = 0{,}002\right) \xrightarrow{Moody} f = 0{,}024 \text{ (turbulent flow).}$$

$$\left(\mathrm{Re}_3 = 4{,}72 \cdot 10^5, \varepsilon/d = 0{,}002\right) \xrightarrow{Moody} f = 0{,}0238 \text{ (absolutely turbulent flow).}$$

11) In a pipe made of cast iron with diameter 8 cm, fluid, with kinematic viscosity $2 \cdot 10^5 \, m^2 / s$, flows at a velocity of 3 m/s. Calculate the friction coefficient *f*.

Solution

– From the relevant table, we get the roughness of the cast iron: ε = 0.026 cm.

– We calculate the relative roughness and the Reynolds number as:

$$\varepsilon/d = (0{,}026cm)/(10cm) = 0{,}0026$$

$$\mathrm{Re} = \frac{V \cdot d}{v} = \frac{(3m/s) \cdot (0{,}1m)}{2 \cdot 10^{-5} m^2 / s} = 1{,}5 \cdot 10^4$$

With these values, we go to the Moody diagram and we have:

$$\left(\mathrm{Re} = 1{,}5 \cdot 10^4, \varepsilon/d = 0{,}0026\right) \longrightarrow f = 0{,}0325 \text{ (Moody).}$$

12) Calculate the type of flow and the friction coefficient using the Moody diagram for a pipe made of cast iron with diameter 6 cm and for a pipe made of stainless steel with diameter 8 cm, if water flows through them at 20°C with velocities of $1, 3, 5, 7 m / \sec$.

$$\left(v_{H2O} = 10^{-6} m^2 / \sec\right)$$

Solution

From the relevant table, we find the pipes' roughnesses as:

$$\varepsilon_{cast} = 0,026cm = 0,00026m,$$

$$\varepsilon_{steel} = 0,0002cm = 0,000002m.$$

For each case, we calculate the relative roughness ε/d and Re.

Thus, we have:

i) $\left(\mathrm{Re} = 6,0 \cdot 10^4, \varepsilon/d = 0,0043\right) \xrightarrow{Moody} f = 0,031$ (turbulent).

ii) $\left(\mathrm{Re} = 1,8 \cdot 10^5, \varepsilon/d = 0,0043\right) \xrightarrow{Moody} f = 0,30$ (turbulent).

iii) $\left(\mathrm{Re} = 3,0 \cdot 10^5, \varepsilon/d = 0,0043\right) \xrightarrow{Moody} f = 0,023$ (absolutely turbulent).

iv) $\left(\mathrm{Re} = 4,2 \cdot 10^5, \varepsilon/d = 0,0043\right) \xrightarrow{Moody} f = 0,023$ (absolutely turbulent).

v) $\left(\mathrm{Re} = 8,0 \cdot 10^4, \varepsilon/d = 0,000025\right) \xrightarrow{Moody} f = 0,019$ (turbulent).

vi) $\left(\mathrm{Re} = 2,4 \cdot 10^5, \varepsilon/d = 0,000025\right) \xrightarrow{Moody} f = 0,0153$ (turbulent).

vii) $\left(\mathrm{Re} = 4,0 \cdot 10^5, \varepsilon/d = 0,000025\right) \xrightarrow{Moody} f = 0,014$ (turbulent).

viii) $\left(\mathrm{Re} = 5,6 \cdot 10^5, \varepsilon/d = 0,000025\right) \xrightarrow{Moody} f = 0,0135$ (turbulent).

From the Moody diagram, we can see that the second pipe is almost smooth.

13) In a horizontal pipe made of commercial steel with length $\ell = 2500m$, fluid with kinematic viscosity $v = 10^{-5} m^3 / \sec$ flows at $Q = 30m^3 / \min$. If the flow's losses are 300 m, calculate the pipe's diameter *d*.

Solution

i) From the continuity and losses relationships for solving *d*, we have:

$$d = \left(\frac{8\ell Q^2}{\pi^2 g h_1} \right)^{0,2} f^{0,2} \text{ with flow losses } h_1 = 300m.$$

ii) Assuming that $f = 0,017$, for $Q = 30\frac{m^3}{\min} = 0,5\frac{m^3}{\sec}$, we have:

$$d = \left(\frac{8 \times 2.500 \times 0,25}{\pi^2 \times 9,81 \times 300} \right)^{0,2} \times (0,017)^{0,2}\ m = 0,3114m.$$

iii) We calculate the Reynolds number:

$$\mathrm{Re} = \frac{Vd}{v} = \frac{4Q}{\pi d v} = \frac{4 \times 0,5}{\pi \times 0,3114 \times 10^{-5}} = 2,03 \times 10^5.$$

iv) We find the relative roughness $\frac{\varepsilon}{d}$ for the commercial steel, where from the table the roughness is given as $\varepsilon = 0,0046cm$. Therefore:

$$\frac{\varepsilon}{d} = \frac{0,0046}{31,14} = 1,477 \times 10^{-4} = 0,0001477.$$

v) With $\mathrm{Re} = 2,05 \times 10^5$ and $\frac{\varepsilon}{d} = 0,0001477$, from the Moody diagram, we find the friction coefficient:

$$f = 0,017.$$

vi) In step ii, we assumed that $f = 0,017$. In step v, we found $f = 0,017$. There is no difference and hence we need not repeat the circle with new value of *f* because the two values coincide. Therefore:

vii) $d = 31cm$

14) Calculate the sum of the coefficients of the local losses in the pipeline shown in the figure below if the pipe has a diameter of 2 in.

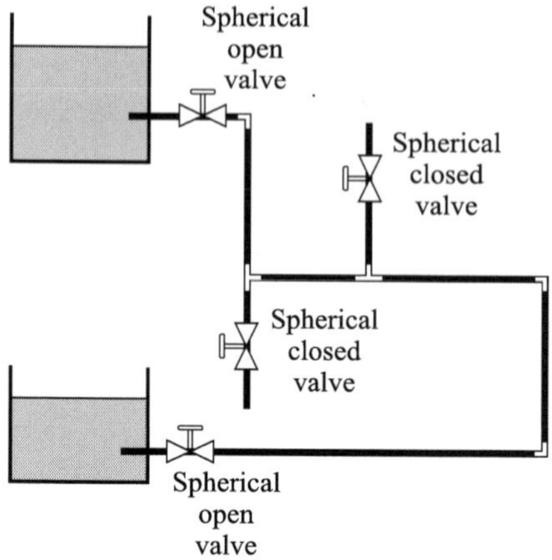

Figure (14)

Solution

We have local disorders of the flow at the inlet from the first tank to the pipe, at the exit from the pipe to the second tank, at the two open spherical valves, at the two tees (rectilinear parts) and at the three (simple) angles 90°. The editing of a table makes the calculation easier.

Kind of local disorder	**Source**	K_i
Entrance to a pipe with extending lips	Table 5.4	0.8
Exit from a pipe to a tank	Table 5.4	1.0
Spherical valves open 2″ (two)	Table 5.5	2 × 6.9
Tee of rectilinear flow	Table 5.6	0.9
Tee of vertical flow	Table 5.6	1.4
Angles 90°, simple (three)	Table 5.4	3 × 1.2
TOTAL		**ΣK = 21.5**

Thus, ΣK=21.5.

15) A pipe made of asphalted cast iron with length $\ell = 50m$ and diameter $d = 2in$ connects two tanks, as shown in Figure (15). The free surfaces of the tanks have a distance of 15 m and roughness of $0,00012$. If the kinematic viscosity of the water is $v = 10^{-6} m^2 / \sec$ and if there are two open spherical screwless valves and two simple bends 90° each in the pipe, calculate the flow rate of the water.

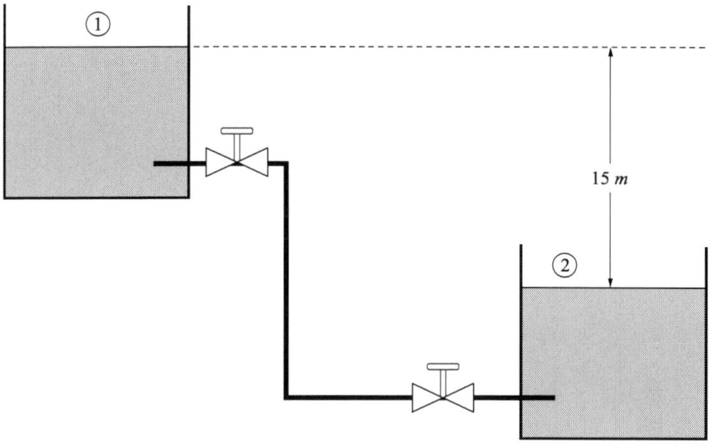

Figure (15)

Solution

We will use the following equations:

$$\text{Total losses } \Sigma h = \left(f \cdot \frac{\ell}{d} + \Sigma K_i \right) \cdot \frac{V_2}{2g} \tag{1}$$

$$\text{Continuity } Q = \frac{\pi \cdot d^2}{4} \cdot V \Rightarrow V = \frac{4 \cdot Q}{\pi \cdot d^2} \tag{2}$$

$$\text{Friction coefficient } f = F\left(\text{Re} = \frac{V \cdot d}{\nu}, \frac{\varepsilon}{d} \right) \tag{3}$$

$$\text{Energy equation } \frac{p_1 - p_2}{\gamma} + \frac{h_1^2 - h_2^2}{2 \cdot g} + (h_1 - h_2) = \Sigma h \tag{4}$$

We can calculate the flow rate from the continuity equation (2) if the velocity has been calculated previously. This can be done using the Darcy–Weisbach equation (1). However, for (1), in addition to the diameter d and the length ℓ of the pipe, which are given, we must know the total losses Σh, the friction coefficient f and the sum of the coefficients of the total losses ΣK_i. The difficulty lies, as we will find out, in calculating the friction coefficient f accurately.

Therefore, the local losses Σh are calculated if we apply the energy equation (4) between the free surfaces of the two tanks.

We know: $y_1 - y_2 = 15m$.

It is: $V_1 = V_2 = 0, p_1 = p_2 = p_{atm} \Rightarrow p_1 - p_2 = 0$.

So, from (4) $\Rightarrow (y_1 - y_2) = \Sigma h \Rightarrow \Sigma h = 15m$.

The tank is considered to be a pipe of infinite diameter and therefore has zero velocity:

$$Q = \frac{\pi \cdot d_1^2}{4} \cdot V_1 = \frac{\pi \cdot d^2}{4} \cdot V \Rightarrow V_1 = \frac{d^2}{d_1^2} \cdot V \Rightarrow V_1 = 0{,}0006 \cdot V_1 \cong 0$$

Then, the *total coefficient of the local losses*, $\mathbf{\Sigma K_i}$, is calculated by taking into account the given data from the problem:

The local disorders from point 1 to point 2 are as follows:

– Inlet in a pipe that extends: $K_1 = 0{,}8$ (Table 5.4)

– Exit from a pipe to a tank: $K_2 = 1{,}0$ (Table 5.4)

– Two (simple) angles $90°$: $K_3 = 2 \cdot 1{,}2$ (Table 5.4)

– Two spherical screwless valves: $K_4 = 2 \cdot 6{,}9$ (Table 5.5)

Adding: $\mathbf{\Sigma K_i = 17{,}6}$.

In order to calculate the friction coefficient *f*, we must know the relative roughness of the pipe ε/d and the Reynolds number Re. From the given data of the exercise and from the data that we have already calculated, we note that we cannot calculate the Reynolds number, because we do not know the flow velocity as the continuity equation (2) has two unknowns: velocity *V* and *Q*.

Thus, although we have *three relationships*, the continuity equation, the losses equation and the Moody diagram, and *three unknowns*, *Q*, velocity *V* and friction coefficient *f*, we *cannot* solve the equations system to calculate the three unknowns, because the third relation is not an equation, but expresses the Moody diagram.

Therefore, we are obliged to follow an *iterative* procedure assuming the asked variables and *correcting* the result each time.

Thus, we assume some value of Q. Assuming Q', we calculate the losses $\Sigma h'$ that correspond to it. Because we know that losses $\Sigma h = 15$, we compare the ones we found with these and we check whether the flow rate we ask is higher or lower than the flow rate Q'. According to the losses equation:

i) if $\Sigma h' > \Sigma h \Rightarrow Q' > Q$;

ii) if $\Sigma h' < \Sigma h \Rightarrow Q' < Q$.

Depending on the result, we proceed with a new assumption, and the whole procedure concludes when we find:

$$\Sigma h' = \Sigma h \Rightarrow Q' = Q.$$

Specifying the whole procedure and using the given data from the problem, we have:

– Assuming that $Q' = 60m^3/h = 60/3600m^3/\sec$

$(2) \Rightarrow V' = 8,22m/\sec$

$\left(\varepsilon/d = 0,00236, \mathrm{Re} = Vd/v = 4,18\cdot 10^5\right) \rightarrow f = 0,025$ (Moody)

$(1) \Rightarrow \Sigma h' = 145,7m > 15m \Rightarrow \Sigma h' > \Sigma h \Rightarrow Q' > Q \Rightarrow \mathbf{Q < 60m^3/h}$.

We repeat by decreasing Q':

– Assuming that $Q' = 20m^3/h = 20/3600m^3/\sec$

$(2) \Rightarrow V' = 2,74m/\sec$

$\left(\varepsilon/d = 0,00236, \mathrm{Re} = Vd/v = 1,39\cdot 10^5\right) \rightarrow f = 0,0255$ (Moody)

$(1) \Rightarrow \Sigma h' = 16,36m > 15m \Rightarrow \Sigma h' > \Sigma h \Rightarrow Q' > Q \Rightarrow \mathbf{Q < 20m^3/h}$.

– Assuming that $Q' = 18m^3 / h = 18 / 3600m^3 / \sec$

$$(2) \Rightarrow V' = 2,74m / \sec$$

$$\left(\varepsilon/d = 0,00236, \mathrm{Re} = Vd/v = 1,25 \cdot 10^5\right) \rightarrow f = 0,0256 \text{ (Moody)}$$

$$(1) \Rightarrow \Sigma h' = 13,29m < 15m \Rightarrow \Sigma h' < \Sigma h \Rightarrow Q' < Q \Rightarrow \mathbf{Q > 18m^3 / h}.$$

We have already started approaching the flow rate: $18m^3 / h < Q < 20m^3 / h$. We continue:

– Assuming that $Q' = 19m^3 / h = 19 / 3600m^3 / \sec$

$$(2) \Rightarrow V' = 2,6m / \sec$$

$$\left(\varepsilon/d = 0,00236, \mathrm{Re} = Vd/v = 1,32 \cdot 10^5\right) \rightarrow f = 0,0256 \text{ (Moody)}$$

$$(1) \Rightarrow \Sigma h' = 14,79m < 15m \Rightarrow \Sigma h' < \Sigma h \Rightarrow Q' < Q \Rightarrow \mathbf{Q > 19m^3 / h}.$$

– Assuming that $Q' = 19,2m^3 / h = 19,2 / 3600m^3 / \sec$

$$(2) \Rightarrow V' = 2,63m / \sec$$

$$\left(\varepsilon/d = 0,00236, \mathrm{Re} = Vd/v = 1,34 \cdot 10^5\right) \rightarrow f = 0,0256 \text{ (Moody)}$$

$$(1) \Rightarrow \Sigma h' = 15,10m > 15m \Rightarrow \Sigma h' > \Sigma h \Rightarrow Q' > Q \Rightarrow \mathbf{Q < 19,2m^3 / h}.$$

Because the value that we have found with the previous repetitions, which is the value $Q \cong 19,2m^3 / h$, satisfies us, we stop here.

If we want higher accuracy, we continue with a smaller value for Q' like $Q' = 19,15m^3 / h$, which gives us $\Sigma h' = 15,02m$ and $Q' = 19,14m^3 / h$, which gives us $\Sigma h' = 15,002m$ and therefore we have finally reached the solution.

$$\Sigma h' = \Sigma h = 15 .$$

Therefore, Q is:

$$Q = 19{,}14 m^3 / h = 0{,}00532 m^3 / \text{sec}$$

From the above procedure, we conclude that the more logical the first random assumption is, the less tests will be needed.

The solving procedure is checked better and exhibited more elegantly, if we make a relevant table with the results of the tests gathered.

Q' (m³/h)	***v' (m/s)***	***Re***	***f***	***Σh'***	***Σh'-Σh***	***Q***
60	8.223	$4.18 \cdot 10^5$	0.0249	145.07	130.07	Q < 60
20	2.741	$1.39 \cdot 10^5$	0.0255	16.36	1.36	Q < 20
18	2.467	$1.25 \cdot 10^5$	0.0256	13.29	−1.71	Q > 18
19	2.604	$1.32 \cdot 10^5$	0.0256	14.79	−0.21	Q > 19
19.2	2.631	$1.34 \cdot 10^5$	0.0256	15.10	0.10	Q < 19.2
19.15	2.625	$1.33 \cdot 10^5$	0.0256	15.02	0.02	Q < 19.15
19.14	2.623	$1.33 \cdot 10^5$	0.0256	15.002	0.00	Q = 19.14

16) Calculate the flow rate of the water of the previous problem with the *assumption of intense turbulence* (AIT).

Solution

i) With the assumption that we are in an area of intense turbulence, from the Moody diagram, knowing the relative roughness ε/d, we calculate the friction coefficient f:

$$(\varepsilon/d = 0{,}00236 \text{AIT}) \rightarrow f = 0{,}0245 \text{ (Moody)}.$$

ii) We calculate the (assumptive) value of the flow rate:

From (1) of problem (15), we have:

$$(15) = \frac{8}{\pi^2 \cdot 9{,}81 \cdot 0{,}0508^4} \cdot \left(0{,}0245 \cdot \frac{50}{0{,}0508} + 17{,}6\right) \cdot Q^2 \Rightarrow$$

$$\Rightarrow Q' = 0{,}005384 m^3 / s = 19{,}38 m^3 / h .$$

iii) We check whether the assumption is valid. From (2) in problem (15), we have:

$V = 2,656m / s$ so:

$\mathrm{Re} = Vd / v = 1,3 \cdot 10^5 \rightarrow f = 0,0256$ (Moody).

If $f' \cong f$, the assumption is verified and therefore, $Q = Q'$.

If $f' < f$, we repeat the calculation with the new value of *f*. In our example, we have a remarkable regression (4.5%), so we continue:

iv) We correct the value of the flow rate, making use of the new value of *f*:

So again from (1) of (15), we have:

$$(15) = \frac{8}{\pi^2 \cdot 9,81 \cdot 0,0508^4} \cdot \left(0,0256 \cdot \frac{50}{0,0508} + 17,6\right) \cdot Q^2 \Rightarrow$$

$$\Rightarrow Q = 0,005317m^3 / s = 19,14m^3 / h .$$

Theoretically, we must once again check the value of *f* and, if needed, correct it. However, in practice, after the first correction, the regression is negligible. Let us carry on with the control:

$$(2) \Rightarrow V = 2,623m / \sec .$$

$$\mathrm{Re} = Vd/v = 1,33 \cdot 10^5 \rightarrow f = 0,0256 \text{ (Moody)}$$

Therefore, the above relation of Q is the one we asked for:

$$Q = 0,005317m^3 / s = 19,14m^3 / h .$$

17) In a pipe made of galvanized iron, which has a length of 20 m, through which water flows at 20°C, we want to have $Q = 60m^3 / h$. If the total losses are 8 m maximum and the total losses coefficient is $\Sigma K = 4$, calculate the minimum diameter that this pipe should have.

Solution

The relationships that we have for the solution of this category problems are:

Total losses equation: $\Sigma h = \left(f \cdot \frac{\ell}{d} + \Sigma K_i \right) \cdot \frac{V^2}{2g}$ (1)

Continuity equation: $Q = \frac{\pi \cdot d^2}{4} \cdot V \Rightarrow V = \frac{4 \cdot Q}{\pi \cdot d^2}$ (2)

Relation of the Moody diagram: $f = F\left(\mathrm{Re} = \frac{V \cdot d}{\nu}, \frac{\varepsilon}{d} \right)$ (3)

Energy equation: $\frac{p_1 - p_2}{\gamma} + \frac{V_1^2 - V_2^2}{2 \cdot g} + (h_1 - h_2) = \Sigma h$. (4)

We note once again that we have three unknowns (V, f, d) and three available relationships (continuity equation, losses equation and Moody diagram). The energy equation cannot be used because an energy value is neither given nor asked.

i) Assuming that $d' = 10cm = 0{,}10m$

(2) $\Rightarrow V' = 2{,}122 m/s$

$\left(\mathrm{Re}' = Vd/\nu = 2{,}1 \cdot 10^5, \varepsilon/d = 0{,}0015 \right) \rightarrow f = 0{,}0227$ (Moody)

(1) $\Rightarrow \Sigma h' = 1{,}96m < 8m \Rightarrow d \leq 10cm$

We try with a smaller diameter:

ii) Assuming that $d' = 5cm = 0{,}05m$

(2) $\Rightarrow V' = 8{,}49 m/s$

$\left(\mathrm{Re}' = 4{,}2 \cdot 10^5, \varepsilon/d = 0{,}003 \right) \rightarrow f = 0{,}0265$ (Moody)

(1) $\Rightarrow \Sigma h' = 53{,}62m > 8m \Rightarrow d > 5cm$

iii) Assuming that $d' = 7cm = 0{,}07m$

$(2) \Rightarrow V' = 4,33m / s$

$\left(\text{Re}' = 3,0 \cdot 10^5, \varepsilon/d = 0,00214\right) \rightarrow f = 0,0244$ (Moody)

$(1) \Rightarrow \Sigma h' = 10,49m > 8m \Rightarrow d > 7cm$

iv) Assuming that $d' = 8cm = 0,08m$

$(2) \Rightarrow V' = 3,32m / s$

$\left(\text{Re}' = 2,7 \cdot 10^5, \varepsilon/d = 0,00187\right) \rightarrow f = 0,0237$ (Moody)

$(1) \Rightarrow \Sigma h' = 5,56m < 8m \Rightarrow d \leq 8cm$

From the tests we have carried out so far, we conclude that the asked diameter lies between 7 and 8 cm: $7cm < d \leq 8cm$. If the data of the problems do not indicate the diameter values between 7 and 8 cm, then we would choose the value as 8 cm (the value $d = 7cm$ is rejected as it does not meet the demands of the problems). However, we continue for a more satisfying approach, for example, 1 mm:

v) Assuming that $d' = 7,5cm = 0,075m$

$(2) \Rightarrow V' = 3,77m / s$

$\left(\text{Re}' = 2,8 \cdot 10^5, \varepsilon/d = 0,002\right) \rightarrow f = 0,024$ (Moody)

$(1) \Rightarrow \Sigma h' = 7,55m < 8m \Rightarrow d \leq 7,5cm$.

vi) Assuming that $d' = 7,4cm = 0,074m$

$(2) \Rightarrow V' = 3,875m / s$

$\left(\text{Re}' = 2,9 \cdot 10^5, \varepsilon/d = 0,00203\right) \rightarrow f = 0,0241$ (Moody)

$(1) \Rightarrow \Sigma h' = 8,05m > 8m \Rightarrow d > 7,4cm$.

So: $7,4cm < d < 7,5cm \Rightarrow d = 7,5m$.

With this diameter, we will get $Q = 60m^3 / h$ and the total losses of the pipeline will be:

$$\Sigma h = 7,55m < 8m$$

As in problem (15), the construction of a table is suggested for better understanding:

d' (cm)	v' (m/s)	ε/d'	Re'	f	$\Sigma h'$ (m)	$\Sigma h'$ - Σh	d (cm)
10	2.122	0.00150	$2.12 \cdot 10^5$	0.0277	1.96	−6.04	$d \leq 10$
5	8.488	0.00300	$4.24 \cdot 10^5$	0.0265	53.62	45.62	$d > 5$
7	4.331	0.00214	$3.03 \cdot 10^5$	0.0244	10.49	2.49	$d > 7$
8	3.316	0.00188	$2.65 \cdot 10^5$	0.0237	5.56	−2.44	$d \leq 8$
7.5	3.773	0.00200	$2.83 \cdot 10^5$	0.0240	7.55	−0.45	$d \leq 7.5$
7.4	3.875	0.00203	$2.87 \cdot 10^5$	0.0241	8.05	0.05	$d > 7.4$

18) Calculate the internal diameter of a pipe made of galvanized iron of length $\ell = 150m$, able to carry $Q = 0,03m^3 / \sec$ water of temperature 15°C when the pressure drop per unit length of the pipe is $2KPa / m$ (e.g. $2000N / m^2 / m$).

Solution

From the relevant tables, we find that water at 15°C has:

Density: $\rho = 999kg / m^3$

Kinematic viscosity: $v = 1,146 \times 10^{-6} m^2 / \sec$

Also, from roughness tables we find that the absolute roughness for pipes made of galvanized iron is:

$$\varepsilon = 0,0152cm \ .$$

From the data of the problem, we have that the losses (pressure drop) are $\frac{\Delta P}{\ell} = 2000 \frac{N / m^2}{m}$, so in length ℓ, we have pressure drop:

$$\Delta P = 2000 \times 150 = 300000 N / m^2.$$

Therefore, the linear losses, given from the *Darcy–Weisbach* relation, will be:

$$h_\ell = \frac{\Delta P}{\rho g} = f \frac{\ell}{d} \frac{V^2}{2g} = 8 \frac{f \ell Q^2}{\pi^2 g d^5}, \tag{1}$$

so by solving for diameter *d*, we have:

$$d = \left[\frac{8 f \ell Q^2 \rho}{\pi^2 \Delta p}\right] \tag{2}$$

or by substituting the values of the parameters in (2), S.I. units, we have:

$$d = \left(\frac{8\times150\times(0,03)^2\times999}{\pi^2\times3\times10^5} f\right)^{0,2} m = 0,2053 f^{0,2} m\,.$$

Therefore: $d = 0,2053\times f^{0,2} m\,.$ (3)

Then, we write Re of the flow compared to the diameter, and we have:

$$\text{Re} = \frac{Vd}{\nu} = \frac{4Q}{\pi d\nu} = \frac{4\times0,3}{\pi\times1,146\times10^{-6}} d^{-1} = 3,33\times10^4 d^{-1} \Rightarrow$$

$\Rightarrow \text{Re} = 3,33\times10^4 d^{-1}$, where *d* is in m. (4)

The diameter of the pipe must have such a value that from the pair $\left(\frac{\varepsilon}{d}, \text{Re}\right)$ and with the assistance of the Moody diagram arises such a value for the friction coefficient *f* that putting it into relation (3) gives us the value of the diameter.

For the solution of this problem, we conduct circles for approaching calculations of *d* as follows:

First circle: we assume a value for d:

$$d = 5cm = 0,05m = 5\times10^{-2} m\,.$$

We calculate $\text{Re} = 3,33\times\left(5\times10^{-2}\right)^{-1}\times10^4 = 6,66\times10^5$

We calculate: $\frac{\varepsilon}{d} = \frac{0,0152}{5} = 0,003\,.$

From the Moody diagram with $\frac{\varepsilon}{d} = 0,003$ and $\text{Re} = 6,66 \times 10^5$, we find that we have:

$$f = 0,0267 .$$

By substituting this value of *f* in relation (3), we find:

$$d = 0,0995m = 9,95cm .$$

We note that this value differs a lot from the one we assumed, so we repeat another approximate circle with a new value of *d*.

Second circle: assume that $d = 9,95cm$.

Then:

$$\begin{cases} \text{Re} = 3,33 \times (0,0995)^{-1} \times 10^4 = 3,35 \times 10^5 , \\ \frac{\varepsilon}{d} = \frac{0,0152}{9,95} = 0,0015. \end{cases}$$

From the Moody diagram, we find that it corresponds to $f = 0,0227$.

So: $d = 0,2053(0,0227)^{0,2} = 0,0963m = 9,63cm$.

This value is also approximately 3 mm different from the one we assumed during the beginning of the second circle, and hence we also conduct a calculative circle.

Third circle: assume that $d = 9,63cm$.

Then:

$$\begin{cases} \text{Rc} = 3,33 \times 10^4 \times (0,0963)^{-1} = 3,46 \times 10^5 , \\ \frac{\varepsilon}{d} = \frac{0,0152}{9,63} = 0,0016. \end{cases}$$

From the Moody diagram, we find that it corresponds to: $f = 0,0227$,

so: $d = 0,2053(0,0227)^{0,2} = 0,0963m = 9,63cm$.

Here we stop the circles of calculations because in this last circle, the value *d* we calculated coincides with the one at the beginning of the (same) circle.

So:

$$d = 9,63cm$$

19) A pipeline that consists of two horizontal, homo-axial steel pipes (class number 40). The nominal diameter of the internal pipe is 1 in and that of the external pipe is 2 in. The pipeline is used for the transportation of benzene and methanol (at 20°C). Benzene flows in the same area between the two pipes, and methanol flows in the small pipe along the same direction and with a constant velocity of 2 m/s. The pressure at the inlet of both pipes is the same. If at some point of the wall of the internal pipe a small crack is created, which stream inserts into the other?

Given: hydraulic diameter of circular rings $d_h = d_2 - d_1$.

Solution

In order to determine which of the two fluids (benzene or methanol) will insert into the other, we must know in which of the two pipes transporting the two fluids the highest pressure drop occurs along the direction of the flow. If it happens in the small pipe, then benzene will insert into methanol, but if it occurs in the cylinder ring, then methanol will insert into benzene.

For a flow in a horizontal pipe of circular cross-sectional area (like the pipe transporting methanol), the pressure drop is calculated from the equation $\left(h_f = \frac{p_1 - p_2}{\gamma} \right)$, which, combined with the *Darcy–Weisbach equation*, is written as:

$$\frac{\Delta p}{\ell} = \frac{f \rho V^2}{2d}, \qquad (1)$$

where $\Delta p / \ell$ is the pressure inclination along the direction of the flow, f is the friction coefficient, d is the internal diameter of the pipe, ρ is the density and V is the average velocity of the fluid. For the calculation of value f of the friction coefficient, we will use the Moody diagram after we calculate the Reynolds number of the flow and the relative roughness of the pipe. So:

$$\mathrm{Re} = \frac{\rho V d}{\mu} = \frac{\left(788 kg/m^3\right)\left(2m/s\right)\left(2,664\times10^{-2} m\right)}{\left(5,98\times10^{-4} Pa\cdot s\right)} = 7,02\times10^4 . \tag{2}$$

The values for the fluid quantities and the values of the internal and external diameters of the two steel pipes are obtained from the relevant tables. The absolute roughness of the pipe transporting methanol is:

$$\frac{\varepsilon}{d} = \frac{(0,046mm)}{(46,64mm)} = 0,0017 . \tag{3}$$

From the Moody diagram, for $\mathrm{Re} = 7,02\times10^4$ and $\varepsilon/d = 0,0017$, we find $f = 0,0255$. Therefore, we can calculate, from equation (1), the inclination of the pressure in the pipe transporting methanol:

$$\frac{\Delta p}{\ell} = \frac{(0,0255)\left(788 kg/m^3\right)\left(2m/s\right)^2}{2\left(2,664\times10^{-2} m\right)} = 1508 Pa/m . \tag{4}$$

For flow in pipes of non-circular cross-sectional area, it is enough to use the hydraulic diameter of the pipe. Therefore, the pressure inclination in the pipe transporting benzene can be calculated from equation (1), which is modified like:

$$\frac{\Delta p}{\ell} = \frac{f\rho V^2}{2d_h} \tag{5}$$

where d_h is the hydraulic diameter of the circular ring:

$$d_h = d_2 - d_1 = (5,25cm) - (3,34cm) = 1,91cm . \tag{6}$$

For the definition of the value of the friction coefficient f, we will use once again the Moody diagram and then calculate (based on the hydraulic diameter d_h) the Reynolds number of the flow:

$$\mathrm{Re}_h = \frac{\rho V d_h}{\mu} = \frac{\left(881 kg/m^3\right)\left(2m/s\right)\left(1,91\times10^{-2} m\right)}{\left(6,51\times10^{-4} Pa\cdot s\right)} = 5,17\times10^4 \tag{7}$$

and the relative roughness of the pipe:

$$\frac{\varepsilon}{d_h} = \frac{(0,046mm)}{(19,1mm)} = 0,0024\ . \tag{8}$$

From the Moody diagram, for $\text{Re} = 5,17 \times 10^4$ and $\varepsilon/d = 0,0024$, we get the value $f = 0,0275$. We can calculate, from equation (5), the pressure drop in the pipe transporting benzene:

$$\frac{\Delta p}{\ell} = \frac{(0,0275)\left(881kg/m^3\right)(2m/s)^2}{2\left(1,91\times 10^{-2}\, m\right)} = 2537\, Pa/m\ . \tag{9}$$

By comparing equations (4) and (9), it arises that the cylindrical pipe experiences a higher pressure drop than the small pipe. Therefore, methanol will insert benzene.

20) In an air-conditioning installation, the rate of airflow under normal circumstances is $0,5m^3/s$. The pipe transporting air is made of commercial steel and has a square cross-sectional area of side 25 cm. Calculate the frictions head of the flow and the pressure drop for a horizontal pipe of length 50 m.

The active diameter is given as $d_{\varepsilon v} = \dfrac{64d_h}{C}$, where C = constant = 56.9.

Solution

For flow in a pipe of non-circular cross-sectional area, the frictions head h_f of the flow is calculated from the generalized Darcy–Weisbach equation:

$$h_f = f\frac{\ell}{d_h}\frac{V^2}{2g}, \tag{1}$$

where f is the friction coefficient, ℓ is the length of the pipe, d_h the hydraulic diameter, g is the acceleration of gravity and V is the average velocity of the fluid. Of these values, we know $\ell = 50m$ and $g = 9,81m/s^2$. The average velocity of the air is calculated from the equation:

$$V = \frac{Q}{S} = \frac{Q}{a^2} = \frac{\left(0,5m^3/s\right)}{(0,25m)^2} = 8m/\text{sec}\ . \tag{2}$$

The hydraulic diameter of a pipe with square cross-sectional area is equal to the length a of its side.

$$d_h = a = 0,25m \,. \tag{3}$$

For the definition of the friction coefficient f, we will use the Moody diagram to calculate the Reynolds number for the relative roughness, based on the active diameter, d_{ac}, of the pipe. The diameter d_{ac} is calculated from the following equation:

$$d_{ac} = \frac{64d_h}{C} = \frac{64(0,25m)}{56,9} = 0,28m \,. \tag{4}$$

The Reynolds number, $\mathrm{Re}_{d_{ac}}$, of the flow is calculated from the equation as:

$$\mathrm{Re}_{d_{ac}} = \frac{\rho V d_{ac}}{\mu} = \frac{\left(1,204kg/m^3\right)\left(8m/s\right)\left(0,28m\right)}{\left(1,82\times10^{-5}\,Pa\cdot s\right)} = 1,58\times10^5 \,. \tag{5}$$

The absolute roughness of the steel pipes is $\varepsilon = 0,046mm$, so the relative roughness of the airway based on the diameter d_{ac} is:

$$\frac{\varepsilon}{d_{ac}} = \frac{(0,046mm)}{(28mm)} = 0,0016 \tag{6}$$

From the Moody diagram, for $\mathrm{Re} = 1,48\times10^5$ and $\varepsilon/d = 0,0016$, we obtain the value $f = 0,024$. Thus, we can calculate the values we ask: the frictions head, h_f, of the flow, from equation (1):

$$h_f = \frac{(0,024)(50m)(8m/s)^2}{2(0,25m)\left(9,81m/s^2\right)} \simeq 15,7m \tag{7}$$

and the relevant pressure drop, Δp:

$$\Delta p = \varepsilon g h_f = \left(1,204kg/m^3\right)\left(9,81m/s^2\right)(15,7m) = 185Pa \,. \tag{8}$$

21) A smooth pipe of square cross-sectional area is given with side $a = 3$ cm and length $\ell = 10m$. Water with kinematic viscosity $\nu = 10^{-6} m^2/\sec$ flows through the pipe into a tank of capacity $\text{v} = 2m^3$. Calculate the losses (in meters of water column) along the pipe when the tank fills (a) in time $t = 10h$ and (b) in time $t = 20\min$.

Solution

a) In order to fill the tank of volume $\text{v} = 2m^3$ in 10 h, the pipe must have flow rate:

$$Q = \frac{\text{v}}{t} = \frac{2}{10 \times 3600} \frac{m^3}{\sec} = 5,556 \times 10^{-5} m/\sec .$$

The Reynolds number of the flow (in a non-cylindrical pipe) will be:

$$\text{Re} = \frac{V(4R_h)}{\nu}$$, where R_h is the hydraulic radius,

which means that:

$$R_h = \frac{S}{P} = \frac{a^2}{4a} = \frac{a}{4} \; (P = \text{perimeter}) \text{ and } \text{v} = \frac{Q}{a^2}$$

so: $$\text{Re} = \frac{Q}{a^2} \cdot \frac{a}{\nu} = \frac{Q}{a\nu} = \frac{5,556 \times 10^{-5}}{3 \times 10^{-2} \times 10^{-6}} = 1852 .$$

We note that Re=1,852 < 2,300, so the flow is *laminar* and the friction coefficient is given by the known relation:

$$f = \frac{64}{R_e} = \frac{64}{1852} = 0,0346 .$$

Losses from the Darcy–Weisbach relation with $d = 4R_h$ in meters of water column are given by:

$$h_\ell = f \frac{\ell}{d} \frac{V^2}{2g} = f \frac{\ell}{4R_h} \cdot \frac{Q^2}{a^4} \cdot \frac{1}{2g} = \frac{1}{2} \frac{f \ell Q^2}{g a^5} \Rightarrow$$

$$h_\ell = \frac{0,0346\times10\times\left(5,55\times10^{-5}\right)^2}{2\times9,81\times\left(3\times10^{-2}\right)^5}m = \frac{4,48}{2}\times10^{-3}m = 2,24\times10^{-3}m\,.$$

Therefore, the losses in m of water column when the tank fills in 10 h are given by:

$$h_\ell = 2,24\times10^{-3}m\,.$$

b) When the tank fills in $t = 20\,\text{min}$:

$$Q = \frac{\text{V}}{t} = \frac{2}{20\times60}\frac{m^3}{\text{sec}} = 1,667\times10^{-3}m/\text{sec}\,,$$

$$\text{so R}_e = \frac{Q}{av} = \frac{1,667\times10^{-3}}{3\times10^{-2}\times10^{-6}} = 55.556\,.$$

This means that the flow is turbulent, as $R_e > 2{,}300$. However, because the pipe has smooth walls and $R_e = 55556 < 10^5$, we can make use of the Blasius type that gives the friction coefficient for pipes with smooth walls because the Reynolds number of the flow is smaller than 10^5. Thus, the Blasius type for $R_e < 10^5$ is:

$$f = \frac{0,316}{R_e^{\,0,25}}$$

And therefore:

$$f = \frac{0,316}{(55.556)^{0,25}} = 0,02058\,,$$

So, the losses are: $h_\ell = \frac{1}{2}\frac{f\ell Q^2}{ga^5} \Rightarrow$

$$h_\ell = \frac{0,02058\times10\times\left(1,667\times10^{-3}\right)^2}{2\times9,81\times\left(3\times10^{-2}\right)^5}m = 1,20m\,.$$

Therefore, the losses in m of water column when the tank fills in 20 min are:

$$h_\ell = 1,20m .$$

22) Crude oil of kinematic viscosity $\nu = 3\times10^{-4} m^2 / \sec$ and relative density $\rho_r = 0,85$ is transferred from a big tank through a horizontal cylindrical pipe made of cast iron with diameter $d = 16''$ to a distance $\ell = 10 km$. Calculate the power of an electric motor in kW and hp that is needed to move the pump of the pair as well as the consumption of electrical power. The following data are also given: pipe's flow rate $Q = 0,7m^3 / \sec$, and coefficients of the electric motor and the pump, $n_K = 0,95$ and $n_A = 0,7$, respectively.

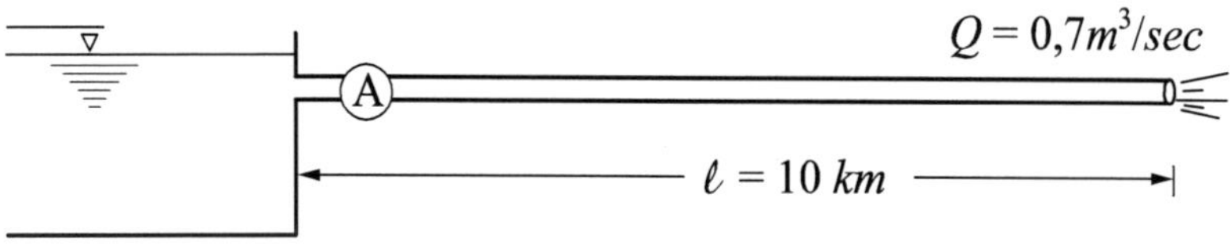

Figure (22)

Solution

From the data of the problem, we find that the Reynolds number of the flow is:

$$R_e = \frac{Vd}{\nu} = \frac{4Q}{\pi d\nu} = \frac{4\times0,7}{\pi\times16\times0,0254\times3\times10^{-4}} = 7.310$$

$$\rho = 1000\rho_{rel} kg / m^3 = 850 kg / m^3 .$$

From the roughness tables, we find for the cast iron:

$$\varepsilon = 0,0102'',$$

so: $\dfrac{\varepsilon}{d} = \dfrac{0,0102''}{16''} = 0,00064$.

From the equation of mechanical energy after simplification of various terms, we finally have that, if:

$$\begin{cases} H_A = \text{energy head offered from the pump to the oil} \\ H_\ell = \text{losses inside the pipe} \end{cases}$$

then: $H_A = H_\ell + \dfrac{V^2}{2g} = f\dfrac{\ell}{d}\dfrac{V^2}{2g} + \dfrac{V^2}{2g} = \left(1 + f\dfrac{\ell}{d}\right)\dfrac{V^2}{2g}$

$$\text{or } H_A = 8\left(1 + f\frac{\ell}{d}\right)\frac{Q^2}{\pi^2 g d^4}. \quad (1)$$

So if,

N_A = power of the exit of the pump (power that offers the pump to the oil

$$= \rho g Q H_A, \quad (2)$$

N_K = exit power of the electric motor $= \dfrac{N_A}{n_A}$

$$\text{or } N_K = \frac{N_A}{n_A} = \frac{\rho g Q H_A}{n_A}, \quad (3)$$

N_{EL} = Electric power consumed by the electric motor

$$= \frac{N_K}{n_K}$$

$$\text{or } N_{EL} = \frac{\rho g Q H_A}{n_A n_K} \quad (4)$$

Therefore, for the calculation of N_K and N_{EL}, we must know H_A from relation (1). The calculation of f in (1) will allow us to calculate H_A afterward.

Therefore, from the Moody diagram, we find what coefficient f corresponds to the pair ($\frac{\varepsilon}{d} = 0,00064$ and $R_e = 7310$). We see that this value is approximately $f = 0,034$.

In this case, we can use the relation:

$$\frac{1}{\sqrt{f}} = 1,74 - 2\ell og\left(2\frac{\varepsilon}{d} + \frac{18,7}{R_e\sqrt{f}}\right), \qquad (5)$$

which is applicable to the value area R_e as the one we have ($R_e = 7310$). However, we note that the values $R_e = 7310$ and $f = 0,034$ satisfy (5). Therefore, we conclude that there is no reason to use (5) because we obtain the solution easily using the Moody diagram. Thus, finally:

$$f = 0,034 .$$

So: $H_A = 8\left(1 + \frac{0,034 \times 10000}{16 \times 0,0254}\right)\frac{(0,7)^2}{\pi^2 \times 9,81 \times (16 \times 0,0254)^4} m \Rightarrow$

$$H_A = 1243,22m$$

Therefore:

$$N_K = \frac{\rho g Q H_A}{n_A} = \frac{850 \times 9,81 \times 0,7 \times 1243,22}{0,7} Watt$$

or $N_K = 10.366,56KW = 13902hp$

And $N_{H\Lambda} = \frac{N_K}{n_K} = \frac{10.366,56}{0,95} KW$

or $N_{EL} = 10.912KW$

23) Calculate the pressure gauge head and the head of the energy line at points A, B, C and D in Figure (23)

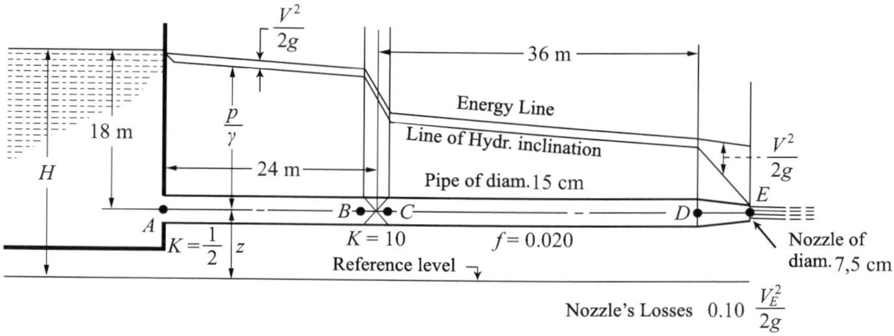

Figure (23)

Solution

If we consider the center of the pipe as the reference level, then both lines will begin at height 18 m.

First, we will solve for the head of the kinetic energy by applying its equation between the tank and point C.

So, we have:

$$18+0+0=\frac{V_E^2}{2g}+0+0+\frac{1}{2}\frac{V^2}{2g}+0,020\cdot\frac{60}{0,85}\cdot\frac{V^2}{2g}+10\frac{V^2}{2g}+0,10\frac{V_E^2}{2g}. \quad (1)$$

From the continuity equation, we have:

$$V_E=4V. \quad (2)$$

Therefore, by substituting (2) into (1) and after simplifications, we have:

$$18+\frac{V^2}{2g}\left(16+\frac{1}{2}+8+10+16\times 0,1\right)=36,1\frac{V^2}{2g}$$

$$\text{and } \frac{V^2}{2g}0,5m \quad (3)$$

a) By now applying the energy equation between the tank and point A, we get:

$$18+0+0=\frac{V^2}{2g}+\frac{p}{\gamma}+z+0,5\frac{V^2}{2g}$$

So, the pressure gauge head is:

$$\left.\frac{p}{\gamma}+z\right|_A=18-1,5\frac{V^2}{2g}-1,5\times 0,5=18,25$$

and the energy head is:

$$\frac{V^2}{2g}+z+\frac{p}{\gamma}=17,5+0,5=17,75m \tag{4}$$

b) For point B:

$$18+0+0=\frac{V^2}{2g}+\frac{p}{\gamma}+z+0,5\frac{V^2}{2g}+0,02\times\frac{24}{0,15}\frac{V^2}{2g}$$

and $\left.\frac{p}{\gamma}+z\right|_B=18-(1,5+3,2)0,5=15,16m$

Thus, the energy line in B lies at the head:

$$15,65+0,5=16,15m \tag{5}$$

c) The line of *hydraulic inclination* goes down the valve and $10V^2/2g$ or 5 m. So, at point C, the *lines of hydraulic inclination and energy* lie at heads 10.65 and 11.15, respectively.

d) At point D, we have:

$$18=\frac{V^2}{2g}+\frac{p}{\gamma}+z+\left(10,5+0,02\times\frac{60}{0,15}\right)\frac{V^2}{2g}$$

and $\left.\frac{p}{\gamma}+z\right|_D=18-19,5\times 0,5=8,25$ (6)

and the energy line has head

$$8,25+0,25=8,75 \tag{7}$$

e) At point E, the *line of hydraulic inclination* lies at point *zero* and the energy line at head:

$$V_E^2 = 16\frac{V^2}{2g} = 16 \times 0,5 = 8m \tag{8}$$

24) A pump of performance 70% with a power of 7.5 KW is connected to a water line that transfers $0,1 m^3/\sec$. The suction pipe diameter is $d_1 = 15cm$ and the discharge pipe diameter is $d_2 = 12cm$. The suction pipe is connected to the pump 1 m lower than the discharge pipe. Calculate the pressure after the pump and the increase in the pressure gauge head because of the pipe, when the suction pressure is $70KN/m^2$.

Solution

i) If E is the energy that is added in $m \cdot N/N$, the power that is attached to the fluid is:

$$Q \cdot \gamma \cdot E = 7.500 \times 0,70 \Rightarrow E = \frac{7.500 \times 0,7}{0,1 \times 9802} = 5.356m \tag{1}$$

If we apply the energy equation for the pump, we get:

$$\frac{V_S^2}{2g} + \frac{p_s}{\gamma} + 0 + 5.356 = \frac{V_a^2}{2g} + \frac{p_a}{\gamma} + 1 \tag{2}$$

where the indicators s and d refer to the suction and the discharge, respectively.

From the continuity equation, we have:

$$V_S = \frac{0,1 \times 4}{0,15^2 \cdot \pi} = 5,66m/\sec \tag{3}$$

$$\text{and } V_a = \frac{0,1 \times 4}{0,12^2 \pi} = 8,84m/\sec \tag{4}$$

By solving (2) for p_d and substituting into (3) and (4), we have:

$$\frac{p_d}{\gamma} = \frac{5,66^2}{2 \times 9.806} + \frac{70.000}{9.8002} - \frac{8,84^2}{2 \times 9,806} - 1 = 9,145m$$

$$p_d = 89,6KN / m^2 \tag{5}$$

ii) The increase of the *pressure gauge head* is:

$$\left(\frac{p_a}{\gamma}+1\right)-\frac{p_a}{\gamma}=9.145+1-\frac{70.000}{9.802}=3.004m \tag{6}$$

25) If we ignore the secondary losses and if we consider the length of the pipe equal to its horizontal distance, calculate the point of the minimum pressure of the pipette in Figure (25).

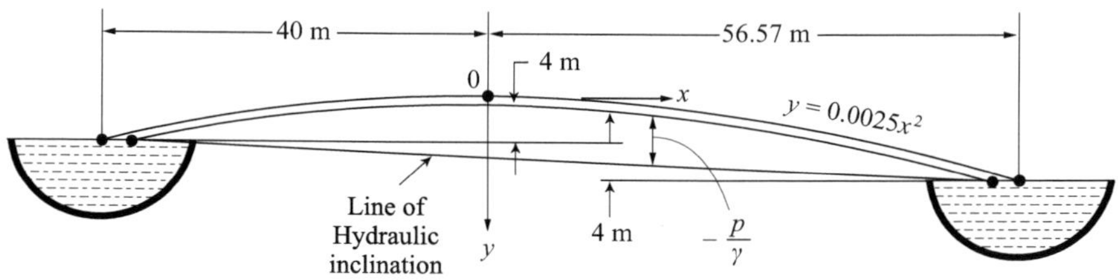

Figure (25)

Solution

When we ignore the secondary losses, usually the head of the kinetic energy $V^2/2g$ is also ignored.

Then, the *line of the hydraulic inclination* is a straight line that connects the two surfaces of the fluid; if we take the coordinates of two points of the line, we have:

$$(x=-40m,\ y=4m) \text{ and } (x=56,57m,\ y=8m) \tag{1}$$

so if we use the equation of a straight line given from the relation:

$$y = ax + b$$

by substituting the values of (1), the equation of the line of the hydraulic inclination is:

$$y = 0,041x + 5.656m$$

The minimum pressure lies at the point where the distance between the line of the hydraulic inclination of the pipe is maximum:

$$\frac{p}{\gamma} = 0,0025x^2 - 0,0414x - 5.656$$

In order to find the minimum $p/8$, we set:

$$\frac{d(p/\gamma)}{dx} = 0$$

which gives $x \simeq 8,28$ and $p/\gamma = -5,827m$ of the fluid that flows.

The minimum point appears there, where the inclinations of the pipe and linear hydraulic inclination are equal.

26) In a horizontal pipe made of cast iron, which is of diameter $d = 12in$, fluid flows with kinematic viscosity $v = 10^{-5} m^2 / \sec$. Calculate the losses in meters of fluid column, per unit length of the pipe when it gives us flow rate $Q = 1,0m^3 / \sec$.

Solution

The linear losses in meters of column are given by the Darcy–Weisbach relation:

$$h_\ell = f\frac{\ell}{d}\frac{V^2}{2g} = \frac{8f\ell Q^2}{\pi^2 gd^5} \qquad (1)$$

So, h_ℓ = losses in meters of column if meter length of the pipe $= \dfrac{8fQ^2}{\pi^2 gd^5}$ (2)

Relation (2) also contains an unknown friction coefficient *f*, which will be calculated from the Moody diagram as follows:

$$R_e = \frac{4Q}{\pi dv} = \frac{4\times 1}{\pi\times(12\times 0,0254)\times 10^{-5}} = 4,18\times 10^5$$

From the roughness tables, we find that the absolute roughness of the pipe made of cast iron is:

$$\varepsilon = 0,00085 ft$$

So, the relative roughness of the pipe is $(1ft = 12in)$

$$\varepsilon / d = \frac{0,00085ft}{1ft} = 0,00085ft$$

Now, we consider the Moody diagram and find that the line which corresponds to $R_e = 4,18 \times 10^5$ cuts the curve of relative roughness $\varepsilon / d = 0,00085ft$. From this point of intersection, we go to the axis of f and read the value $f = 0,0195$.

So, returning to relation (1), we have:

$$\frac{h_\ell}{\ell} = \frac{8fQ^2}{\pi^2 g d^5} = \frac{8 \times 0,0195 \times 1^2}{\pi^2 \times 9,81 \times (12 \times 0,0254)^5} = 0,6125m/m$$

So, $\frac{h_\ell}{\ell} = 0,6125m$ column/m pipe.

27) Two tanks A and B with free surface height difference $h = 6m$ connected to two pipes in a row have the following parameters. Pipe (1) has length $\ell_1 = 300m$, diameter $d_1 = 0,6m$ and roughness $\varepsilon_1 = 0,0015m$ and pipe (2) has length $\ell_2 = 240m$, diameter $d_2 = 0,9m$ and roughness $\varepsilon_2 = 0,0003m$. If the kinematic viscosity of the flow $v = 9 \times 10^{-7} m^2/\sec$, calculate the flow rate of the fluid from inside the system (Figure (27)) Given: $K = 0,5$.

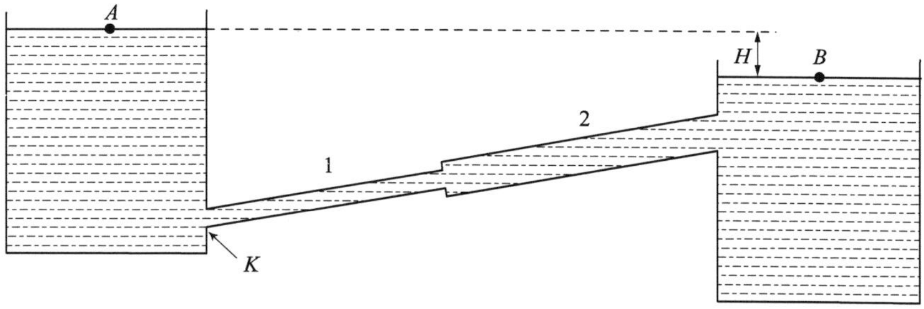

Figure (27)

Solution

By substituting in the energy equation, we have:

$$6=\frac{V_1^2}{2g}\left[0,5+f_1\left(\frac{300}{0,6}\right)+\left(1-\left(\frac{0,6}{0,8}\right)^2\right)^2+f_2\left(\frac{240}{0,9}\right)+\left(\frac{0,6}{0,9}\right)^4+\left(\frac{0,6}{0,9}\right)^4\right] \quad (1)$$

and after simplification, we have:

$$6=\frac{V_1^2}{2g}(1,01+500f_1+52,6f_2) \quad (2)$$

Now, the relative roughness of the two pipes will be $\varepsilon_1/d_1=0,0025$ and $\varepsilon_2/d_2=0,00033$ and from the Moody diagram for the area of the absolute turbulence, we have:

$$f_1=0,025 \text{ and } f_2=0,015 \quad (3)$$

By substituting (3) into (2) and solving for V_1, we have:

$$V_1=2,87m/\sec \text{ and } V_2=1,28m/\sec$$

By substituting now the values V_1 and V_2 into the equations of the Reynolds number, we have:

$$\mathrm{Re}_1 - \frac{2.87\times 0,6}{9\times 10^{-7}}=1.913.000$$

$$\mathrm{Re}_2=\frac{1.28\times 0,9}{9\times 10^{-7}}=1.280.000$$

So, from the Moody diagram once again, we have:

$$f_1=0,025 \text{ and } f_2=0,016$$

Solving again for V_1 from (2), we have $V_1=2,86m/\sec$

So:

$$Q = \frac{2,86 \times \pi \times 0,6^2}{4} = 0,81 m^3 / \sec$$

28) Solve the previous problem with the method of equivalent lengths.

Solution

Two pipe systems are called *equivalent* when the *same head losses* correspond to the *same flow rate* for both systems.

So, it must be applicable that:

$$h_{f1} = h_{f2} \text{ and } Q_1 = Q_2$$

Therefore, for the given data of the problem, we will express the secondary losses K_1 in connection with the equivalent lengths.

So, for pipe (1), we have:

$$K_1 = 0,5\left[1 + \left(\frac{0,6}{0,9}\right)^2\right] = 0.809$$

and $\ell_{\ell 1} = \dfrac{K_1 d_1}{f_1} = \dfrac{0.809 \times 0.6}{0,025} = 19,4m$

And for pipe (2), we have:

$$K_2 = 1 \text{ and } \ell_{\ell 2} = \frac{K_2 d_2}{f_2} = \frac{1 \times 0.9}{0,015} = 60m$$

The values f_1 and f_2 are approximately chosen for the area of the absolute turbulence.

This problem is now simplified in $319,4m$ of pipe with diameter 0.6 m and in 300 m of pipe with diameter 0.9 m. If we express the pipe with diameter 0.9 m with a pipe of equivalent length with diameter 0.6 from the equation of the equivalent lengths:

$$\ell_1 = \ell_2 \frac{f_1}{f_2}\left(\frac{d_1}{d_2}\right)^5,$$

we get $\ell_1 = 300\dfrac{0,015}{0,025}\left(\dfrac{0,6}{0,9}\right)^5 = 23,7m$

Adding this length to the pipe with diameter 0.6 m, we have to calculate the flow rate of the pipe with length $319,4+23,7=343,1m$ with diameter $d_1=0,6$ and $\varepsilon=0,0015m$ for head losses $h_f=6m$, so we have:

$$6 = f\frac{343,1}{0,6}\frac{V^2}{2g}$$

With $f=0,025,\ 0,9m=2,87m/\sec$, the Reynolds number gives:

$$R_e = 2,87\times\frac{0,6}{9\times10^{-7}} = 1,913.000$$

and for $\varepsilon/d=0,0025,\ f=0,025$ and

$$Q = \frac{2,87\times\pi\times0,62}{4} = 0,811m^3/\sec$$

29) A system of kerosene transportation at 11°C is shown in Figure (29):

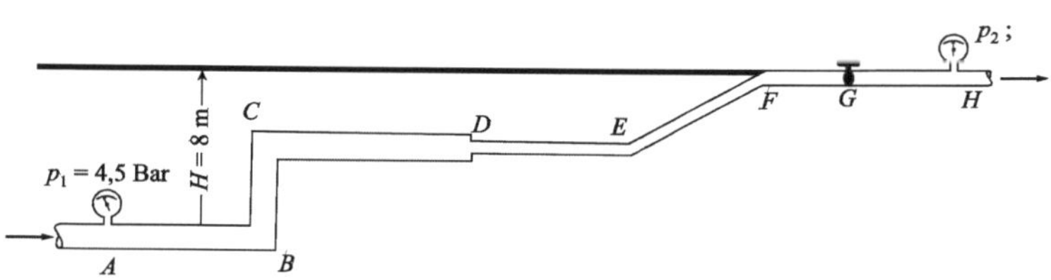

Figure (29)

In the above system, pipe ABCD has length $\ell=45m$ and diameter $d_1=4''$. Angles B and C are standard 90° angles, and angles E and F are standard 45° angles. The pipe DEFGH has total length $\ell_2=80m$ and diameter $d_2=2''$. Calculate the pressure p_2 that the manometer must show if the flow rate of the system is

$Q = 140 g.p.m$ (imperial gallons), the pipes are drawn and the valve at G has $K_G = 0,2$.

Solution

From the diagrams (or tables) of kinematic viscosity, we find that, for kerosene at 11°C:

$$v = 3\times10^{-5}\, ft^2/\sec = 3\times 0,0929\times10^{-5}\, m^2/\sec = 2,787\times10^{-6}\, m^2/\sec$$

$$\mu = 5\times10^{-5}\frac{lb_f - \sec}{ft^2} = 5\times10^{-5}\times 47,899\frac{N-\sec}{m^2} = 2,395\times10^{-4}\frac{kg}{m-\sec}$$

because: $1 ft^2/\sec = 0,0929 m^2/\sec$

$$1\frac{lb_f - \sec}{ft^2} = 47,899\frac{N-\sec}{m^2} = 47,899\frac{kg}{m-\sec}$$

So: $\rho = \dfrac{\mu}{v} = \dfrac{2,395\times10^{-4}}{2,787\times10^{-6}} = 859,33 kg/m^3$

Also from Figure 5.10 we have that the absolute roughness for a drawn tube is $\varepsilon = 0,00006''$.

So: $\dfrac{\varepsilon}{d_1} = \dfrac{0,00006''}{4''} = 0,000015$

and: $\dfrac{\varepsilon}{d_2} = \dfrac{0,00006''}{2''} = 0,00003$

From Tables 5.3 and 5.4 of the local losses, we find that the coefficients K of the local losses are:

$$K_B = K_C = 0,9 K_E = K_F = 0,35,\ K_D = 0,33$$

From the equation of the mechanical energy between the positions A and H, we have:

$$p_A + h_A + \frac{V_A^2}{2g} = p_H + h_H + \frac{V_H^2}{2g} + H_\ell$$

$$\frac{p_A - p_H}{\gamma} = \frac{V_H^2}{2g} - \frac{V_A^2}{2g} + h_H - h_A + H_\ell \qquad (1)$$

where $p_A = p_1 = 4,5\text{Bars} = 4,5\times10^5 N/m^2$

$$p_H = p_2 = ?$$

$$V_A = V_1 = \frac{4Q}{\pi d_1^2} = \frac{4\times140\times4,5\times10^{-3}}{60\times\pi\times(4\times0,0254)^2}\frac{m}{\sec} = 1,295\frac{m}{\sec}$$

$$V_H = V_2 = \frac{4Q}{\pi d_2^2} = \frac{4\times140\times4,5\times10^{-3}}{60\times\pi\times(2\times0,0254)^2}\frac{m}{\sec} = 5,181\frac{m}{\sec}$$

$$h_H - h_A = H = 8m$$

$H_\ell =$ total losses= $h_{\ell1} + h_{\ell2} + h_\ell$

$$h_{\ell1} = f_1 + \frac{\ell_1}{d_1}\frac{V_1^2}{2g} = \frac{45\times(1,295)^2}{2\times9,81\times(4\times0,0254)} f_1 = 37,858 f_1$$

$$h_{\ell2} = f_2 + \frac{\ell_2}{d_2}\frac{V_2^2}{2g} = \frac{80\times(5,181)^2}{2\times9,81\times(2\times0,0254)} f_2 = 2154,539 f_2$$

$$h'_\ell = K_B\frac{V_1^2}{2g} + K_C\frac{V_1^2}{2g} + K_D\frac{V_2^2}{2g} + K_E\frac{V_2^2}{2g} + K_F\frac{V_2^2}{2g} + K_G\frac{V_2^2}{2g}$$

or $h'_\ell = (K_B + K_C)\frac{V_1^2}{2g} + (K_D + K_E + K_F + K_G)\frac{V_2^2}{2g} =$

$$= \frac{1,8\times(1,295)^2}{2\times9,81}m + \frac{1,23\times(5.181)^2}{2\times9,81}m = 1,837m$$

or $h'_\ell = 1,837m \simeq 1,84m$

Then, we calculate the friction coefficients f_1 and f_2 by examining each pipe separately.

Pipe 1

$$\begin{cases} R_{e1} = \dfrac{V_1 d_1}{\nu} = \dfrac{1,295 \times 4 \times 0,0254}{2,787 \times 10^{-6}} = 4,7 \times 10^4 \\ \dfrac{\varepsilon}{d_1} = 0,000015 \end{cases}$$

So, from the Moody diagram, we have that in the pair of these values corresponds the friction coefficient $f_1 = 0,021$.

Pipe 2

$$\begin{cases} R_{e2} = \dfrac{V_2 d_2}{\nu} = \dfrac{5,181 \times 2 \times 0,0254}{2,787 \times 10^{-6}} = 9,44 \times 10^4 \\ \dfrac{\varepsilon}{d_2} = 0,00003 \end{cases}$$

From the Moody diagram, we find that the friction coefficient $f_2 = 0,0183$ corresponds in this pair.

So: $h_{\ell 1} = 37,858 f_1 = 0,81m$

$$h_{\ell 2} = 2154,539 f_2 = 39,42m$$

Therefore: $H_\ell = (0,81 + 39,42 + 1,84)m = 42,07m$

$$H_\ell = 42,07m \qquad (2)$$

From relation (1), we have that:

$$\frac{p_1 - p_2}{\rho g} = \frac{V_2^2 - V_1^2}{2g} + H + H_\ell$$

or $p_1 - p_2 = -\frac{1}{2}\rho\left(V_1^2 - V_2^2\right) + \left(H + H_\ell\right)\rho g$

or $p_2 = p_1 + \frac{1}{2}\left(V_1^2 - V_2^2\right)\rho - \left(H + H_\ell\right)\rho g$ (3)

By substituting numerical values in (3), we have:

$$p_2 = 4{,}5\times10^5\, N/m^2 - 0{,}5\times859{,}33\times\left[(5{,}181)^2 - (1{,}295)^2\right]N/m^2 -$$

$$-\left(8+42{,}07\right)\times859{,}33\times9{,}81 N/m^2 \Rightarrow$$

$$\Rightarrow p_2 = 1{,}71\times10^4\, N/m^2 = 0{,}17\,Bar$$

30) In the horizontal hydraulic circuit shown in Figure (30), calculate flow rates Q_1 and Q_2 of the pipes if the fluid that flows is water at temperature 20°C, the tubes are made of cast iron and the local losses are negligible.

Solution

For water at $20°C: v = 1{,}011\times10^{-6}\, m^2/\sec$

Also, we know that for cast iron, $\varepsilon_1 = \varepsilon_2 = 0{,}00085\, ft$

$$Q = 0{,}1 m^3/\sec = Q_1 + Q_2 \qquad (1)$$

$$V_1 = \frac{4Q_1}{\pi d_1^2} \text{ and } V_2 = \frac{4Q_2}{\pi d_2^2}$$

Because in a system of parallel pipes the losses of all the pipes, regardless of their lengths and diameters, are all equal, then:

$$h_{\ell 1} = h_{\ell 2}$$

where: $h_{\ell 1} =$ total losses in pipe 1 $= h_{\ell 1} + h_{\ell' 2}$ and

$h_{\ell 2} =$ total losses in pipe 2 $= h_{\ell 2} + h_{\ell 2}$

so: $h_{\ell 1} = h_{\ell 2}$

or $f_1 \frac{\ell_1}{d_1}\frac{V_1^2}{2g} = f_2 \frac{\ell_2}{d_2}\frac{V_2^2}{2g}$

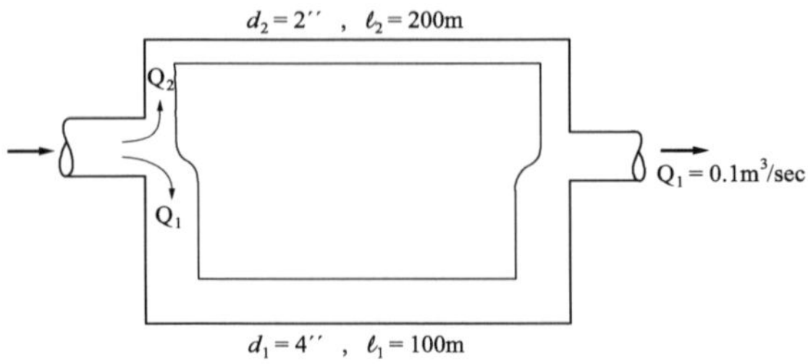

Figure (30)

$$\text{or } \frac{8f_1\ell_1}{\pi^2 d_1^5 g}Q_1^2 = \frac{8f_2\ell_2}{\pi^2 d_2^5 g}Q_2^2$$

$$\text{or } Q_1 = Q_2\left[\frac{f_2}{f_1}\frac{\ell_2}{\ell_1}\left(\frac{d_1}{d_2}\right)^5\right]^{1/2} = Q_2\left[\frac{f_2}{f_1}\frac{200}{100}\left(\frac{4}{2}\right)^5\right]^{1/2}$$

$$Q_1 = 8Q_2\left(\frac{f_2}{f_1}\right)^{1/2} \tag{2}$$

Combining (1) and (2), we have:

$$Q = Q_2\left[8\left(\frac{f_2}{f_1}\right)^{1/2} + 1\right] \tag{3}$$

Coefficients f_1 and f_2 are calculated as follows:

First circle

For pipe (1):

$$\frac{\varepsilon_1}{d_1} = \frac{0{,}00085 \times 12''}{4''} = 0{,}00255$$

From the Moody diagram, we know that in the case where the flow was developed turbulent, and for

$$\frac{\varepsilon_1}{d_1} = 0{,}00255\text{, it would correspond to } f_1 = 0{,}025 \tag{4}$$

For pipe (2):

$$\frac{\varepsilon_2}{d_2} = \frac{0,00085 \times 12''}{2''} = 0,0051$$

From the Moody diagram as in pipe (1), we find that in the case where the flow was developed turbulent, and for

$$\frac{\varepsilon_2}{d_2} = 0,0051 \text{, it corresponds to } f_2 = 0,03 \qquad (5)$$

By substituting (4) and (5) into (3), we find:

$$Q_2 = \frac{Q}{8\left(\frac{f_2}{f_1}\right)^{1/2} + 1} = \frac{0,1}{8\left(\frac{3}{2,5}\right)^{1/2} + 1} m^3 / \sec$$

or $Q_2 = 0,0102 m^3 / \sec$

so $Q_1 = 0,0898 m^3 / \sec$

$$R_{e2} = \frac{4Q_2}{\pi d_2 v} = \frac{4 \times 0,0102}{\pi \times 2 \times 0,0254 \times 1,011 \times 10^{-6}} = 2,53 \times 10^5$$

$$R_{e1} = \frac{4Q_1}{\pi d_1 v} = \frac{4 \times 0,0898}{\pi \times 4 \times 0,0254 \times 1,011 \times 10^{-6}} = 1,11 \times 10^6$$

So, now using the Moody diagram, we find:

$$\left\{ \begin{aligned} \frac{\varepsilon_1}{d_1} &= 0,00255 \\ R_{e1} &= 1,11 \times 10^6 \end{aligned} \right\} \Rightarrow f_1 = 0,0251 \text{ (Moody)}$$

and

$$\left\{ \begin{aligned} \frac{\varepsilon_2}{d_2} &= 0,0051 \\ R_{e2} &= 2,53 \times 10^5 \end{aligned} \right\} \Rightarrow f_2 = 0,031 \text{ (Moody)}$$

Because the calculated f_1 and f_2 are a little different from the values presumed initially, we repeat a new circle.

Second circle

With $f_1 = 0,0251$ and $f_2 = 0,031$,

relation (3) becomes:

$$Q_2 = \frac{Q}{8\left(\frac{f_2}{f_1}\right)^{1/2} + 1} = \frac{0,1}{8\left(\frac{3,1}{2,51}\right)^{1/2} + 1} m^3 / \text{sec}$$

or $Q_2 = 0,0101 m^3 / \text{sec}$

so $Q_1 = 0,0899 m^3 / \text{sec}$

Therefore:

$$R_{e1} = \frac{4Q_1}{\pi d_1 v} = \frac{4 \times 0,0898}{\pi \times 4 \times 0,0254 \times 1,011 \times 10^{-6}} = 1,11 \times 10^6$$

$$R_{e2} = \frac{4Q_2}{\pi d_2 v} = \frac{4 \times 0,0101}{\pi \times 2 \times 0,0254 \times 1,011 \times 10^{-6}} = 2,50 \times 10^5$$

so:

$$\left\{ \begin{aligned} \frac{\varepsilon_1}{d_1} &= 0,00255 \\ R_{e1} &= 1,11 \times 10^6 \end{aligned} \right\} \Rightarrow f_1 = 0,0251 \text{ (Moody)}$$

$$\left\{ \begin{aligned} \frac{\varepsilon_2}{d_2} &= 0,0051 \\ R_{e2} &= 2,51 \times 10^5 \end{aligned} \right\} \Rightarrow f_2 = 0,031 \text{ (Moody)}$$

We note that the new values coincide with those presumed at the beginning of the second circle, so we stop.

So:

$$Q_1 = 0,0899 m^3 / \text{sec}$$

$$Q_2 = 0,0101 m^3 / \text{sec}$$

31) Water at temperature 20°C flows from a big tank through pipes, as shown in Figure (31). Calculate (a) the flow rate of each pipe and (b) the pressure at points (2) and (3).

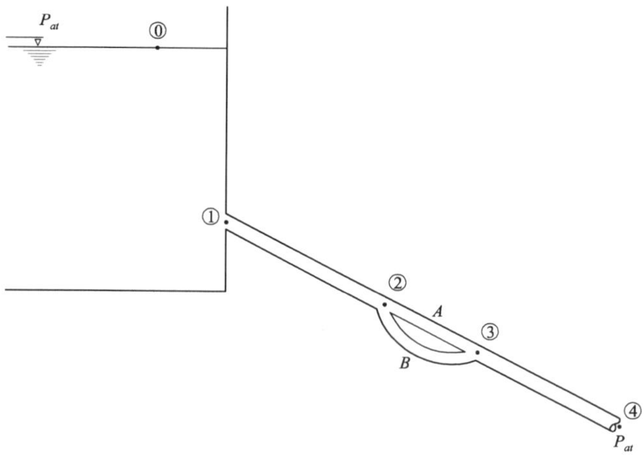

Figure (31)

Given:

Height of point (0): 25 m

Height of point (1): 18 m

Height of point (2): 9 m

Height of point (3): 5 m

Height of point (4): 0 m

$\nu =$ kinematic viscosity $10^{-6} m^2 / \text{sec}$, $\rho = 998 kg / m^3$

For pipe A: $\ell_A = 18m$, $d_A = 1''$, $f_A = 0,015$

For pipe B: $\ell_B = 25m,\ d_B = 0,5'',\ f_B = 0,030$

For pipe 1–2: $\ell_{12} = 36m,\ d_{12} = 1,5'',\ f_{12} = 0,025$

For pipe 3–4: $\ell_{34} = 18m,\ d_{34} = 1,5'',\ f_{34} = 0,025$

Solution

Equations of mechanical energy between positions (0)-(1), (1)-(2), $(2)_A$-$(3)_A$, $(2)_B$-$(3)_B$ and (3)-(4) are:

$$\frac{p_0}{\gamma} + h_0 + \frac{V_0^2}{2g} = \frac{p_1}{\gamma} + h_1 + \frac{V_1^2}{2g} \tag{1}$$

$$\frac{p_1}{\gamma} + h_1 + \frac{V_1^2}{2g} = \frac{p_2}{\gamma} + h_2 + \frac{V_2^2}{2g} + h_{\ell 12} \tag{2}$$

$$\frac{p_2}{\gamma} + h_2 + \frac{V_2^2}{2g} = \frac{p_3}{\gamma} + h_3 + \frac{V_3^2}{2g} + h_{\ell_A} \tag{3}$$

$$\frac{p_2}{\gamma} + h_2 + \frac{V_2^2}{2g} = \frac{p_3}{\gamma} + h_3 + \frac{V_3^2}{2g} + h_{\ell_B} \tag{4}$$

$$\frac{p_3}{\gamma} + h_3 + \frac{V_3^2}{2g} = \frac{p_4}{\gamma} + h_4 + \frac{V_4^2}{2g} + h_{\ell_{34}} \tag{5}$$

where:

$$h_{\ell_{12}} = f_{12}\frac{\ell_{12}}{d_{12}}\frac{V_1^2}{2g} = f_{12}\frac{\ell_{12}}{d_{12}}\frac{V_2^2}{2g} \text{ because } V_1 = V_2$$

$$h_{\ell_A} = f_A\frac{\ell_A}{d_A}\frac{V_A^2}{2g}$$

$$h_{\ell_B} = f_B\frac{\ell_B}{d_B}\frac{V_B^2}{2g}$$

$$p_0 = p_4 = p_{at}$$

$$h_{\ell_{3\,4}} = f_{3\,4}\frac{\ell_{3\,4}}{d_{3\,4}}\frac{V_4^2}{2g} = f_{3\,4}\frac{\ell_{3\,4}}{d_{3\,4}}\frac{V_3^2}{2g} \text{ because } V_3 = V_4.$$

Comparing (3) and (4), it must be applicable that:

$$h_{\ell_A} = h_{\ell_B}$$

$$\text{or } f_A\frac{\ell_A}{d_A}\frac{V_A^2}{2g} = f_B\frac{\ell_B}{d_B}\frac{V_B^2}{2g}$$

$$\text{or } V_A = \left[\frac{f_B}{f_A}\frac{\ell_B}{\ell_A}\frac{d_A}{d_B}\right]^{1/2} V_B = \left[\frac{0,030}{0,025}\times\frac{25}{18}\times\frac{1}{0,5}\right]^{1/2} V_B$$

$$\text{or } V_A = 1,826V_B \tag{6}$$

From the continuity equation of the flow, we also have:

$$Q = V_1\frac{\pi d_{1\,2}^2}{4} = V_A\frac{\pi d_A^2}{4} + V_B\frac{\pi d_B^2}{4} = V_4\frac{\pi d_{3\,4}^2}{4}$$

$$\text{so } V_A = V_4\left(\frac{d_{3\,4}}{d_A}\right)^2 - V_B\left(\frac{d_B}{d_A}\right)^2 = 2,25V_4 - 0,25V_B$$

$$\text{or } V_A = 2,25V_4 - 0,25V_B \tag{7}$$

From (6) and (7), we find:

$$1,826V_B = 2,25V_4 - 0,2V_B$$

$$V_B = \frac{2,25}{2,076}V_4 = 1,084V_4$$

$$\text{so } V_B = 1,084V_4 \tag{8}$$

$$\text{and } V_A = 1,979V_4 \tag{9}$$

Adding relationships (1), (2), (3) and (5), we find:

$$\frac{p_0}{\gamma}+h_0+\frac{V_0^2}{2g}=\frac{p_4}{\gamma}+h_4+\frac{V_4^2}{2g}+h_{\ell_2}+h_{\ell_A}+h_{\ell_{34}}$$

or $h_0-h_4=\frac{V_4^2}{2g}+f_{12}\frac{\ell_{12}}{d_{12}}\frac{V_1^2}{2g}+f_A\frac{\ell_A}{d_A}\frac{V_A^2}{2g}+f_{34}\frac{\ell_{34}}{d_{34}}\frac{V_4^2}{2g}$

and because $d_{12}=d_{34}\Rightarrow V_1=V_4$, using (9), we have:

$$h_0-h_4=\frac{V_4^2}{2g}+f_{12}\frac{\ell_{12}}{d_{12}}\frac{V_4^2}{2g}+f_A\frac{\ell_A}{d_A}(1{,}979)^2\frac{V_4^2}{2g}+f_{34}\frac{\ell_{34}}{d_{34}}\frac{V_4^2}{2g}\Rightarrow$$

$$V_4=\left[\frac{(h_0-h_4)2g}{1+f_{12}\frac{\ell_{12}}{d_{12}}+f_A\frac{\ell_A}{d_A}(1{,}979)^2+f_{34}\frac{\ell_{34}}{d_{34}}}\right]^{1/2}\Rightarrow$$

$$V_4=\left[\frac{25\times2\times9{,}81}{1+\frac{0{,}025\times36}{1{,}5\times0{,}0254}+\frac{0{,}015\times18\times(1{,}979)^2}{1\times0{,}0254}+\frac{0{,}025\times18}{1{,}5\times0{,}0254}}\right]^{1/2} m/\sec\Rightarrow$$

$$V_4=2{,}507 m/\sec \tag{10}$$

so

$$\text{(a) } Q_{12}=Q_{34}=V_4\frac{\pi d_{34}^2}{4}=2{,}507\times\frac{\pi}{4}\times(1{,}5\times0{,}0254)^2\ m^3/\sec\Rightarrow$$

or $Q_{12}=Q_{34}=2{,}858\times10^{-3} m^3/\sec$.

$$Q_A = V_A \frac{\pi d_A^2}{4} = 1{,}979 V_4 \frac{\pi d_A^2}{4} = \frac{1{,}979 \times 2{,}507 \times \pi \times (0{,}0254)^2}{4} \frac{m^3}{\sec} \Rightarrow$$

or $Q_A = 2{,}514 \times 10^{-3} m^3 / \sec$

$$Q_B = 0{,}344 \times 10^{-3} m^3 / \sec$$

(b) From relation (5), we have:

$$\frac{p_3}{\gamma} + h_3 = \frac{p_4}{\gamma} + h_4 + h_{\ell_{3\,4}}, \text{ because } V_3 = V_4$$

or $\dfrac{p_3 - p_4}{\gamma} = h_4 - h_3 + h_{\ell_{3\,4}}$,

where $h_4 - h_3 = -5m$

and $h_{\ell_{3\,4}} = f_{3\,4} \dfrac{\ell_{3\,4}}{d_{3\,4}} \dfrac{V_4^2}{2g} = \dfrac{0{,}025 \times 18}{1{,}5 \times 0{,}0254} \times \dfrac{(2{,}507)^2}{2 \times 9{,}81} m \Rightarrow$

or $h_{\ell_{3\,4}} = 3{,}784m$.

so $\dfrac{p_3 - p_4}{\gamma} = -5m + 3{,}784m = -1{,}216m$

and because $p_4 = p_{at}$.

Then, the differential pressure at position (2) is:

$p_3 = -1{,}216m$ water column (depression).

From relation (4), we have:

$$\frac{p_2}{\gamma} = \frac{p_3}{\gamma} + h_3 - h_2 + h_{\ell_B}, \text{ because } V_2 = V_3 \text{ as } d_{1\,2} = d_{3\,4}$$

or

$$\frac{p_2}{\gamma} = -1,216m + 5m - 9m + f_B \frac{\ell_B}{d_B}\frac{V_B^2}{2g} =$$

$$= -1,216m - 4m + \frac{0,030 \times 25}{0,5 \times 0,0254} \times \frac{(1,084 \times 2,507)^2}{2 \times 9,81} m$$

$$\frac{p_2}{\gamma} = -5,216m + 22,229m = 17,013m$$

So: $p_2 = 17,013m$ water column.

32) For the system in Figure (32), calculate Q_1, Q_2 and Q_3 of the three pipes. The solution is considered satisfied when the flow rate to branch D and the flow rate from branch D are up to the second decimal digit.

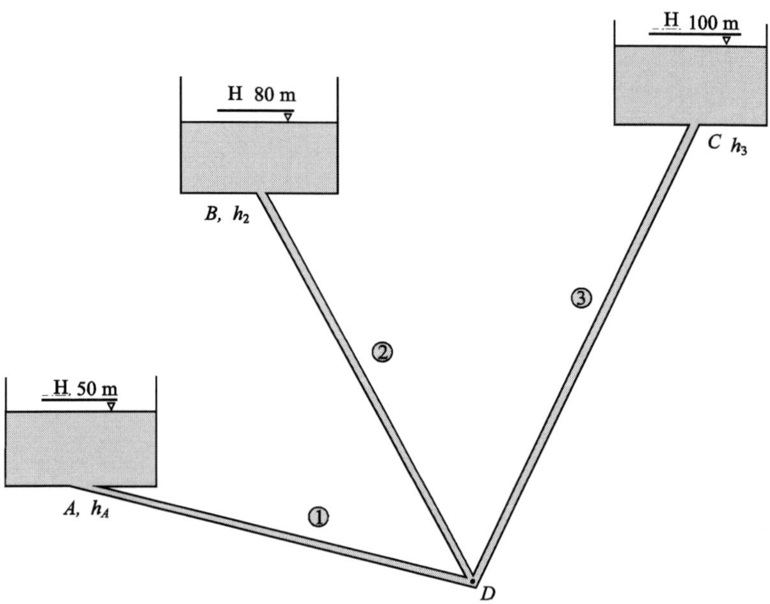

Figure (32)

The following data are given:

Water at temperature $12°C \Rightarrow v = 1,244 \times 10^{-6} m^2 / \sec$

$$d_1 = 12'', \ \ell_1 = 2000m, \ \varepsilon_1 = 0,00048 ft, \ h_1 = 50m$$

$$d_2 = 8'', \ \ell_2 = 1000m, \ \varepsilon_2 = 0,00048 ft, \ h_2 = 80m$$

$$d_3 = 8'', \ \ell_3 = 3000m, \ \varepsilon_3 = 0,00048 ft, \ h_3 = 100m$$

Solution

Assume that $h_A =$ height of the cross-sectional area of the pressure gauge of the lines of the three pipes.

Thus, if $h_A > h_2$, then we have: $Q_3 = Q_1 + Q_2$. However, if $h_A < h_2$, then we have: $Q_3 + Q_2 = Q_1$.

Generally, the solution is found after a number of circles of approaching solutions by applying the following thought: if the flow rate to branch D is found to be higher than the flow rate from D at the end of a circle, then we proceed with a new circle, although the new presumed value of height h_A is higher than that of the previous circle. However, if the flow rate to D is found to be smaller than the flow rate from D at the end of a circle, we proceed with a new circle of calculations by choosing a smaller h_A. We stop the circles of the approaching solutions when we finally we find that the flow rate to and from D agree, here, to the second decimal digit.

First circle: assume that $h_A = 75m$

Then:

i) For pipe (1):

$$h_{\ell_1} = |h_A - h_1| = f_1 \frac{\ell_1}{d_1} \frac{V_1^2}{2g} \Rightarrow V_1 = \left[\frac{2 h_{\ell_1} g d_1}{f_1 \ell_1} \right]^{1/2}$$

and because $h_{\ell 1} = 75m - 50m = 25m$

then: $V_1 = \left[\dfrac{2\times 25\times 9,81\times 12\times 0,0254}{2000}\right]^{1/2} \times f_1^{-1/2} = 0,2734 f_1^{1/2}$ in $m/\sec$

$$\frac{\varepsilon_1}{d_1} = \frac{0,00048\times 12''}{12''} = 0,00048$$

Assuming that $f_1 = 0,017$, then:

$$V_1 = 0,2734(0,017)^{-1/2}\ m/\sec = 2,0969 m/\sec$$

so: $R_{e1} = \dfrac{V_1 d_1}{\nu} = \dfrac{2,0969\times 12\times 0,254}{1,244\times 10^{-6}} = 5,14\times 10^5$

Thus, with $\left\{\dfrac{\varepsilon_1}{d_1} = 0,00048,\ R_{e1} = 5,14\times 10^5\right\} \Rightarrow f_1 = 0,0175$ (Moody)

so $V_1 = 0,2734(0,0175)^{-1/2} m/\sec = 2,0667 m/\sec$

and $R_{e1} = \dfrac{V_1 d_1}{\nu} = \dfrac{2,0667\times 12\times 0,0254}{1,244\times 10^{-6}} = 5,06\times 10^5$

Thus, with $\left\{\dfrac{\varepsilon_1}{d_1} = 0,00048,\ R_{e1} = 5,06\times 10^5\right\} \Rightarrow f_1 = 0,0175$ (Moody)

Therefore: $V_1 = 2,0667 m/\sec$ and $Q_1 = V_1 \dfrac{\pi d_1^2}{4} \Rightarrow$

$$\Rightarrow Q_1 = \frac{2,0667\times \pi\times (12\times 0,0254)^2}{4} m^3/\sec = 0,1508 m^3/\sec$$

ii) For pipe (2):

$$h_{\ell_2} = |h_A - h_2| = f_2 \frac{\ell_2}{d_2}\frac{V_2^2}{2g} \Rightarrow V_2 = \left[\frac{2h_{\ell_2} g d_2}{f_2 \ell_2}\right]^{1/2}.$$

Thus, with $h_{\ell 2} = |75 - 80|\, m = 5m$

then $V_2 = \left[\dfrac{2\times5\times9,81\times8\times0,0254}{1000}\right]^{1/2} \times f_2^{-1/2} = 0,1412 f_2^{1/2}$ in $m/\sec$

$$\frac{\varepsilon}{d_2} = \frac{0,00048\times12''}{8''} = 0,00072$$

Assuming that $f_2 = 0,019$, then:

$$V_2 = 0,1412(0,019)^{-1/2}\ m/\sec = 1,0244 m/\sec$$

so, $R_{e2} = \dfrac{V_2 d_2}{v} = \dfrac{1,0244\times8\times0,254}{1,244\times10^{-6}} = 1,67\times10^5$

Thus, with $\left\{\dfrac{\varepsilon}{d_2} = 0,00072,\ R_{e2} = 1,67\times10^5\right\} \Rightarrow f_2 = 0,02$ (Moody)

then, $V_2 = 0,1412(0,02)^{-1/2} = 0,9984 m/\sec$

So $R_{e2} = \dfrac{V_2 d_2}{v} = \dfrac{0,9984\times8\times0,0254}{1,244\times10^{-6}} = 1,63\times10^5$

Thus, with $\left\{\dfrac{\varepsilon}{d_2} = 0,00072,\ R_{e2} = 1,63\times10^5\right\} \Rightarrow f_2 = 0,023$ (Moody)

Then $V_2 = 0,1412(0,023)^{\ 1/2} = 0,9310 m/\sec$

Therefore, $R_{e2} = \dfrac{V_2 d_2}{v} = \dfrac{0,9310\times8\times0,0254}{1,244\times10^{-6}} = 1,52\times10^5$

Thus, with $\left\{\dfrac{\varepsilon}{d_2} = 0,00072,\ R_{e2} = 1,52\times10^5\right\} \Rightarrow f_2 = 0,023$ (Moody)

So, $V_2 = 0,931 m/\sec$

And $\Rightarrow Q_2 = V_2 \dfrac{\pi d_2^2}{4} \Rightarrow \dfrac{0,931\times\pi\times(8\times0,0254)^2}{4} m^3/\sec = 0,0302 m^3/\sec$

iii) For pipe (3):

$$h_{\ell_3} = |h_A - h_3| = f_3 \frac{\ell_3}{d_3}\frac{V_3^2}{2g} \Rightarrow V_3 = \left[\frac{2h_{\ell_3} g d_3}{f_3 \ell_3}\right]^{1/2}.$$

Thus, with $h_{\ell 3} = |75-100| m = 25m$:

$$V_3 = \left[\frac{2\times 25\times 9,81\times 8\times 0,0254}{3000}\right]^{1/2} \times f_3^{-1/2} = 0,1823 f_3^{1/2} \text{ in } m/\sec$$

$$\frac{\varepsilon}{d_3} = \frac{0,00048\times 12''}{8''} = 0,00072$$

Assuming that $f_3 = 0,019$

then, $V_3 = 0,1823 f_3^{-1/2} = 0,1823(0,019)^{-1/2} = 1,3225 m/\sec$

$$R_{e3} = \frac{V_3 d_3}{\nu} = \frac{1,3225\times 8\times 0,254}{1,244\times 10^{-6}} = 2,16\times 10^5$$

with $\left\{\frac{\varepsilon}{d_3} = 0,00072,\ R_{e3} = 2,16\times 10^5\right\} \Rightarrow f_3 = 0,0197$ (Moody)

Then, $V_3 = 0,1823(0,0197)^{-1/2} = 1,2988 m/\sec$

Therefore, $R_{e3} = \frac{V_3 d_3}{\nu} = \frac{1,2988\times 8\times 0,0254}{1,244\times 10^{-6}} = 2,12\times 10^5$

Thus, with $\left\{\frac{\varepsilon}{d_3} = 0,00072,\ R_{e3} = 2,12\times 10^5\right\} \Rightarrow f_3 = 0,0197$ (Moody)

so, $V_3 = 1,2988 m/\sec$ and

$$\Rightarrow Q_3 = V_3 \frac{\pi d_3^2}{4} \Rightarrow \frac{1,2988\times \pi\times (8\times 0,0254)^2}{4} m^3/\sec = 0,0421 m^3/\sec$$

Therefore, from the results of the first circle, in which we assumed that $h_A = 75m < h_2 = 80m$ and:

$$Q_1 = 0,1508 m^3 / \sec$$

$$Q_2 = 0,0302 m^3 / \sec$$

$$Q_3 = 0,0421 m^3 / \sec$$

Thus, it always has to be $Q_3 + Q_2 = Q_1$.

However, we can see that $Q_2 + Q_3 = 0,0723 m^3 / \sec < Q_1 = 0,1508 m^3 / \sec$

Therefore, we perform another circle choosing for h_A a value smaller than the first one.

Second circle: assume that $h_A = 65m$

i) For pipe (1):

$$h_{\ell_1} = 65 - 50 = 15m$$

$$V_1 = \left[\frac{2h\ell_1 g d_1}{f_1 \ell_1}\right]^{1/2} = \left[\frac{2\times 15\times 9,81\times 12\times 0,0254}{2000}\right]^{1/2} \times f_1^{-1/2} = 0,2118 f_1^{-1/2}$$

$$\frac{\varepsilon}{d_1} - \frac{0,00048\times 12''}{12''} = 0,00048$$

Assuming that $f_1 = 0,017$

then $V_1 = 0,2118(0,017)^{-1/2} = 1,6243 m / \sec$

and $R_{e1} = \frac{V_1 d_1}{\nu} = \frac{1,6243\times 12\times 0,0254}{1,244\times 10^{-6}} = 3,98\times 10^5$

Thus, with $\left\{\frac{\varepsilon_1}{d_1} = 0,00048,\ R_{e1} = 3,98\times 10^5\right\} \Rightarrow f_1 = 0,018$ (Moody) then $V_1 = 0,2118(0,018)^{-1/2} = 1,5787 m/\sec$

$$R_{e1} = \frac{V_1 d_1}{\nu} = \frac{1,5787 \times 12 \times 0,0254}{1,244 \times 10^{-6}} = 3,87 \times 10^5$$

Thus, with $\left\{ \frac{\varepsilon_1}{d_1} = 0,00048,\ R_{e1} = 3,87 \times 10^5 \right\} \Rightarrow f_1 = 0,018$ (Moody)

ii) For pipe (2):

$$h_{\ell_2} = 80 - 65 = 15m$$

$$V_2 = \left[\frac{2h\ell_2 g d_2}{f_2 \ell_2} \right]^{1/2} = \left[\frac{2 \times 15 \times 9,81 \times 8 \times 0,0254}{1000} \right]^{1/2} \times f_2^{-1/2} \Rightarrow$$

$$\Rightarrow V_2 = 0,2445 f_2^{-1/2} \text{ in m/sec}$$

$$\frac{\varepsilon}{d_2} = \frac{0,00048 \times 12''}{8''} = 0,00072$$

Assuming that $f_2 = 0,02$,

then $V_2 = 0,2445(0,02)^{-1/2} = 1,7289 m/\sec$

$$R_{e2} = \frac{V_2 d_2}{\nu} = \frac{1,7289 \times 8 \times 0,0254}{1,244 \times 10^{-6}} = 2,82 \times 10^5$$

with $\left\{ \frac{\varepsilon}{d_2} = 0,00072,\ R_{e2} = 2,82 \times 10^5 \right\} \Rightarrow f_2 = 0,0195$ (Moody)

So $V_2 = 0,2445(0,0195)^{-1/2} = 1,7509 m/\sec$

and $R_{e2} = \frac{V_2 d_2}{\nu} = \frac{1,7509 \times 8 \times 0,0254}{1,244 \times 10^{-6}} = 2,86 \times 10^5$

so, with $\left\{ \frac{\varepsilon}{d_2} = 0,000722,\ R_{e2} = 2,86 \times 10^5 \right\} \Rightarrow f_2 = 0,0195 \text{ (Moody)}$

iii) For pipe (3):

$$h_{\ell_3} = 100 - 65 = 35m\ , \frac{\varepsilon}{d_3} = \frac{0,00048 \times 12''}{8''} = 0,00072$$

$$V_3 = \left[\frac{2h\ell_3 g d_3}{f_3 \ell_3} \right]^{1/2} = \left[\frac{2 \times 35 \times 9,81 \times 8 \times 0,0254}{3000} \right]^{1/2} \times f_3^{-1/2} = 0,2157 f_3^{-1/2} m/\sec$$

Assuming that $f_3 = 0,019$

then $V_3 = 0,2157(0,019)^{-1/2} m/\sec = 1,564 m/\sec$

and $R_{e3} = \frac{V_3 d_3}{v} = \frac{1,5646 \times 8 \times 0,0254}{1,244 \times 10^{-6}} = 2,56 \times 10^5$

Thus, with $\left\{ \frac{\varepsilon}{d_3} = 0,00072,\ R_{e3} = 2,56 \times 10^5 \right\} \Rightarrow f_3 = 0,0195 \text{ (Moody)}$

then $V_3 = 0,2157(0,0195)^{-1/2} m/\sec = 1,5447 m/\sec$

and $R_{e3} = \frac{V_3 d_3}{v} = \frac{1,5447 \times 8 \times 0,0254}{1,244 \times 10^{-6}} = 2,52 \times 10^5$

Thus, with $\left\{ \frac{\varepsilon}{d_3} = 0,00072,\ R_{e3} = 2,52 \times 10^5 \right\} \Rightarrow f_3 = 0,0195 \text{ (Moody)}$

So, the second circle gives us the results:

$$Q_1 = V_1 \frac{\pi d_1^2}{4} = \frac{1,5787 \times \pi \times (12 \times 0,0254)^2}{4} \frac{m^3}{\sec} = 0,115 \frac{m^3}{\sec}$$

$$Q_2 = V_2 \frac{\pi d_2^2}{4} = \frac{1,7509 \times \pi \times (8 \times 0,0254)^2}{4} \frac{m^3}{\sec} = 0,057 \frac{m^3}{\sec}$$

$$Q_3 = V_3 \frac{\pi d_3^2}{4} = \frac{1,5447 \times \pi \times (8 \times 0,0254)^2}{4} \frac{m^3}{\sec} = 0,050 \frac{m^3}{\sec}$$

The following relation must always be satisfied:

$$Q_2 + Q_3 = Q_1$$

The results of the second circle give:

$$Q_2 + Q_3 = 0,107 m^3 / \sec < Q_1 = 0,115 m^3 / \sec$$

So, we perform the third circle as well by choosing a value for h_A a little smaller than the one we assumed at the beginning of the second circle.

Third circle: assume that $h_A = 63,8m$

Then:

i) For pipe (1):

$$h_{\ell 1} = |h_A - h_1| = 13,8m \,, \quad \frac{\varepsilon}{d_1} = \frac{0,00048 \times 12''}{12''} = 0,00048$$

$$V_1 = \left[\frac{2h\ell_1 g d_1}{f_1 \ell_1} \right]^{1/2} = \left[\frac{2 \times 13,8 \times 9,81 \times 12 \times 0,0254}{2000} \right]^{1/2} \times f_1^{-1/2} = 0,2031 f_1^{-1/2}$$

Assuming that $f_1 = 0,018$, then:

$$V_1 = 0,2031(0,018)^{-1/2} m / \sec = 1,5141 m / \sec$$

$$R_{e1} = \frac{V_1 d_1}{\nu} = \frac{1,5141 \times 12 \times 0,0254}{1,244 \times 10^{-6}} = 3,71 \times 10^5$$

Thus, with $\left\{ \frac{\varepsilon}{d_1} = 0,00048, \; R_{e1} = 3,71 \times 10^5 \right\} \Rightarrow f_1 = 0,018$ (Moody)

ii) For pipe (2):

$$h_{\ell 2} = |h_A - h_2| = 16,2m\,,\ \frac{\varepsilon}{d_2} = \frac{0,00048\times 12''}{8''} = 0,00072$$

$$V_2 = \left[\frac{2h\ell_2 g d_2}{f_2 \ell_2}\right]^{1/2} = \left[\frac{2\times 16,2\times 9,81\times 8\times 0,0254}{1000}\right]^{1/2} \times f_2^{-1/2} = 0,254 f_2^{-1/2}$$

Assuming that $f_2 = 0,0195$

then $V_2 = 0,2541(0,0195)^{-1/2}\, m/\sec = 1,8199 m/\sec$

and $R_{e2} = \dfrac{V_2 d_2}{\nu} = \dfrac{1,8199\times 8\times 0,0254}{1,244\times 10^{-6}} = 2,97\times 10^5$

with $\left\{\dfrac{\varepsilon}{d_2} = 0,00072,\ R_{e2} = 2,97\times 10^5\right\} \Rightarrow f_2 = 0,0195$ (Moody)

iii) For pipe (3):

$$h_{\ell 3} = |h_A - h_3| = 36,2m\,,\ \frac{\varepsilon}{d_3} = \frac{0,00048\times 12''}{8''} = 0,00072$$

$$V_3 = \left[\frac{2h\ell_3 g d_3}{f_3 \ell_3}\right]^{1/2} = \left[\frac{2\times 36,2\times 9,81\times 8\times 0,0254}{3000}\right]^{1/2} \times f_3^{-1/2} = 0,2193 f_3^{-1/2}$$

Assuming that $f_3 = 0,0195$

then $V_3 = 0,2193(0,0195)^{-1/2} = 1,5707 m/\sec$

$$R_{e3} = \frac{V_3 d_3}{\nu} = \frac{1,5707\times 8\times 0,0254}{1,244\times 10^{-6}} = 2,57\times 10^5$$

with $\left\{\dfrac{\varepsilon}{d_3} = 0,00072,\ R_{e3} = 2,57\times 10^5\right\} \Rightarrow f_3 = 0,0195$ (Moody)

So, the results of the third circle are:

$$Q_1 = V_1 \frac{\pi d_1^2}{4} = \frac{1,5141 \times \pi \times (12 \times 0,0254)^2}{4} \frac{m^3}{\sec} = 0,110 m^3 / \sec$$

$$Q_2 = V_2 \frac{\pi d_2^2}{4} = \frac{1,8199 \times \pi \times (8 \times 0,0254)^2}{4} \frac{m^3}{\sec} = 0,059 m^3 / \sec$$

$$Q_3 = V_3 \frac{\pi d_3^2}{4} = \frac{1,5707 \times \pi \times (8 \times 0,0254)^2}{4} \frac{m^3}{\sec} = 0,051 m^3 / \sec$$

So, we see that:

$$Q_2 + Q_3 = 0,059 + 0,051 = 0,110 m^3 / \sec = Q_1 = 0,110 m^3 / \sec$$

This means that we have agreement to the third decimal digit, so the flow rates will be:

$Q_1 = 0,110 m^3 / \sec$ to D

$Q_2 = 0,059 m^3 / \sec$ from D

$Q_3 = 0,051 m^3 / \sec$ from D

5.15. Problems to be solved

1) Define the type of the flow in a pipe of $18''$ when (a) water with $v = 10^{-6} m/\sec$ flows with velocity $V = 1,2 m / \sec$ and (b) crude oil with $v = 2,3 \times 10^{-4} m^2 / \sec$ flows with the same velocity.

2) Crude oil of kinematic viscosity $3 \times 10^{-4} m^2 / \sec$ and relative density 0.85 is transferred through a horizontal cylindrical pipe of diameter $16''$ made of cast iron to a distance of 10 km. Calculate the power of an electric motor in kW that will move a pump of distribution coefficient 0.7 and flow rate $0,1 m^3 / \sec$.

3) Water with kinematic viscosity $v = 1,7 \times 10^{-6} m^2 / \sec$ flows in a pipe of length 350 m. Compare the pressure drops between pipes made of (a) cast iron and (b) commercial steel. The specific weight of water is given as $\gamma = 995 Kp / m^3$.

4) Oil SAE 10 flows in a pipe of diameter 2 in. Calculate the maximum velocity for which the flow remains laminar if the temperature of the oil is: a) 20°C and b) 60°C.

5) Through the pipe, shown in the figure below, of diameter 6 cm, glycerin flows at 20°C and 100 l/min flow rate. The pressures at points A and B were measured 2.1 and $3,7bar$, respectively. Determine the direction of fluid flow and calculate the Reynolds number, the head losses and the length AB (secondary losses negligible).

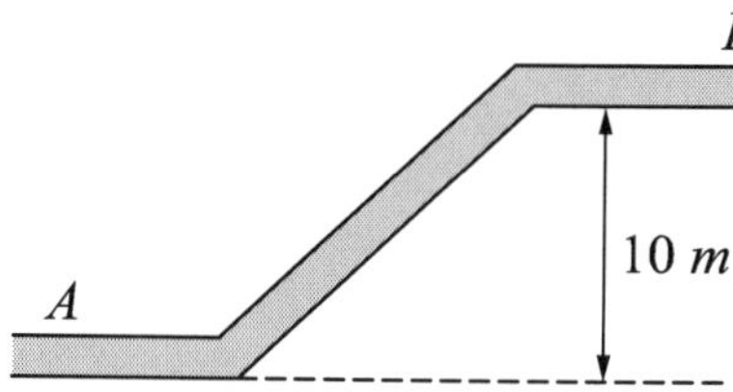

6) In a horizontal pipe of diameter 5 mm, oil flows with density $950kg/m^3$. When the flow rate is $1,2l/\min$, the pressure drop is $380kPa$ per meter of pipe. Calculate the kinematic viscosity of the oil and the Reynolds number.

7) Oil SAE $10, 20°C$, of density $870kg/m^3$, flows through a rectilinear pipe of diameter 2 cm, which has a 30° inclination to the horizontal level. If the pressure along the pipe remains constant, determine the direction of the flow and calculate the head of the losses per meter of the pipe, the Reynolds number and the flow rate of the oil (in m^3/h).

8) In a horizontal cylindrical pipe, made of wrought iron, of length 800 m and diameter 3 in, water flows at 20°C. If the flow rate is 60 m^3/h, calculate the head losses and pressure drop.

9) In a horizontal cylindrical pipe, made of cast iron, of length 500 m and diameter 18 cm, kerosene flows at 20°C with density $800kg/m^3$ and flow rate $400m^3/h$. Calculate the head of the losses and the power that attribute the used pumps at the flow system.

10) Crude oil with kinematic viscosity $7 \cdot 10^{-4} m^2/\sec$ and relative density 0.86 is transferred through a cylindrical pipe made of cast iron of diameter 15 in for a distance of 5 km with flow rate $0,1m^3/\sec$. Calculate the energy losses in kW at (a) $\Sigma K = 25$ and (b) $\Sigma K = 0$.

11) A pipe made of commercial steel of length 500 m and diameter 7 cm is put on a sloping ground with a constant inclination of 6° to the horizontal level. The pipe is connected to a water tank, which lies at a higher point, with its surface 5 m above the entrance of the pipe. Calculate the flow rate of the water.

12) The pipe in the figure below is made of asphalted cast iron of diameter 6 cm. The turbine Tα that is inserted absorbs from the flow system power 2 kW. Calculate the water flow rate if the total coefficient of the local losses is equal to 10.

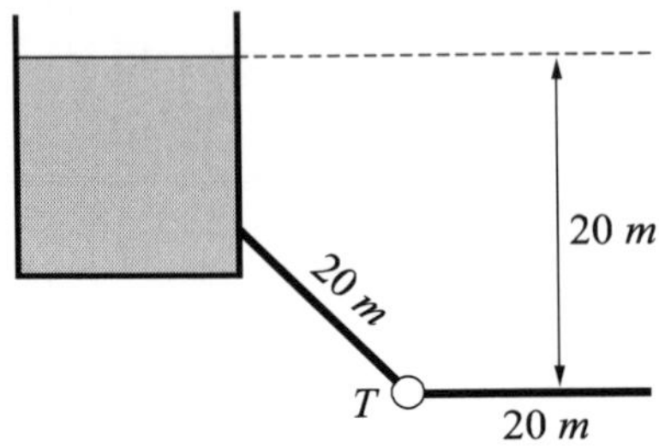

13) A 12 m^3 water tank is supplied from a horizontal pipe made of commercial steel of length 20 cm and diameter 1 in. Calculate the time needed to fill the tank if the pressure of the net is 30 m of water column and the secondary losses are negligible.

14) A pipe made of commercial steel of length 120 m is going to transfer gasoline at 20°C from one tank to another, whose free surface lies 8 m below the first. What diameter of the pipe will we choose if we want the flow rate to be at least $30m^3/h$? $(\Sigma K = 0)$

15) Pump A′ supplies energy head $h_A = 50m$ to water that is transferred through a pipe of the same diameter from point (1) to point (2), which is 35 m above. If the pressure at (1) is 60 cmHg and the losses from (1) to (2) is $h_\ell = 8\frac{V^2}{2g}$, calculate the flow rate. Given: $p_{atm} = 76cmHg$, $\gamma = 1000Kp/m^3$ and $\gamma_{Hg} = 13.600Kp/m^3$.

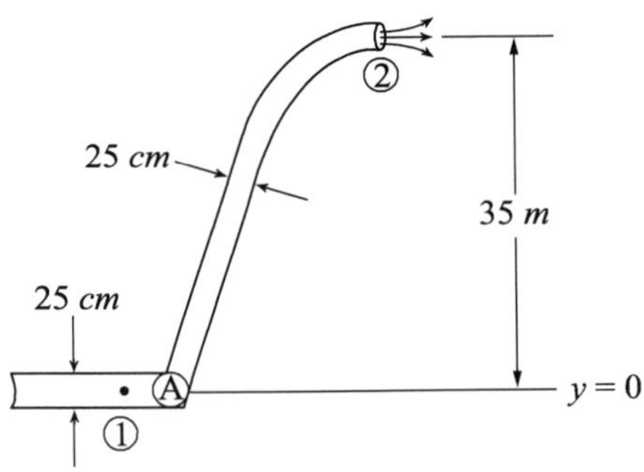

16) Calculate the time t required to fill water in a small pool of capacity 48 m^3 using a plastic tube of length ℓ = 25 m and diameter d = 2. It is given that the roughness of the tube is ε=0.00008 and the water temperature is 15°C, which gives to the adaptation hydrant of the tube 30 m water column differential pressure.

17) A transportation system is shown in the below figure. Calculate p_2 if, in this system, the pipe ABCD has length 100 ft and diameter 3. Angles B and C are standard 90° angles, and angles E and F standard 45° angles $(k = 0,35)$. Use the other values k from the relevant table because the valve at G has k = 0.2, and the flow rate is $15\,ft^3\,/\min$ of water with kinematic viscosity $v = 1,41\times10^{-5}\,ft^2\,/\sec$. The total length of the pipe DEFGH is 1,500 ft and its diameter is 2″.

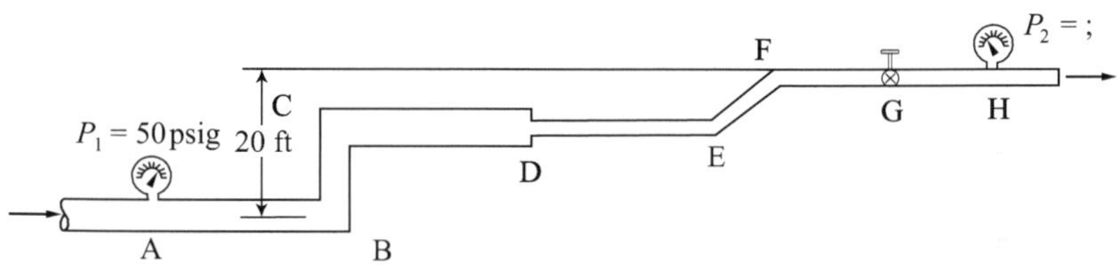

18) (a) Calculate the flow rate of water in the system shown in the figure below. It is given that the offered power to the pump is 3 hp, and the coefficient is $n = 0,8$. (b) Calculate the pressure at the inlet of the pump.

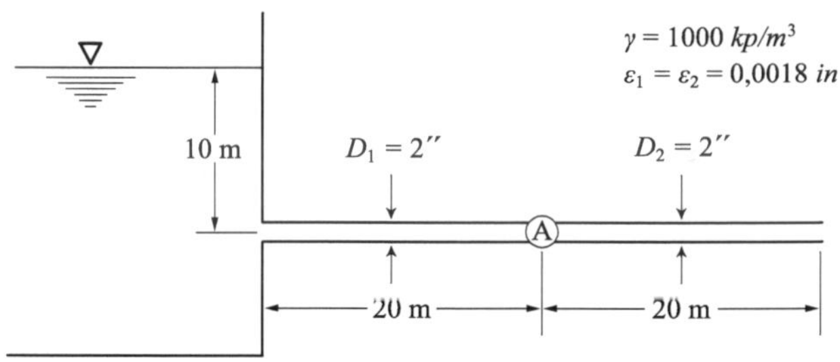

19) Water flows from a tank through pipelines, as shown in the figure below. Calculate (a) the flow rate at each tube and (b) the pressure at points 2 and 3.

Given:

Height of point (0): 25 m

Height of point (1): 18 m

Height of point (2): 9 m $\nu = 10^{-6} m^2 / \sec$

Height of point (3): 5 m

Height of point (4): 0 m

For pipe A: $\ell_A = 18m,\ d_A = 1'',\ f_A = 0{,}015$

For pipe B: $\ell_B = 25m,\ d_B = 0{,}5''\ f_B = 0{,}030$

For pipe (1)–(2): $\ell_{12} = 36m,\ d_{12} = 1{,}5'',\ f_{12} = 0{,}025$

For pipe (3)–(4): $\ell_{34} = 18m,\ d_{34} = 1{,}5'',\ f_{34} = 0{,}025$

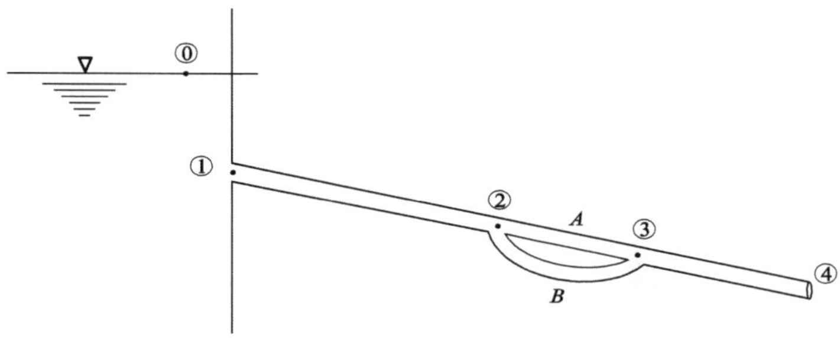

20) Kerosene at 20°C is transferred through a steel pipe (series number 40) to a distance of 500 m at a flow rate of 0.03 m^3/s. The size of the pipe has been chosen in such a way that the head of the frictions of the flow does not surmount the 25 m. What is the nominal diameter of the pipe?

21) For a tube of diameter $d = 12''$ and length l = 100 m, calculate (a) the shear stress τ_0 on the walls of the pipe when the losses during the flow come up to 5 m of the fluid's column that flows and (b) the average flow velocity V. It is given that the fluid is water at temperature 90°C and the friction coefficient of the flow is f = 0.04.

22) Two tanks with altitude range 12 m are connected to two pipes made of galvanized iron, connected in a row, with lengths 100 and 120 m and diameters 15 and 10 cm, respectively. If the local losses are considered to be negligible, calculate the flow rate of the water to the tank that lies below. What energy variations do we notice?

23) The two water tanks, shown in the figure below, must be connected to pipes AB and BC. Pipe BC is of length 40 m and diameter 10 cm. Calculate (in mm) the minimum diameter of pipe AB of length 20 m, so that the flow rate is at least

60 m^3/h. Both pipes are made of asphalted cast iron, with coefficients of local losses $K_{AB} = 0{,}9$ and $K_{BC} = 4{,}5$.

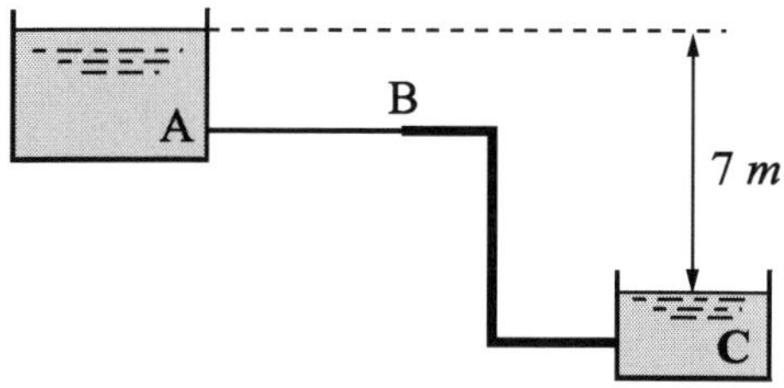

24) A pipe made of commercial steel with diameter $d = 0{,}5''$ and length $\ell = 15m$ is used for emptying an open tank that contains oil of density $\rho = 800 Kg / m^3$ and viscosity $\mu = 0{,}1$ *poise*.

Calculate the flow rate when the free level of the oil in the tank (*H*) is 3 m above the free edge (outflow) of the pipe. It is given that the part of the pipe extends inside the tank and that there is an open valve in the pipe.

25) Calculate the size (power – operation head) of a pump able to ensure $Q_1 = Q_2 = 0{,}14 m^3 / \sec$ of water at 10°C from each one of the two parallel pipes in the following figure. It is given that the pipes are made of commercial steel (common steel), the level that contains the two pipes are horizontal and the local losses can be considered to be negligible.

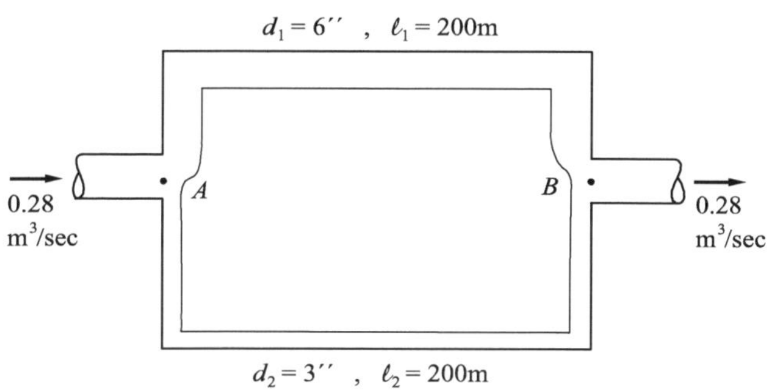

26) Three bronze pipes (ε = 0.0002 m) connected in parallel have lengths 400, 300 and 500 ft and diameters 2, 3 and 4 in. If the total flow rate of the crude oil is 0.6 ft^3/s, calculate the losses and the flow rate in each pipe (in m^3/h). The relative density of crude oil is 0.86, $v = 8 \cdot 10^{-5}$ m^2/s secondary losses are negligible.

The pipes connected in parallel, as shown in the figure below, are made of galvanized iron with $\ell_1 = 60m,\ \ell_2 = 55m,\ d_1 = 5cm,\ d_2 = 4cm$, and they transfer fluid of relative density 0.9 and kinematic viscosity $4{,}4 \cdot 10^{-6}\, m^2/s$. In pipe (2), there is a pump, which has a coefficient of local losses $\mathrm{K}_p = 2.5$. If the pipe is out of order and the total flow rate is 14 m^3/h, calculate the flow rates of the pipes and the pressure drop from A to B.

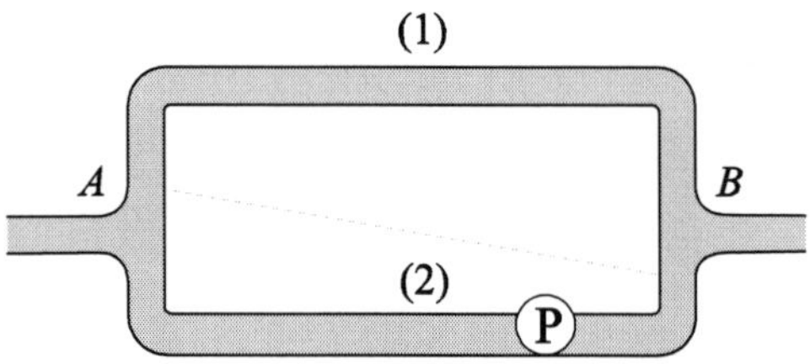

27) The pipes shown in the figure are made of asphalted cast iron with:

$\ell_1 = 250m,\ d_1 = 10cm$

$\ell_2 = 150m,\ d_2 = 12cm$

$\ell_3 = 200m,\ d_3 = 8cm$

$\ell_4 = 150m,\ d_4 = 15cm$

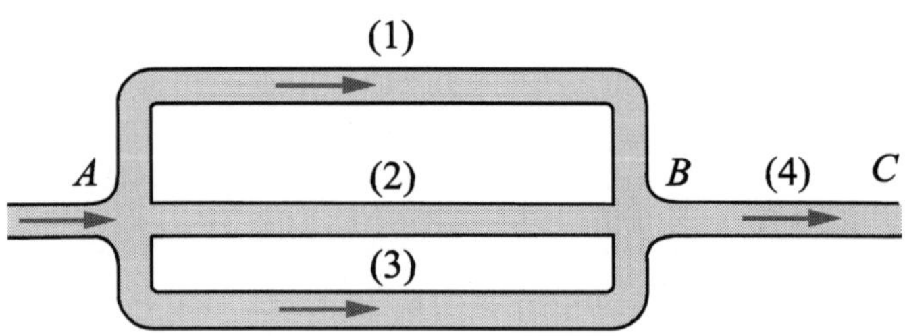

Considering that the secondary losses are negligible, calculate the head of the losses from A to C and the flow rates in the three parallel pipes if $Q_4 = 250m^3/h$.

28) In the branch of the figure below, the pipes are made of asphalted cast iron with:

$\ell_1 = 200m \quad d_1 = 20cm$

$\ell_2 = 100m \quad d_2 = 14cm$

$\ell_3 = 150m \quad d_3 = 12cm$

If the water inflow in tank C is 150 m^3/h, calculate the altitude range of A from B and Q_1 and Q_2.

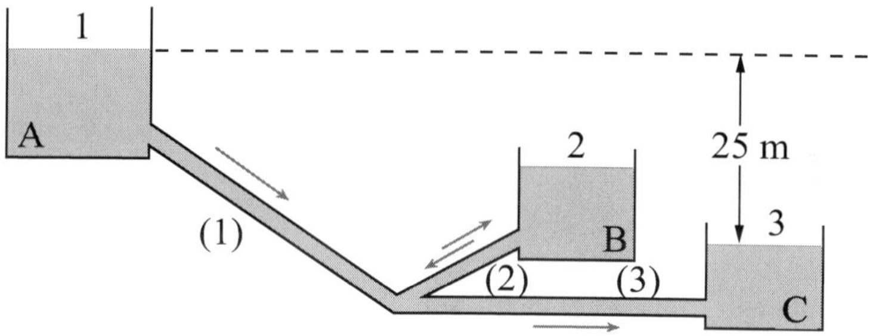

29) Three tanks A, B and C with altitudes of their free surfaces h_1 = 25 m, h_2 = 75 m and h_3 = 60 m are connected to pipes (1), (2) and (3) made of cast iron, as shown in the figure below. The diameters of the pipes are $d_1 = d_2 = d_3 = 8$ in and the lengths are $\ell_1 = 95m$, $\ell_2 = 125m$ and ℓ_3 =160 m. Calculate the flow rates of the three tanks.

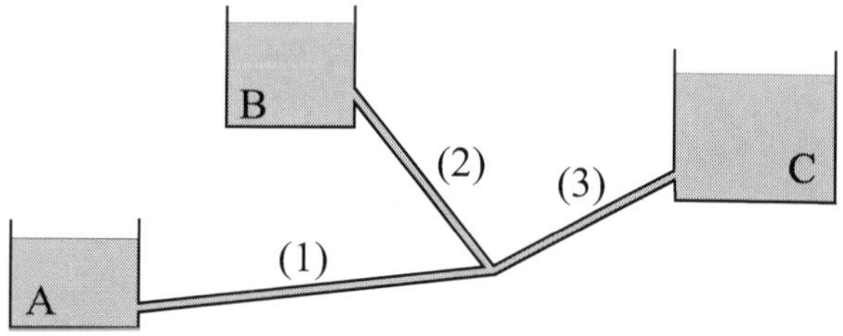

30) The pipes shown in the figure below are made of asphalted cast iron with:

$\ell_1 = 100m \quad d_1 = 8cm$

$\ell_2 = 60m \quad d_2 = 6cm$

$\ell_3 = 80m \quad d_3 = 8cm$

If we have atmospheric pressure on the surface of the tanks B and C and the flow rate to C is 60 m^3/h, calculate pressure p_A and the flow rates in the two other tanks.

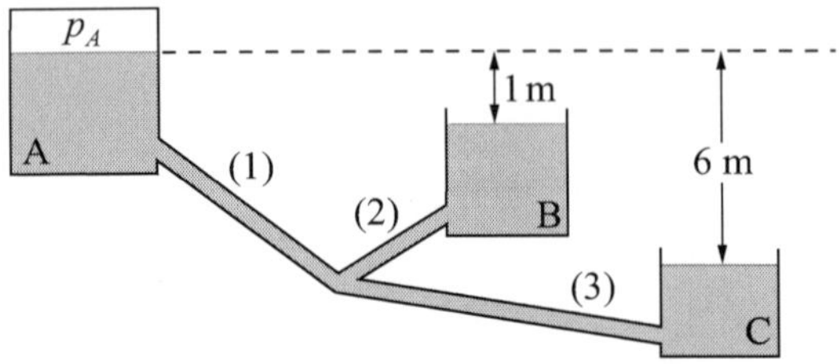

31) Pipes shown in the figure below are made of galvanized iron with:

$\ell_1 = 30m \quad d_1 = 5cm \quad \Sigma K_1 = 3$

$\ell_2 = 20m \quad d_2 = 5cm \quad \Sigma K_2 = 2$

$\ell_3 = 25m \quad d_3 = 7cm \quad \Sigma K_3 = 5$

If water flows out at a flow rate of 20 m^3/h from tank *B*, calculate the flow rates of the two other tanks and the altitude range between A and B.

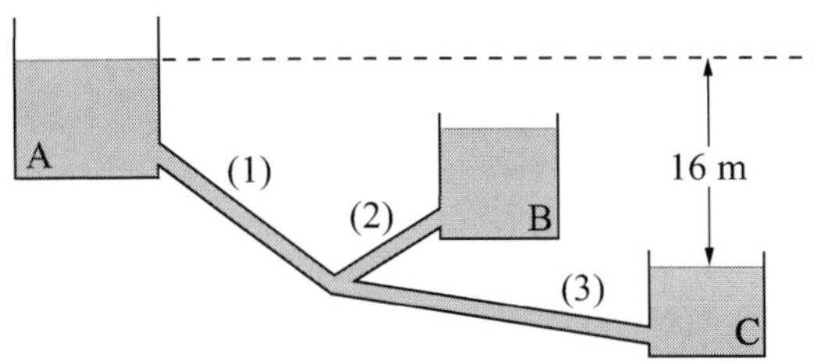

Appendices

Appendix 1

Symbols and Units

Symbol	Unit	Symbol name
S	m^2	Area
A	N	Lift
b	m	Base (length)
b	m	Wall thickness
b	m	Blade thickness
Br	–	Brinkman number
c	J/kg.K	Specific temperature
c_p	J/kg.K	Specific temperature under constant pressure
c_v	J/kg.K	Specific temperature under constant volume
C	–	Hazen–Williams roughness coefficient
C_D	–	Drag coefficient
d	m	Internal diameter
d_h	m	Equivalent or hydraulic diameter

d_x, d_y	m	Axis distance
D	m	External diameter
D	m	Blade diameter
E	J	Energy
E_δ	J	Dynamic energy
E_κ	J	Kinetic energy
E	W	Energy flow
Eu	–	Euler number
f	–	Friction coefficient
f_f	–	Fanning friction coefficient
F	N	Force
D	N	Drag force
F_κ	N	Centrifugal force
Fr	–	Froude number
g	m/s^2	Acceleration of gravity
h	m	Height
h	m	Head loss
K	–	Minor loss coefficient
H	m	Total head loss
H_m	m	Manometric head
P	$kg\text{-}m^2$	Moment of inertia
P_x	m^4	Moment of inertia on x-axis
J	kg-m/s	Momentum

K	Pa	Bulk modulus
ℓ_1	m	Length
ℓ_{eq}	m	Equivalent length
j	kg-m^2/s	Angular momentum
m	kg	Mass
q	kg/s	Mass flow rate
P	Nm	Torque
M	–	Mach number
n	s^{-1}	Frequency
Nu	–	Nusselt number
p	Pa	Pressure
p_{atm}	Pa	Atmospheric pressure
p_{ax}	Pa	Relative pressure
p_w	Pa	Vapor pressure
P	W	Power
P_r	–	Prandtl number
Q	m^3/s	Flow rate (volumetric)
Q_c	m^3/s	Internal flow rate of a pump
r	m	Radius
R	J/kg.K	Gas constant
Re	–	Reynolds number
T	K	Temperature

T	s	Period
t	s	Time
u	m/s	Local velocity
V	m/s	Velocity (total)
v	m^3	Volume
h	m	Height
δ	m	Boundary layer thickness
We	–	Weber number
K	–	Compressibility coefficient
α	m/s^2	Acceleration
α_κ	m/s^2	Centrifugal acceleration
β	Pa^{-1}	Compressibility
W	N	Weight
W_φ	N	Apparent weight
γ	N/m^3	Specific weight
γ_{re}	–	Relative specific weight
ε	m	Roughness
n	–	Efficiency
η_h	–	Hydraulic efficiency
η_m	–	Mechanical efficiency
θ,ω	rad	Angle
μ	kg/ms	Viscosity (dynamic)
ν	m^2/s	Kinetic viscosity

Π	–	Dimensionless parameter
ρ	kg/m^3	Density
ρ_{re}	–	Relative density
σ	N/m	Surface stress
σ	–	Cavitation coefficient
Σh	m	Total head loss
τ	Pa	Shear stress
U	m^3/kg	Specific volume
ω	rad/s	Angular velocity
Ω	$kg \cdot m/s$	Thrust
x,y,z	m	Cartesian coordinates

Appendix 2

Tables and Diagrams of Natural Values

Gases	c_p KJ/KgK	$k = c_p / c_v$	Fluids	c_p KJ/KgK
Air	1,005	1,400	Water	4,18
Nitrogen	1,04	1,400	Methanol	2,55
Oxygen	0,92	1,395	Ethanol	2,46
Carbon dioxide	0,88	1,289	Acetone	2,13
Chlorine	0,48	1,324	Acetic acid	2,05
Hydrogen	14,31	1,405	Benzene	1,74
Methane	2,26	1,304	Glycerin	2,62
Ethane	1,72	1,187	Gasoline	2,22
Propane	1,63	1,127	Kerosene	2,09
Ethylene	1,54	1,240	Castrol oil	1,97
Hydrazine	1,65	1,195	Olive oil	1,97
Ammonia	2,09	1,304	Mercury5	0,14

Table A2.1. *Specific temperatures (in regular conditions) of gases and fluids*

T (°C)	C_p (K/KgK)
0	4,192
10	4,181
20	4,177
30	4,178
40	4,180
50	4,184
60	4,190
70	4,197
80	4,206
90	4,216
100	4,230

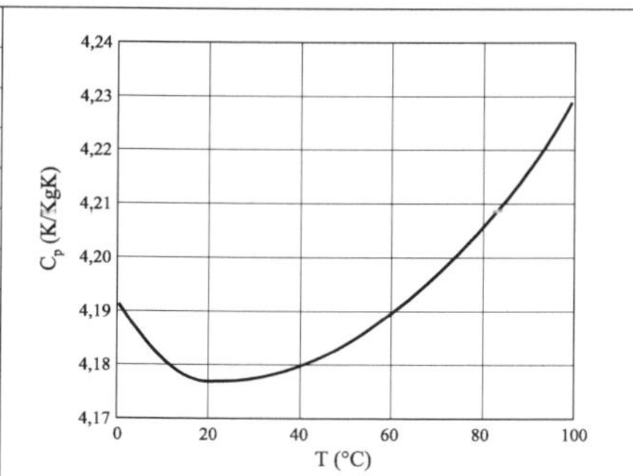

Table A2.2. *Specific water temperature from 0°C to 100°C (p = 101.3 kPa)*

Relations between c_p, c_v, k, R		$c_p - c_v = R$ $c_p / c_v = k$	$R = R_w/m_r$ $R_w = 8{,}314$ J/KmolK
State equation for ideal gases		$p \cdot \upsilon = R \cdot T \Rightarrow p = \rho \cdot R \cdot T$	
Variation of internal energy and enthalpy		$\Delta u = c_u \cdot \Delta T$ $\Delta h = c_p \cdot \Delta T$	
Process	**Relationp, u, T** $(\rho = 1/\upsilon)$	**Closed System** $q - w = \Delta u$	**Diagram** $p - \upsilon$
General	$p_1 \cdot \upsilon_1^n = p_2 \cdot \upsilon_2^n$ $\frac{T_1}{T_2} = \left(\frac{p_1}{p_2}\right)^{\frac{n-1}{n}} = \left(\frac{\upsilon_2}{\upsilon_1}\right)^{n-1}$	$w = \frac{R \cdot T_1}{n-1} \cdot \left[1 - \left(\frac{p_2}{p_1}\right)^{\frac{n-1}{n}}\right]$	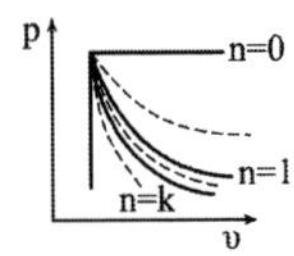
n = 0 **Isobaric** **(p = constant)**	$p_1 = p_2 = p$ $\frac{\upsilon_1}{\upsilon_2} = \frac{T_1}{T_2}$	$w = p \cdot (\upsilon_2 - \upsilon_1)$ $q = \Delta h = c_p \cdot (T_2 - T_1)$	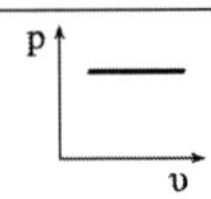
n = 1 **Isochoric** **(u = constant)**	$\upsilon_1 = \upsilon_2 = \upsilon$ $\frac{p_1}{p_2} = \frac{T_1}{T_2}$	$w = 0$ $q = \Delta u = c_v \cdot (T_2 - T_1)$	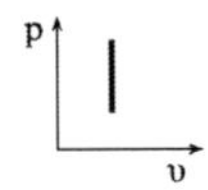
n = 1 **Isothermal** **(T=constant)**	$T_1 = T_2 = T$ $\frac{\upsilon_1}{\upsilon_2} = \frac{p_2}{p_1}$	$q = w = R \cdot T \cdot \ln \frac{p_1}{p_2}$	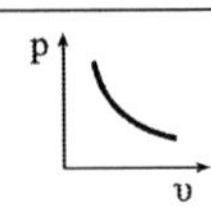
n = k **Adiabatic** **(q = 0)**	$p_1 \cdot \upsilon_1^k = p_2 \cdot \upsilon_2^k$ $\frac{T_1}{T_2} = \left(\frac{p_1}{p_2}\right)^{\frac{k-1}{k}} = \left(\frac{\upsilon_2}{\upsilon_1}\right)^{k-1}$	$w = -\Delta u = c_\upsilon \cdot (T_1 - T_2)$	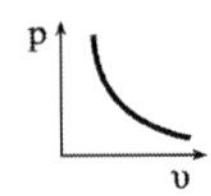

Table A2.3. *Characteristics of gases*

Fluid	**θ**	***ρ*** **kg/m³**	**ν** **m²/s**	**σ** **N/m**	**β** **Kpa⁻¹**
Water	0	1000	$1{,}79\cdot10^{-06}$	$7.56\cdot10^{-02}$	$5{,}03\cdot10^{-07}$
	3.98	1000	$1{,}57\cdot10^{-06}$		
	10	1000	$1{,}31\cdot10^{-06}$	$7{,}42\cdot10^{-02}$	$4{,}72\cdot10^{-07}$
	20	998	$1\cdot10^{-06}$	$7.28\cdot10^{-02}$	$4{,}52\cdot10^{-07}$
	30	996	$8{,}06\cdot10^{-07}$	$2{,}26\cdot10^{-02}$	$3{,}40\cdot10^{-07}$
	40	992	$6{,}58\cdot10^{-07}$	$6{,}96\cdot10^{-02}$	$4{,}37\cdot10^{-07}$
	50	988	$5{,}57\cdot10^{-07}$	$6{,}79\cdot10^{-02}$	$4{,}37\cdot10^{-07}$
	60	983	$4{,}78\cdot10^{-07}$	$6{,}62\cdot10^{-02}$	4,38· IO*^{7}
	70	978	$4{,}16\cdot10^{-07}$	$6{,}64\cdot10^{-02}$	$4{,}46\cdot10^{-07}$
	80	972	$3{,}67\cdot10^{-07}$	$6.26\cdot10^{-02}$	$4{,}55\cdot10^{-07}$
	90	965	$3{,}28\cdot10^{-07}$		$4{,}67\cdot10^{-07}$
	100	958	$2{,}96\cdot10^{-07}$	$5{,}89\cdot10^{-02}$	$4.83\cdot10^{-07}$
Ethanol	0	806	$2{,}2\cdot10^{-06}$	$2.41\cdot10^{-02}$	$9{,}80\cdot10^{-07}$
	20	789	$1{,}52\cdot10^{-06}$	$2.28\cdot10^{-02}$	$1{,}11\cdot10^{-06}$
	40	772	$1{,}08\cdot10^{-06}$		$1{,}27\cdot10^{-06}$
	60	754	0,785 'IO*^{6}		$1.47\cdot10^{-06}$
Ethylene Glycol	20	1117	$19.1\cdot10^{-06}$	$4.84\cdot10^{-02}$	
	40	1110	$8{,}3\cdot10^{-06}$		
	60	1090	$4.54\cdot10^{-06}$		
	80	1070	$2{,}82\cdot10^{-06}$		
	100	1060	$1{,}88\cdot10^{-06}$		
Benzene	0	900	$1{,}01\cdot10^{-06}$	$3{,}02\cdot10^{-02}$	$8.13\cdot10^{-07}$
	20	881	$0{,}742\cdot10^{-06}$	$2.88\cdot10^{-02}$	$7.14\cdot10^{-07}$
	40	858	$0.586\cdot10^{-06}$		$1{,}1\cdot10^{-06}$
	60	836	$0{,}469\ 10^{-06}$		$1{,}28\ 10^{-06}$
	80	815	$0{,}404\cdot10^{-06}$		$1{,}54\cdot10^{-06}$
Kerosene	-18	841	$8{,}4\cdot10^{-06}$		
	20	814	$2{,}37\cdot10^{-06}$	$2.90\cdot10^{-02}$	
Methanol	0	810	$1{,}01\cdot10^{-06}$	$2.45\cdot10^{-02}$	$1{,}07\cdot10^{-06}$
	10	801			$1.14\cdot10^{-06}$
	20	791	$0{,}756\cdot10^{-06}$	$2.26\cdot10^{-02}$	$1{,}22\cdot10^{-06}$

Fluid	θ	ρ kg/m³	ν m²/s	σ N/m	β Kpa⁻¹
Methanol	30	783	$0{,}651 \cdot 10^{-06}$		$1{,}29 \cdot 10^{-06}$
	40	774	$0{,}581 \cdot 10^{-06}$		$1{,}38 \cdot 10^{-06}$
	50	765	$0{,}518 \cdot 10^{-06}$		$1{,}47 \cdot 10^{-06}$
Octane (regular)	0	718	$0{,}983 \cdot 10^{-06}$		$1 \cdot 10^{-06}$
	20	703	$0{,}772 \cdot 10^{-06}$		
	25				$1{,}2 \cdot 10^{-06}$
	40	686	$0{,}631 \cdot 10^{-06}$		$1{,}34 \cdot 10^{-06}$
Mercury	0	13600	$0{,}124 \cdot 10^{-06}$		$4 \cdot 10^{-08}$
	4	13590			
	20	13550	$0.115 \cdot 10^{-06}$	$48{,}4 \cdot 10^{-02}$	$4 \cdot 10^{-08}$
	40	13500	$0{,}107 \cdot 10^{-06}$		
	60	13450	$0{,}102 \cdot 10^{-06}$		
	80	13400	$0{,}0970 \cdot 10^{-06}$		
	100	13350	$0{,}0929 \cdot 10^{-06}$		

Table A2.4. *Fluid properties (density ρ, kinematic viscosity ν, surface stress σ and compressibility β) at various temperatures*

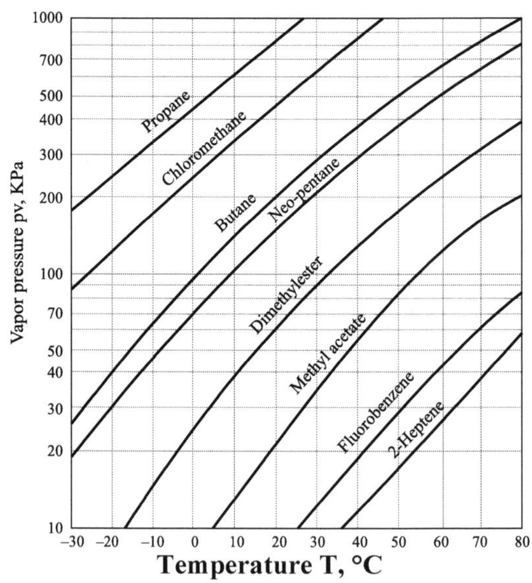

Figure A2.1. *Vapor pressure of volatile materials*

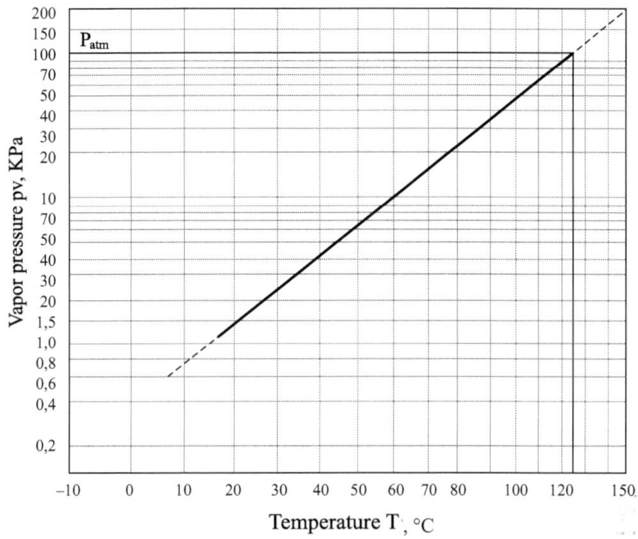

Figure A2.2. *Octane vapor pressure*

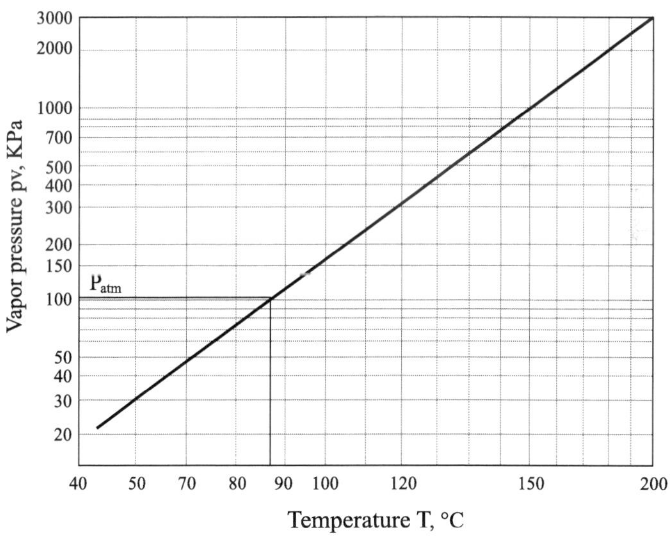

Figure A2.3. *Ethanol vapor pressure*

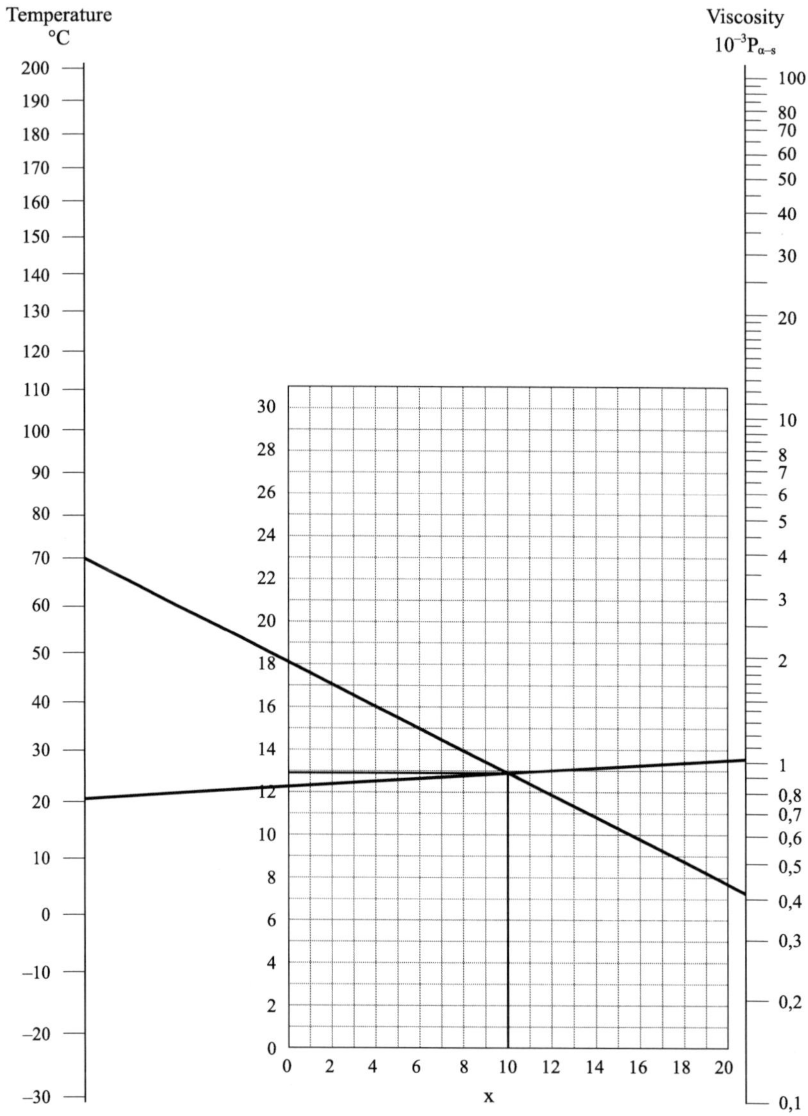

Figure A2.4. *Nomograph of viscosity calculation*

Fluid	X	Y
Ethanol, 100% Ethanol, 40%	**10,5** **6,5**	**13,8** **16,6**
Ethylene glycol	**6,0**	**23,6**
Acetaldehyde	**15,2**	**4,8**
Acetone, 100% Acetone, 35%	**14,5** **7,9**	**7,2** **15,0**
Ammonia, 100% Ammonia, 26%	**12,6** **10,1**	**2,0** **13,9**
Aniline	**8,1**	**18,7**
Benzene	**12,5**	**10,9**
Glycerin, 100% Glycerin, 50%	**2.0** **6,9**	**30,0** **19,6**
Nitrogen dioxide	**12,9**	**8,6**
Carbon dioxide	**11,6**	**0,3**
Sulphur dioxide	**15,2**	**7,1**
Dichloromethane	**14,6**	**8,9**
Hexane	**14,7**	**7,0**
Heptane	**14,1**	**8,4**
Sulphuric acid, 98% Sulphuric acid, 60%	**7,0** **10,2**	**24,8** **21,3**
Cyclohexane	**9,8**	**12,9**
Kerosene	**10,2**	**16,9**
Methanol.90% Methanol, 40%	**12,3** **7,8**	**11,8** **15,5**
Water	**10,2**	**13,0**
Nitric acid	**12,8**	**13,8**
Xylene	**13,9**	**10,6**
Octane	**13,7**	**10,0**
Pentane	**14,9**	**5,2**
Carbon tetrachloride	**12,7**	**13,1**
Toluene	**13,7**	**10,4**
Mercury	**18,4**	**16,4**
Sodium hydroxide, 50%	**3,2**	**25,8**
Hydrochloric acid, 31,5%	**13,0**	**16,6**
Phosphoric acid	**12,8**	**13,8**
Chlorobenzene	**12,3**	**12,4**
Chloroform	**14,4**	**10,2**
Freon-1 1	**14,4**	**9,0**
Freon -12	**16,8**	**5,6**
Freon -21	**15,7**	**7,5**
Freon -22	**17,2**	**4,7**
Freon -113	**12,5**	**11,4**

Table A2.5. *Values of X and Y used in the nomograph of viscosity*

The various measurement grades of viscosity (which constitutes the most important natural quality of lubricants), such as SAE, SUS, ANSI, BSI, °E and RI, led to an international agreement and the establishment of the ISO scales in 1975. With the standard ISO 3448:1992 (Industrial liquid lubricants – ISO viscosity classification), a system is established to classify the viscosity for industrial fluid lubricants and relevant fluids, including liquid petroleum used as a lubricant, hydraulic fluids and for other applications. The following table presents the viscosity grades (ISO VG) and their corresponding SI units (m^2/s).

ISO VG	Viscosity Grade Ranges ($10^{-7}m^2/s$ in 40°C)		
	Average Value	Minimum (–10%)	Maximum (+10%)
2	2,2	1.98	2.42
3	3,2	2.88	3.52
5	4,6	4.14	5.06
7	6,8	6.12	7.48
10	10	9.00	11.0
15	15	13.5	16.5
22	22	19.8	24.2
32	32	28.8	35.2
46	46	41.4	50.6
68	68	61.2	74.8
100	100	90.0	110
150	150	135	165
220	220	198	242
320	320	288	352
460	460	414	506
680	680	612	748
1000	1000	900	1100
1500	1500	1350	1650

NOTE.– The ISO viscosity grades are the same as those of the ASTM and BSI. The viscosities in ISO grades are measured at 40°C, whereas in ASTM and BSI, they are measured at 100°F (37.8°C). The lubricants of the ASTM or BSI grade are slightly more viscous than those of the ISO grade.

Table A2.6. *Viscosity grades (VG) ISO*

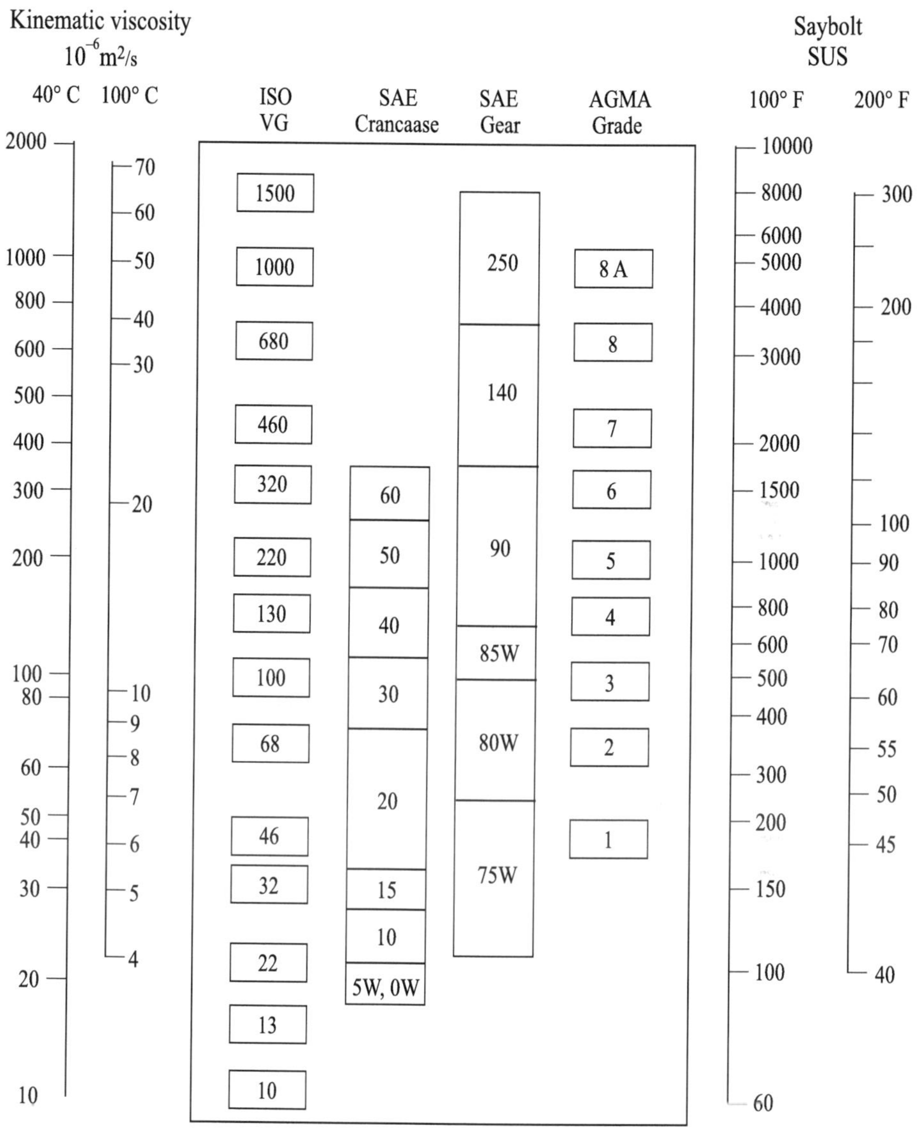

Table A2.7. *Correspondence of the scales of lubricants' viscosity*

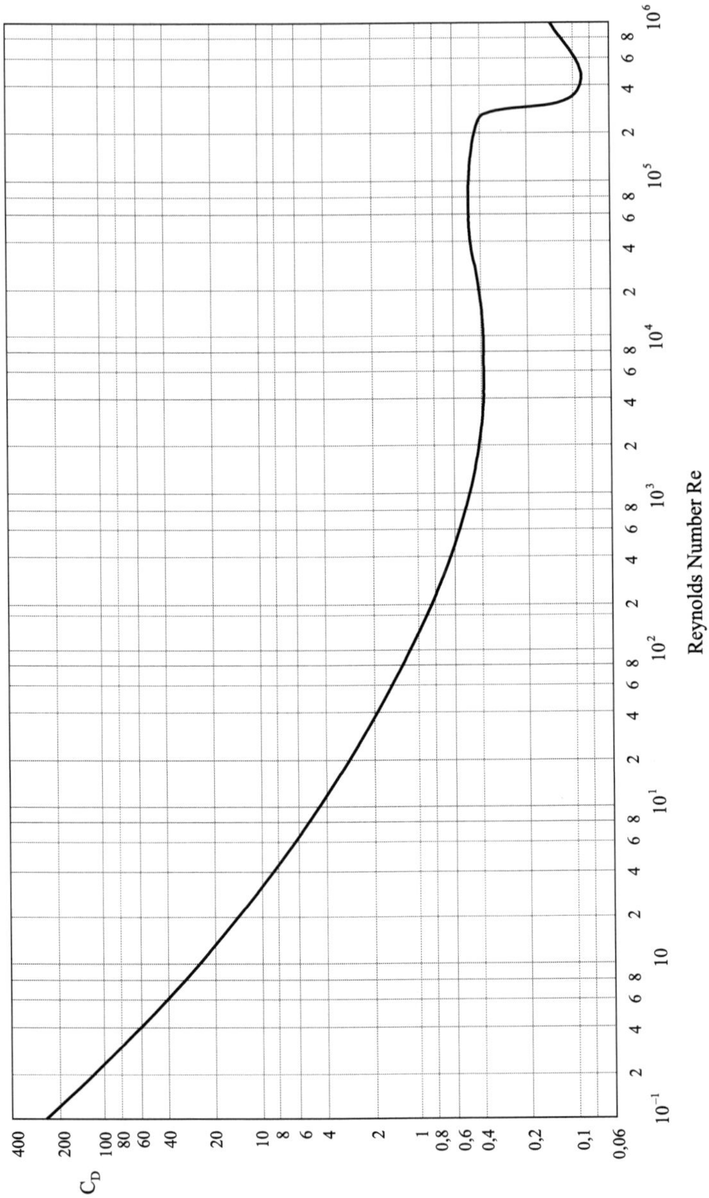

Figure A2.5. *Drag coefficient of a sphere*

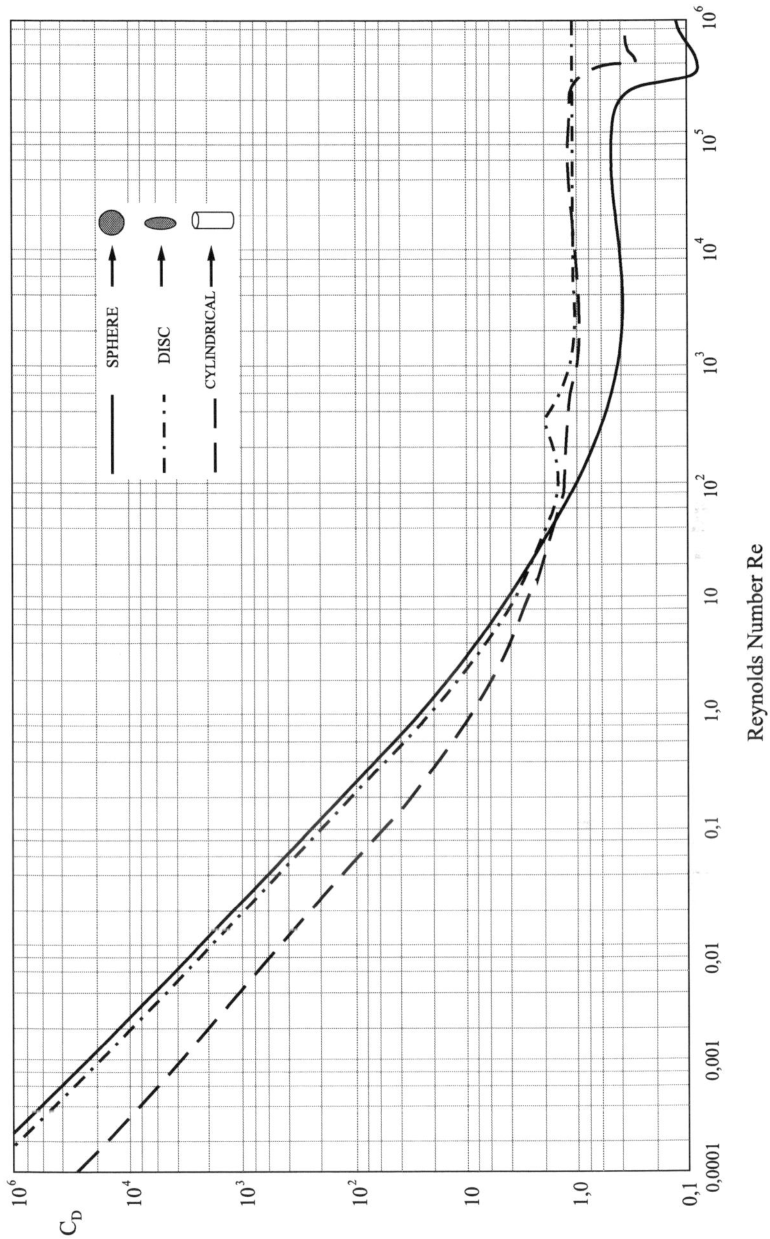

Figure A2.6. *Drag coefficient of a sphere, disk and cylinder*

Material resistance in corrosion

MATERIAL \ FLUID	Sulphuric acid 10%	Sulphuric acid 30%	Nitric acid 10%	Nitric acid 20%	Diluted hydrochloric acid	Hydrochloric acid 20%	Hydrochloric acid 30%	Hydrofluoric acid 30%	Phosphoric acid 10%	Formic acid 10-85%	Acetic acid 20%	Oxalic acid 10%
METALS												
Aluminum	U	U	U	U	U	U	U	U	53	113	102	59
Bronze	U	U	U	U	U	U	75	75	U	113	U	113
Carbon steel	U	U	U	U	U	U	U	U	U	U	U	U
Copper	U	U	U	U	U	U	75	75	U	113	U	113
Ductile iron, Pearlitic												
Hasterlloy C	113	102	113	102	97	81	113	113	113	113	164	113
Inconel	U	U	47	41	47	41	U	U	108	108	47	113
Monel	41	41	U	U	47	41	219	219	41	113	113	113
Nickel	41	41	U	U	47	41	91	91	41	113	47	53
Stainless Steel 304	U	U	175	164	U	U	U	U	103	119	119	U
Stainless Steel 316	U	U	113	159	U	U	U	U	159	219	219	191
PLASTIC - RUBBER												
ABS	75	47	75	69	47	47	U	U	75	U	69	53
BUtyl	97	97	86	86	64	U	191	191	81	81	80	103
CPVC	97	97	97	86	97	97	U	U	97	75	97	103
EPDM	75	75	30	30	164	53	30	30	164	164	75	164
FEP	219	219	219	219	219	219	219	219	219	219	219	219
FKM	191	191	219	219	191	191	113	113	219	103	108	219
HDPE	75	75	75	75	86	75	75	75	75	75	75	75
Neoprene	108	108	U	U	81	97	108	108	108	86	86	108
N-RUbber	80	80	U	U	75	81	53	53	81	U	41	81
PFA	136	136	108	108	137	137	108	108	108	108	108	
PP	108	108	108	75	119	119	97	97	137	113	119	81
PTFE	258	258	258	258	258	258	258	258	258	258	258	258
PVC Type 2	75	75	75	75	75	75	69	69	75	47	75	75
PVDC	64	41	69	81	97	97	86	86	97	81	64	91
PVDF	136	119	108	97	153	153	142	142	153	137	164	81
SBR Styrene	U	U	U	U			U	U	108		U	
Epoxy resins	75	64	75	53	103	108	U	U	86	35	58	147
Furan resins	136	136	41	υ	142	142	U	U	137	142	136	108
Polyester resins	119	119	81	53	103	103	47	47	119	81	108	119
Vinyl resins	108	97	81	81	125	119	U	U	108	53	108	108

Table A2.8. *Resistance in acid*

MATERIAL \ FLUID	Ammonia (anhydrous)	Ammonium hydroxide 10%	Ferric chloride, 50%	Ferrous sulphate	Calcium chloride	Calcium hydroxide 10%	Cupric sulphate	Hydrogen peroxide 30%	Potassium peroxide 5%	Potassium permanganate 10-%	Sodium bicarbonate 20%	Sodium carbonate	Sodium chloride	Caustic soda	Zinc chloride
METALS															
Aluminum	97	191	U	U	30	41	U	191	U	113	80	U	U	U	U
Bronze	41	U	U	U			U	U	47	108	47	53	113	102	U
Carbon steel	219	113	U	U	30	41	U	U	113	41	53	64	86	113	U
Copper	41	U	U	41	30	113	U	U	53	41	41	53	113	113	U
Ductile iron, Pearlitic											45		97	65	U
Hastelloy C	313	113	113	81	108	91	113	53	113	113	113	113	113	124	136
Inconel	313	47	41	U	30	113	47	75	113	113	113	113	113	164	41
Monel	313	U	U	41	113	113	47	30	113	113	113	113	113	191	108
Nickel	47	U	U	U	30	113	47		113	113	113	113	113	113	108
Stainless Steel 304	264	113	U	41	80	113	113	113	164	113	136	113	113	113	U
Stainless Steel 1316	313	113	U	108	75	113	219	219	191	190	191	191	191	191	108
PLASTIC – RUBBER															
ABS	U	41		75	75		75	U	75	U	75	75	75	75	75
BUtyl	U	102	86	103	102	102	102	U	97	69	97	97	97	97	102
CPVC	97	108	97	97	97	91	113	97	97	102	113	113	113	102	97
EPDM	164	113	164	153	113	113	164	53	164	113	164	164	164	164	164
FEP	219	219	219	219	219	219	219	219	219	219	219	219	219	219	219
FKM	U	102	219	219	158	164	219	191	175	86	219	102	219	30	219
HDPE	75	75	75		75	75	75	75	75	75	75	75	75	75	75
Neopreme	108	105	86	108	108	119	108	U	108	53	108	108	108	108	86
N-RUbber	U	41	81	81	80	108	80	U	53	U	80	97	69	80	80
PFA	108	153	108	108	108	108	108	108	108	108	108	108	108	136	108
PP	119	119	113	108	119	108	108	53	113	80	119	119	119	119	108
PTFE	258	258	258	258	258	258	258	258	258	258	258	258	258	258	258
PVC Type 2	47	75		75	75		75	U	75	75	75	75	75	75	75
PVDC		U	75	81	97	86	97	64	53	69	97	97	97	47	91
PVDF	153	153	153	153	153	147	153	137	113	153	153	153	153	113	142
SBR Styrene	108		108			108	108		U				108		
Epoxy resins	U	102	137	108	108	113	113	75	108	80	136	164	113	102	136
Furan resins	142	97	131	142	142	119	142	U	136	142	142	142	142	U	142
Polyester resins	119	75	119	119	119	97	119	47	80	113	86	86	119	69	136
Vinyl resins	119	81	113	108	97	97	131	91	80	119	108	97	97	91	97

Table A2.9. *Resistance in inorganic substances*

MATERIAL \ FLUID	Acetone	Benzene	Alcohol	Ester	Ether	Ethylene glycol	Dilute formaldehyde	Glycol	Heptane	Hexane	Oil and fats	Toluene	Xylene
METALS													
Aluminum	275	113	113		47	53		41	53	41	81	113	108
Bronze	219	219	219	219	108	186	81	53	191	191	81	191	136
Carbon steel	164	75	131		108	53		41	191	191	81	191	108
Copper	75	53	53		41	53			41			113	108
Ductile iron, Pearlitic						164							
Hastelloy C	108	113	113		108	314	113		108	137	137	113	164
Inconel	102	113	41		47	113	113	53	108			113	108
Monel	102	113	113		47	113	113	53	108	53		113	54
Nickel	102	113	108		41	113	113		113	41	30	113	108
Stainless Steel 304	102	125	108		108	113	313	53	137	137	81	113	108
Stainless Steel 316	219	219	108	219	107	186	125	41	191	137	137	191	108
PLASTIC - RUBBER													
ABS	U	U	64		U	75	53	75	69	U	75	U	U
BUtyl	86	U	103		U	103		81		U		U	U
CPVC	U	U	97	U	U	113	75	97	97	81	113	U	U
EPDM	164	U	159			164	75	164	U	U		U	U
FEP	219	219	219	219	219	219	219	219	219	219	219	219	242
FKM	U	219	191		U	219	125	219	191	225	97	219	219
HDPE	64	U	75	41	U	75	75	75	59	41	U	U	U
Neopreme	U	U	108		U	86	75	86	108	108	41	U	U
N-RUbber	U	U	81		U	81		64	U	U		U	U
PFA	108	108	108		108	108	108	108	108	108	108	113	108
PP	119	75	97		U	125	108	81	41	59	97	30	30
PTFE	258	258	258	259	259	258	164	258	258	258	258	258	258
PVC Type 2	U	U	75	U	U	75	75	75	75	35	75	U	U
PVDC	47	41	81	41		97	75		81	81	81	43	U
PVDF	U	80	153	91	64	153	64	153	153	153	159	113	113
SBR Styrene	108	U	108			108	108		U	U		U	U
Epoxy resins	58	97	81	86	47	164	59	164	81	97		80	75
Furan resins	108	142	142	137	47	142	86		113	81	137	142	142
Polyester resins	U	U	47			119	41	119	108	47	119	U	47
Vinyl resins	U	U	53	81	97	113	81	113	113	86	113	64	75

Table A2.10. *Resistance in organic substances*

MATERIAL \ FLUID	Crude Oil	Diesel Fuels	Fuel Oil	Gasohol	Gasoline, Refined	Gasoline, unleaded	Jet Fuel JP-4	Kerosene	Lubricating Oil	Machine Oil	Motor Oil	Naphtha	Petroleum Oils, Refined	Transformer Oil
METALS														
Aluminum	53	47	75	81	113	113	91	91	81			97	47	41
Bronze	53	47	191	81	108	191	219	191			53	219	41	47
Carbon steel	53	102	108	81	108	191	91	191	81	113	137	47		41
Copper	41		41	81	47	47		47	47		81	47	47	
Ductile iron, Pearlitic														
Hastelloy C	47	108	108	81	108	175	53	113		113		108		47
Inconel			75			41	47	47			47	81		
Monel	164		97	81	53	53	47	91	53		47	64	47	47
Nickel			97		53	53	41	113				64		47
Stainless Steel 304	113	47	137		147	41	53	219	81	113	137	137	41	47
Stainless Steel 316	113	47	86	81	113	41	219	219	81	113	137	113	41	47
PLASTIC - RUBBER														
ABS	47			U	U	U		47	53		47	75		
BUtyl	U		U				U	U	U	U		U		U
CPVC	113	53			81	υ	108	97	97	97	97	75	97	97
EPDM	U	U	U	U		U	U	U	U	U	U	U	U	U
FEP	219	219	219		219	219	219	219	219	75	219	219	219	219
FKM	164	219	214	47	97	97	219	219	219	108	103	219	103	219
HDPE	64	64	108		U	75		41	U			41	41	75
Neoprene	U	41	108		47	108	U	108	108	108		U	53	69
N-RUbber	U	U	U		U	U	U	U	U			U	U	U
PFA	108	108	108		108	108	108	108	108		108	108		108
PP	80	53	91	U	U	U	35	47	35	59	U	59	81	58
PTFE	258	258	258	108	258	258	258	258	258	258	258	258	258	258
PVC Type 2	75		75	75	U		75	75	75	75	75	75		
PVDC	80	64	64		47	81	41	64	64			81		
PVDF	153	153	153	153		153	137	142	153	108	137	153	142	
SBR Styrene	U	108			U	U	U	U	U			U	U	U
Epoxy resins	164	137	137		81	137	81	137	125		41	119		124
Furan resins		137	137		142	153	75	137				142		
Polyester resins	119	108	41		41	47	41	81				81		119
Vinyl resins	136	119	108		97	53	97	147	108		137	113	108	164

Table A2.11. *Resistance in fuels – lubricants*

MATERIAL \ FLUID	Seawater	Drinking water	Purified water	Distilled water	Chlorinated water	Salts water	Sewage	Detergents	Soap solutions
METALS									
Aluminum	53	113	97	U	41	U			164
Bronze	136	113		108	U	136	47		191
Carbon steel	47		U	U	U	41	47		91
Copper	41	113		47	U	41	47	30	41
Ductile iron, Pearlitic	47	45				47			
Hastelloy C	313	113	108	313	113	164			47
Inconel	41		75	30	47	41			47
Monel	136	113		U	U	136			53
Nickel	47		108	41	U	41			75
Stainless Steel 304	41	113	242	136	U	41	47	97	47
Stainless Steel 316	136	113	242	136	U	136	47	97	47
PLASTIC -RUBBER									
ABS	47	41	75	75	75	75	41		
BUtyl			75		U	102			80
CPVC	97	113	97	97	113	97	97	86	97
EPDM	136	136	136	164	30	136	113	158	169
FEP	219		219	219	219	219	219	219	219
FKM	102	164	102	102	102	102	102	219	219
HDPE	75		75	75	75	75	75	75	75
Neoprene	113	97	113	108	U	113	86	86	108
N-RUbber			80	80	80	80			80
PFA	108		108	108		108	108	108	113
PP	119	97	119	119	75	119	119	80	97
PTFE	258	219	258	258	258	258	258	258	258
PVC Type 2	75	75	75	75	75	75	75	75	41
PVDC	91	91	91	91	97	97	91		97
PVDF	153	153	153	153	119	153	136		53
SBR Styrene	108		113	108		108		108	108
Epoxy resins	164		136	113	U	113		136	
Furan resins	136		136	108	142				
Polyester resins	119	113	86	108	119	97			41
Vinyl resins	97	113	113	113	97	97		64	75

Table A2.12. *Resistance in water and water solutions*

The values mentioned in Tables A2.8–A2.12 correspond to the maximum temperature (°C) up to which the material shows sufficient resistance to corrosion. For metals, this means that the corrosion rate is lower than 1.27 mm per year.

Metals and alloy	M.O. (Volt)	min	max
Magnesium (Mg) and alloy	-1,62	-1,63	-1,60
Zinc, Zn	-1,01	-1,03	-0,98
Aluminum alloy	-0,88	-1,00	-0,76
Cadmium, Cd	-0,73	-0,76	-0,70
Mild Steel	-0,66	-0,71	-0,60
Cast iron	-0,66	-0,72	-0,60
Wrought iron	-0,66	-0,71	-0,60
Plain steel	-0,65	-0,70	-0,60
Stainless steel, Type 410 (13% Cr)	-0,52	-0,58	-0,46
Stainless steel, Type 304 (18 Cr -8 Ni)(*)	-0,52	-0,58	-0,46
Nickel, Ni	-0,52	-0,58	-0,46
Stainless steel, Type 304 (18 Cr -8 Ni -3 Mo)	-0,49	-0,54	-0,43
Inconel (78 Ni-13.5 Cr-6 Fe)	-0,41	-0,46	-0,35
Aluminum Bronze (92 Cu-8 AI)	-0,37	-0,42	-0,31
Navel Brass (60 Cu-39 Zn)	-0,35	-0,40	-0,30
Yellow Brass (65 Cu-35 Zn)	-0,35	-0,40	-0,30
Red Brass (85 Cu-15 Zn)	-0^5	-0,40	-0,30
Muntz Metal (alpha-beta brass, 60 Cu-40 Zu)	-0,35	-0,40	-0,30
Copper, Cu	-0,32	-0,36	-0,28
Tin, Sn	-032	-0,33	-0,31
Aluminum Brass (76 Cu-22 Zn-2 AI)	-0,32	-0,36	-0,28
50-50 Lead/Tin Solder	-0,31	-0,35	-0.26
Stainless steel , Type 410 (13 Cr) (passive)	-031	-0,35	-0,26
Admiralty Brass (71 Cu-28Zn 1 Sn)	-0,30	-0,34	-0,25
Manganese Bronze (58.5 Cu-39 Zn-1 Sn-1 Fe-0.3 Mn)	-0,29	-0,33	-0,25
Bronze Composition G (88 Cu-2 Zn-10 Sn)	-0,28	-0,31	-0,24
Bronze Composition M (88 Cu-3 Zn-6.5 Sn- 1.5 Pb)	-0,28	-0,31	-0,24
Silicon Bronze (96 Cu, Max-Fe-Zn-Si-Mn-Sn)	-0,26	-0,27	-0,24
90 Cu-10 Ni	-0,25	-0,28	-0,21
75 Cu-20 Ni-5 Zn	-0,22	-0,25	-0,19
Lead, Pb	-0,22	-0,25	-0,19
70 Cu-30 Ni	-0,20	-0,22	-0,18
Inconel (78 Ni-13.5 Cr-6 Fe) (Passive)	-0,16	-0,17	-0,14
Stainless steel 17-4 PH	-0,15	-0,20	-0,10
Nickel 200	-0,15	-0,20	-0,10
Silver, Ag	-0,12	-0,14	-0,09
Monel 400 (70 Ni-30 Cu), K-500	-0,09	-0,14	-0,04
Stainless steel , Type 304 (18-8), (passive)	-0,08	-0,10	-0,05
Stainless steel , Type 316 (18-8, 3 Mo) (passive)	-0,05	-0,10	0,00
Titanium	0,01	-0,05	+0,06
Inconel 625	0,03	-0,04	+0,10
Hastelloy C	0,03	-0,04	+0,10
Platinum, Pt	0,22	+0,18	+0,25
Graphite	0,25	+0,20	+0,30

Table A2.13. *Galvanic metals series in sea water that is in motion*

Metal	Atmospheric air	Neutral and Alkaline solutions			Acetate solutions		
		Sodium hydroxide, 5%	Fresh water	Seawater	Sulphuric acid, 5% (non-oxidizing)	Acetic acid, 5% (non-oxidizing)	Nitric acid, 5% (oxidative)
Aluminum	0-0,5	13.000	0,1	1-50	8-100	0,5-5	15-80
Copper alloy	0-0,2	2-5	0-1	0,2-15	2-50	2-15	150-1500
Gold	<0,1	<0,1	<0,1	<0,1	<0,1	<0,1	<0,1
Iron	1-8	0-0,2	0,1-10	0,1-10	15-400	10-400	1000-10000
Lead	0-0,2	5-500	0,1-2	0,2-15	0-2	10-150	100-6000
Molybdenum	<0,1	<0.1	<0,1	<0,1	0-0,2	<0,1	high
Nickel alloy	0-0,2	0-0.2	0-0,2	0-1	2-35	2-10	0,1-1500
Platinum	<0,1	<0,1	<0,1	<0,1	<0,1	<0,1	<0,1
Silver	<0,1	<0,1	<0,1	<0,1	0-1	<0,1	high
Stainless steel	0-0,2	0-0,2	0-0.2	0-200	0-100	0-0,5	0-2
Tantalum	<0,1	<1	<0,1	<0,1	<0,1	<0,1	<0,1
Tin	0-0,2	5-20	0-0,5	0,1	2-500	2-500	100-400
Titanium	<0,1	<0,2	<0,1	<0,1	10-100	<0,1	0,1-1
Zinc	0-0,5	15-200	0,5-10	0,5-10	high	600-800	high
Zirconium	<0,1	<0,1	<0,1	<0,1	<0,5	<0,1	<0,1

NOTE.– The corrosion velocities (in mills/year, 1 mill = 0.1 in) refer to temperatures up to 100°C.

Table A2.14. *Corrosion rate of metals*

Appendix 3

Symbols and Basic Conversion Factors

Symbols

meter, m

kilogram, kg

second, s

kelvin, K

foot, ft

poundforce, Ib or lb_f

poundmass, lb_m

degree Rankine, °R

newton, N

atmosphere, atm

Conversions

1 ft = 0.3048

1 ft^2 = 0.092903 m^2

1 ft^3 = 0.0283168 m^3

1 slug = 14.594 kg

1 slug = 32.2 lb_m

1 lb_m = 0.4536 kg

1 lb = 4.448 N

1 atm = 2116 lb/ft^2 = 1.01×10^5 N/m^2

1 K = 1.8°R

Bibliography

[ATH 88] ATHANASIADIS N.A., TOUZOPOULOS D., TSANGARIS S., *Solved Examples in Fluid Mechanics*, Symeon, Athens, 1988.

[ATH 89] ATHANASIADIS N.A., *Fluid Mechanics*, Symeon, Athens, 1989.

[BAN 05] BANSAL K.R., *A Textbook of Fluid Mechanics and Hydraulic Machines*, Firewall Media, Delhi, 2005.

[CEN 16] CENGEL Y.A., *Thermal – Fluid Sciences*, MacGraw-Hill Higher Education, New York, 2016.

[CIM 04] CIMBALA J., CENGEL Y.A., *Fluid Mechanics: Fundamentals and Applications*, McGraw Hill Higher Education, New York, 2004.

[DAU 97] DAUGERTY R.L., FRANZINI J.B., *Fluid Mechanics and Technical Applications I and II*, Plaisio, Athens, 1997.

[DIM 97] DIMITRIOU I., *Fluid Mechanics I and II*, NTUA, Athens, 1997.

[EVE 92] EVETT J.B., LIU C., *2500 Solved Problems in Fluid Mechanics and Hydraulics*, McGraw Hill, New York, 1992.

[FOX 94] FOX A., MCDONALD T., *Introduction to Fluid Mechanics*, John Wiley & Sons, Hoboken, 1994.

[FRA 76] FRANCIS J.R.D., *Fluid Mechanics for Engineering Students*, Edward Arnold, London, 1976.

[GEO 81] GEORGANTOPOULOS G., *Aerodynamics I*, Symeon, Dekeleia, 1981.

[GEO 04] GEORGANTOPOULOU C., Numerical simulation and estimation of flows using adaptive Cartesian grids, PhD Thesis, National Technical University of Athens, Athens, 2004.

[GEO 16] GEORGANTOPOULOU C., GEORGANTOPOULOS G., *Fluid Mechanics and Hydraulic Applications*, Tsotras, Athens, 2016.

[GRA 89] GRANET I., *Fluid Mechanics*, Prentice Hall, Upper Saddle River, 1989.

[HOL 72] HOLMAN J.P., *Heat Transfer*, McGraw-Hill Company, New York, 1972.

[KAF 83] KAFOUSIA N., GEORGANTOPOULOS G., *Fluid Mechanics Notes*, HAFA, Dekeleia, 1983.

[KOR 80] KORONAKIS P.S.P., *Solved Problems in Fluid Mechanics*, Ivos, Athens, 1980.

[KOR 05] KORONAKIS P.S.P., *Fluid Mechanics*, Ivos, Athens, 2005.

[KUN 02] KUNDU R., COHEL I., *Fluid Mechanics*, Academic Press, 2002.

[MAV 99] MAVROMATAKIS A.K., *Physics Vol II*, Evgenidio, Athens, 1999.

[NAK 99] NAKAYAMA Y., *Introduction to Fluid Mechanics*, Butterworth-Heinemann, Oxford, 1999.

[NOU 72] NOUTSOPOULOS I.K., *Hydraulic Lessons*, Athens, 1972.

[PAN 08] PANTZALIS N., *Fluid Mechanics I and II*, Evdenidio, Athens, 2008.

[PAP 78] PAPANIKAS D.G., *Applied Fluid Mechanics*, Patra, UOP, 1978.

[PAP 95] PAPANTONIS D.E., *Hydraulic Machines – Pumps – Turbines*, Symeon, Athens, 1995.

[PAP 02] PAPAIOANNOU A.T.H., *Fluid Mechanics*, NTUA, Athens, 2002.

[POT 09] POTTER M.C., *Fluid Mechanics Demystifield*, McGraw Hill, New York, 2009.

[SCH 10] SCHOBEIRI M.T., *Fluid Mechanics for Engineers*, Springer-Verlag, Berlin, 2010.

[TER 97] TERZIDIS G.L., *Applied Hydraulics*, Zisis, Thessaloniki, 1997.

[TSA 15] TSANGARIS S., *Fluid Mechanics – Theory and Examples*, Tsotras, Athens, 2015.

[WHI 99] WHITE F.W., *Fluid Mechanics*, McGraw Hill Company, New York, 1999.

[YAM 08] YAMAGUCHI H., *Engineering Fluid Mechanics*, Springer Science & Business Media, Berlin, 2008.

Index

A, B, C

E, F, H

I, L, M

N, P, R

S, T, V

Other titles from

iSTE

in

Science, Society and New Technologies

2018

BARTHES Angela, CHAMPOLLION Pierre, ALPE Yves
Evolutions of the Complex Relationship Between Education and Territories (Education Set - Volume 1)

2017

ANICHINI Giulia, CARRARO Flavia, GESLIN Philippe, GUILLE-ESCURET Georges
Technicity vs Scientificity – Complementarities and Rivalries (Social Interdisciplinarity Set – Volume 2)

DUGUÉ Bernard
Information and the World Stage – From Philosophy to Science, the World of Forms and Communications (Engineering, Energy and Architecture Set – Volume 1)

GESLIN Philippe
Inside Anthropotechnology – User and Culture Centered Experience (Social Interdisciplinarity Set – Volume 1)

GORIA Stéphane
Methods and Tools for Creative Competitive Intelligence

KEMBELLEC Gérald, BROUDOUS EVELYNE
Reading and Writing Knowledge in Scientific Communities– Digital Humanities and Knowledge Construction

MAESSCHALCK Marc
Reflexive Governance for Research and Innovative Knowledge (Responsible Research and Innovation Set - Volume 6)

PARK Sejin, GUILLE-ESCURET Georges
Sociobiology vs Socioecology – Consequences of an Unraveling Debate (Interdisciplinarity between Biological Sciences and Social Sciences Set - Volume 1)

PELLÉ Sophie
Business, Innovation and Responsibility (Responsible Research and Innovation Set - Volume 7)

2016

BRONNER Gérald
Belief and Misbelief Asymmetry on the Internet

EL FALLAH SEGHROUCHNI Amal, ISHIKAWA Fuyuki, HÉRAULT Laurent, TOKUDA Hideyuki
Enablers for Smart Cities

GIANNI Robert
Responsibility and Freedom (Responsible Research and Innovation Set - Volume 2)

GRUNWALD Armin
The Hermeneutic Side of Responsible Research and Innovation (Responsible Research and Innovation Set - Volume 5)

LAGRANA Fernando
E-mail and Behavioral Changes – Uses and Misuses of Electronic Communications

LENOIR Virgil Cristian
Ethical Efficiency – Responsibility and Contingency
(Responsible Research and Innovation Set - Volume 1)

MAESSCHALCK Marc
Reflexive Governance for Research and Innovative Knowledge
(Responsible Research and Innovation Set - Volume 6)

PELLÉ Sophie, REBER Bernard
From Ethical Review to Responsible Research and Innovation
(Responsible Research and Innovation Set - Volume 3)

REBER Bernard
Precautionary Principle, Pluralism and Deliberation – Sciences and Ethics
(Responsible Research and Innovation Set - Volume 4)

VENTRE Daniel
Information Warfare – 2nd edition

Printed and bound by CPI Group (UK) Ltd, Croydon, CR0 4YY